Communications
in Computer and Information Science   898

Commenced Publication in 2007
Founding and Former Series Editors:
Phoebe Chen, Alfredo Cuzzocrea, Xiaoyong Du, Orhun Kara, Ting Liu,
Dominik Ślęzak, and Xiaokang Yang

More information about this series at http://www.springer.com/series/7899

Juan Antonio Lossio-Ventura
Denisse Muñante · Hugo Alatrista-Salas (Eds.)

Information Management and Big Data

5th International Conference, SIMBig 2018
Lima, Peru, September 3–5, 2018
Proceedings

 Springer

Editors
Juan Antonio Lossio-Ventura (iD)
Biomedical Informatics,
College of Medicine
University of Florida
Gainesville, FL, USA

Hugo Alatrista-Salas (iD)
Facultad de Ingeniería
University of the Pacific
Jesús María, Lima, Peru

Denisse Muñante (iD)
Fondazione Bruno Kessler
Trento, Italy

ISSN 1865-0929   ISSN 1865-0937 (electronic)
Communications in Computer and Information Science
ISBN 978-3-030-11679-8   ISBN 978-3-030-11680-4 (eBook)
https://doi.org/10.1007/978-3-030-11680-4

Library of Congress Control Number: 2018968324

This Springer imprint is published by the registered company Springer Nature Switzerland AG
The registered company address is: Gewerbestrasse 11, 6330 Cham, Switzerland

Preface

Today, data scientists use the term "big data" to describe the exponential growth and availability of data, which could be structured and unstructured. Big data has taken place over the past 20 years and will stay with us for the next few years. In a recent study, Cisco Systems predicted that in 2021, over 27.1 billion electronic devices would be connected to the Internet. From these devices, smartphones only will generate — on average — 14.9 gigabytes of data per month. This massive amount of data contains, possibly, patterns of behavior that can be exploited by organizations to take decisions or to design public policies with a high impact. In this context, the term big data not only concerns the storage, the management, and the analysis of a large amount of data but also is associated with the ability to implement new algorithms, to propose new techniques, to apply new strategies, and to find new real-world applications.

Several domains, including biomedicine, life sciences, and scientific research, have been affected by big data. For instance, social networks such as Facebook, Twitter, and LinkedIn generate masses of data, which are available to be accessed by other applications. Therefore, there is a need to understand and exploit all kinds of data (structured and unstructured). This process can be performed with data science, which encompasses methods of data mining, natural language processing, the Semantic Web, and statistics, among others, which allow us to gain new insights from data.

SIMBig (International Conference on Information Management and Big Data) seeks to present new methods of the fields related to data science to assess large volumes of data. SIMBig aims to bring together main — national and international — actors in the decision-making field to state in new technologies dedicated to handling amounts of data. Besides, the conference is a convivial place where these actors can present their innovative contributions and receive feedback from the experts.

This book collects three invited talks and 30 accepted contributions from 101 submitted papers belonging to the following four special tracks:

- ANLP - Track on Applied Natural Language Processing
- DISE - Track on Data-Driven Software Engineering
- SNMAM - Track on Social Network and Media Analysis and Mining
- PSCBig - Track on Privacy and Security Challenges on Big Data

SIMBig 2018 had six keynote speakers who are experts in the main topics of the conference. The conference started with the talk given by our invited speaker Andrei Broder. Broder, a distinguished scientist at Google, reviewed the evolution of Internet search engines, highlighting that his success and survival is centered on three axes: scalability, speed, and functionality. Broder said that the last big step of Google is the jump from the search drawer to the personal assistant. He also stressed that the technology of Internet search is based on the user experience and that the life cycle of innovations is very short, since "the new" quickly becomes "normal." "If you looked at the technology of the search when the autocomplete was introduced in the query box, this was a breakthrough; people were surprised to be able to type only a couple of letters and get what they wanted. Now, if you have a query box and do not auto-complete, the user is annoyed. Surely very soon if you have to write will generate a total annoyance, because users will want to have voice recognition," said Broder exemplifying the virtual circle of innovation.

Furthermore, Lise Getoor from the University of California presented a talk entitled "Effectively Exploiting Structure in Data Science Problems," which is summarized as: "Our ability to collect, manipulate, analyze, and act on vast amounts of data is having a profound impact on all aspects of society. Much of this data is heterogeneous in nature and interlinked in a myriad of complex ways. From information integration to scientific discovery to computational social science, we need machine learning methods that are able to exploit both the inherent uncertainty and the innate structure in a domain. Statistical relational learning (SRL) is a machine learning subfield that builds on principles from probability theory and statistics to address uncertainty while incorporating tools from knowledge representation and logic to represent structure." Getoor overviewed her recent work on probabilistic soft logic (PSL), an SRL framework for large-scale collective, probabilistic reasoning in relational domains. PSL provides a tractable approach to probabilistic inference. She showed the theoretical foundations for PSL, which connect work from the theoretical computer science community on randomized algorithms with work from the probabilistic graphical model's community on local consistency relaxations with work from the AI community on soft logic. She also described several successful applications of PSL to problems from computational social science (stance in online forums, social trust, latent political groups, cyberbullying) and data integration (entity resolution and knowledge graph construction). Getoor closed up with a discussion of responsible data science, which requires understanding both the inherent structure in the domain, the structure and potential bias in the data, and the potential implications and feedback loops in the algorithms.

Later, Jian Pei, vice-president of JD.com and professor at the Simon Fraser University, gave a talk about data mining and logistic. The talk is summarized as: "The future of retail is in breaking the limitations of the current business models in customer connection, logistics and stores. The key to breaking those limitations is the integrated smart supply chain, which covers smart consumption, smart logistics, and smart supply. The foundation of a smart supply chain is intelligent big data science. Through a series of examples, we discuss how smart supply chain techniques and data science can enable the future of retail. Specifically, we demonstrate how big data and AI techniques together can deepen our understanding of customers, create new convenience and efficiency in retail scenarios, minimize the cost store operation, and shorten the path from customers to manufacturers".

Finally, the remaining three invited talks were presented in the form of short papers in this book.

To share the new analysis methods for managing large volumes of data, we encouraged participation from researchers in all fields related to big data, data science, data mining, natural language processing, and Semantic Web, but also multilingual text processing, biomedical NLP, data-driven software engineering, and data privacy.

Topics of interest to SIMBig included: data science, big data, data mining, natural language processing, bio NLP, text mining, information retrieval, machine learning,

Semantic Web, ontologies, IoT, privacy on social networks, Web mining, knowledge representation and linked open data, social networks, social Web, and Web science, information visualization, OLAP, data warehousing, business intelligence, spatiotemporal data, health care, agent-based systems, data-driven security, and privacy.

SIMBig is positioning itself as one of the most important conferences in South America on issues related to information management and Big Data. Two Springer CCIS books were published in the context of the SIMBig conference. The first publication compiles the selected papers from the SIMBig 2015 and SIMBig 2016 conferences [1]. The second one presents the best papers of the SIMBig 2017 conference [2].

References

1. J. A. Lossio-Ventura and H. Alatrista-Salas, editors. *Information Management and Big Data - Second Annual International Symposium, SIMBig 2015, Cusco, Peru, September 2–4, 2015, and Third Annual International Symposium, SIMBig 2016, Cusco, Peru, September 1–3, 2016, Revised Selected Papers*, volume 656 of *Communications in Computer and Information Science*. Springer, 2017.
2. J. A. Lossio-Ventura and H. Alatrista-Salas, editors. *Information Management and Big Data - 4th Annual International Symposium, SIMBig 2017, Lima, Peru, September 4–6, 2017, Revised Selected Papers*, volume 795 of *Communications in Computer and Information Science*. Springer, 2018.

January 2019   Juan Antonio Lossio-Ventura
                                                        Hugo Alatrista-Salas

Organization

SIMBig 2018: Organizing Committee

General Organizers

Juan Antonio Lossio-Ventura	University of Florida, USA
Hugo Alatrista-Salas	Universidad del Pacífico, Peru

Local Organizers

Michelle Rodriguez Serra	Universidad del Pacífico, Peru
Cristhian Ganvini Valcarcel	Universidad Andina del Cusco, Peru
Pilar Hidalgo Leon	Universidad del Pacífico, Peru

SNMAM Track Organizers

Jorge Valverde-Rebaza	Visibilia, Brazil
Alneu de Abdrade Lopes	University of São Paulo, Brazil

DISE Track Organizers

Denisse Muñante Arzapalo	Fondazione Bruno Kessler (FBK), Italy
Nelly Condori Fernandez	Universidade da Coruna, Spain
Carlos Gavidia Calderon	University College London, UK

ANLP Track Organizers

Marco Antonio Sobrevilla-Cabezudo	University of São Paulo, Brazil
Félix Arturo Oncevay-Marcos	Pontificia Universidad Católica del Perú, Peru
Armando Fermin Perez	Universidad Nacional Mayor de San Marcos, Peru

PSCBᵢɢ Track Organizers

Ali Tekeoglu	SUNY Polytechnic Institute, USA
Miguel Nuñez-del-Prado	Universidad del Pacífico, Peru

SIMBig 2018: Program Committee

SIMBig Program Committee

Nathalie Abadie	COGIT IGN, France
Amine Abdaoui	Stack-Labs, France
César Antonio Aguilar	Pontifica Universidad Católica de Chile, Chile
Frank D. J. Aguilar	University of São Paulo, Brazil

Amanda Hicks	University of Florida, USA
William Hogan	University of Florida, USA
Ian Horrocks	Oxford University, UK
Diana Inkpen	University of Ottawa, Canada
Clement Jonquet	LIRMM - University of Montpellier, France
Alípio Jorge	University of Porto, Portugal
Georgios Kontonatsios	Edge Hill University, UK
Yannis Korkontzelos	Edge Hill University, UK
Ravi Kumar	Google, USA
Nikolaos Lagos	NAVER Labs, France
Juan Guillermo Lazo Lazo	Universidad del Pacífico, Peru
Ulf Leser	Humboldt University of Berlin, Germany
Jose Leomar Todesco	Universidade Federal de Santa Catarina, Brazil
Christian Libaque-Saenz	Universidad del Pacífico, Peru
Cédric López	Emvista, France
Franco Luque	Universidad Nacional de Córdoba and CONICET, Argentina
Maysa Macedo	IBM Research, Brazil
Sabrine Mallek	Higher Institute of Management, France
Florent Massegliaó	Zenith - Inria, France
Héctor Andrés Melgar Sasieta	Pontificia Universidad Católica del Perú, Peru
Claudio Miceli	Federal University of Rio de Janeiro, Brazil
André Miralles	SISO - Irstea, France
Giovanni Montana	University of Warwick, UK
Nils Murrugarra-Llerena	University of Pittsburgh, USA
Mark A. Musen	Stanford University, USA
Jordi Nin	BBVA Data & Analytics and Universidad de Barcelona, Spain
Miguel Nuñez-del-Prado-Cortéz	Universidad del Pacífico, Peru
José Eduardo Ochoa Luna	Catholic University San Pablo, Peru
Maciej Ogrodniczuk	Institute of Computer Science, Polish Academy of Sciences, Poland
Jonice Oliveira	Federal University of Rio de Janeiro, Brazil
Thomas Opitz	INRA, France
José Manuel Perea-Ortega	University of Extremadura, Spain
Jessica Pinaire	LIRMM - University of Montpellier, France
Yoann Pitarch	IRIT, France
Bianca Pereira	National University of Ireland, Galway, Ireland
Jorge Poco	San Pablo Catholic University, Peru
Pascal Poncelet	LIRMM - University of Montpellier, France
Mattia Prosperi	University of Florida, USA
Marcos Quiles	Federal University of São Paulo, Brazil
Julien Rabatel	Catholic University of Leuven, Belgium
José-Luis Redondo-García	Amazon, UK

Mathieu Roche	Cirad - TETIS, France
Nancy Rodriguez	LIRMM, CNRS - University of Montpellier, France
Fatiha Saïs	University of Paris-Sud 11, France
Arnaud Sallaberry	LIRMM - Paul Valéry University, France
Nazha Selmaoui-Folcher	PPME - University of New Caledonia, New Caledonia
Matthew Shardlow	University of Manchester, UK
Selja Seppälä	University College Cork, Ireland
Gerardo Sierra-Martínez	Universidad Autónoma de México, Mexico
Newton Spolaor	State University of Western Paraná, Brazil
Claude Tadonki	MINES ParisTech - PSL Research University, France
Alvaro Talavera López	Universidad del Pacífico, Peru
Andon Tchechmedjiev	LIRMM - University of Montpellier, France
Maguelonne Teisseire	Irstea, TETIS, France
Paul Thompson	University of Manchester, UK
Thibaut Thonet	University of Grenoble Alpes, France
Camilo Thorne	Institut für Maschinelle Sprachverarbeitung, University of Stuttgart, Germany
Ilaria Tiddi	Open University, UK
Juan Manuel Torres	University of Avignon, France
Turki Turki	King Abdulaziz University, Saudi Arabia
Willy Ugarte	University of Applied Sciences, Peru
Carlos Vázquez	École de technologie supérieure, Canada
Didier Vega	Universidade de São Paulo, Brazil
Julien Velcin	ERIC Lab - University of Lyon 2, France
Maria-Esther Vidal	Universidad Simón Bolívar, Venezuela
Boris Villazon-Terrazas	Fujitsu Laboratories of Europe, Spain
Sebastian Walter	Semalytix GmbH, Germany
Florence Wang	LIRMM, France
Guo Yi	University of Florida, USA
Osmar Zaïane	University of Alberta, Canada
Amrapali Zaveri	Dumontier Lab, Stanford University, USA
He Zhe	Florida State University, USA
Pierre Zweigenbaum	LIMSI-CNRS, France

SNMAM Program Committee

Nazli Bagherzadeh Karimi	National University of Ireland Galway, Ireland
Lilian Berton	University of Santa Catarina State, Brazil
Ricardo Campos	Polytechnic Institute of Tomar and LIAAD/INESC TEC, Portugal
Alexandre Donizeti	Federal University of ABC, Brazil
Brett Drury	Scicrop, Brazil
Thiago de Paulo Faleiros	University of Brasilia, Brazil
Huei Diana Lee	Western Paraná State University, Brazil
Sabrine Mallek	Institut Supérieur de Gestion de Tunis, Tunisia
Ricardo Marcacini	Federal University of Mato Grosso do Sul, Brazil

Nils Murrugarra-Llerena	University of Pittsburgh, USA
Fabricio Olivetti de França	Federal University of ABC, Brazil
Pascal Poncelet	LIRMM - University of Montpellier, France
Ronaldo C. Prati	Federal University of ABC, Brazil
Mathieu Roche	Cirad - TETIS, France
Luca Rossi	Aston University, UK
Rafael Santos	National Institute for Space Research (INPE), Brazil
Pedro Shiguihara Juárez	Universidad Peruana de Ciencias Aplicadas, Peru
Newton Spolaor	State University of Western Paraná, Brazil
Victor Stroele	Federal University of Juiz de Fora, Brazil

ANLP Program Committee

Fernando Emilio Alva Manchego	University of Sheffield, UK
Fernando Antônio Asevedo Nóbrega	University of São Paulo, Brazil
Leandro Borges dos Santos	University of São Paulo, Brazil
Shay Cohen	University of Edinburgh, UK
Paula Christina Figueira Cardoso	Federal University of Lavras, Brazil
Nathan Siegle Hartmman	University of São Paulo, Brazil
Roque Enrique López Condori	Institute for Research in Computer Science and Automation, France
Shashi Narayan	University of Edinburgh, UK
Thiago Alexandre Salgueiro Pardo	University of São Paulo, Brazil
Márcio de Souza Dias	Federal University of Goiás, Brazil
Francis M. Tyers	UiT Norgga árktalaš universitehta, Norway

DISE Program Committee

Nour Ali	Brunel University London, UK
Joao Araujo	Universidade Nova de Lisboa, Portugal
Yudith Cardinale	Universidad Simon Bolivar, Venezuela
Alejandro Catala	University of Twente, The Netherlands
Vanea Chiprianov	Université de Pau et des Pays de l'Adour, France
Davide Fucci	University of Hamburg, Germany
David Gomez-Jauregui	ESTIA, France
Vincent Lalanne	Université de Pau et des Pays de l'Adour, France
Cristian Lopez	Universidad de la Salle, Peru
Nazim Madhavji	University of Western Ontario, Canada
Itzel Morales Ramirez	Infotec, Mexico
Manuel Munier	Université de Pau et des Pays de l'Adour, France
Oscar Pastor	Universitat Politècnica de València, Spain
Dietmar Pfahl	University of Tartu, Estonia

Jose Antonio Pow Sang	Pontificia Universidad Católica del Perú, Peru
Daniel Rodriguez	Universidad de Alcalá, Spain
Lizeth Tapia	University of Oslo, Norway
Yuanyuan Zhang	University College London, UK

PSCBig Program Committee

Nihat Altiparmak	University of Louisville, USA
Hisham Kholidy	SUNY Polytechnic Institute, USA
John Marsh	SUNY Polytechnic Institute, USA
Ali Tosun	University of Texas at San Antonio, USA

SIMBig 2018: Organizing Institutions and Sponsors

Organizing Institutions

Universidad del Pacífico, Peru[1]
University of Florida, USA[2]

Collaborating Institutions

Springer, Germany[3]
Universidad Andina del Cusco, Peru[4]

Sponsoring Institutions

North American Chapter of the ACL, USA[5]
Telefónica del Perú, Peru[6]
BBVA, Peru[7]

SNMAM Organizing Institutions

Visibilia, Brazil[8]
Labóratorio de Intêligencia Computacional, ICMC, USP, Brazil[9]

---

[1] http://www.up.edu.pe/.
[2] http://www.ufl.edu/.
[3] http://www.springer.com/la/.
[4] http://www.uandina.edu.pe/.
[5] http://naacl.org.
[6] https://www.telefonica.com.pe/.
[7] https://www.bbvacontinental.pe.
[8] http://visibilia.net.br.
[9] http://labic.icmc.usp.br/.

ANLP Organizing Institutions

Universidad Nacional Mayor de San Marcos, Peru[10]
IA Labs, Pontificia Universidad Católica del Perú, Peru[11]
Labóratorio de Intêligencia Computacional, ICMC, USP, Brazil[12]

DISE Organizing Institutions

Fondazione Bruno Kessler, Italy[13]
Universidade da Coruna, Spain[14]
University College London, UK[15]

PSCBig Organizing Institutions

SUNY Polytechnic Institute, USA[16]
Universidad del Pacífico, Peru[17]

[10] http://www.unmsm.edu.pe/.
[11] http://ia.inf.pucp.edu.pe.
[12] http://labic.icmc.usp.br/.
[13] https://www.fbk.eu/en/.
[14] https://www.udc.es.
[15] https://www.ucl.ac.uk.
[16] https://sunypoly.edu.
[17] http://www.up.edu.pe/.

Contents

Clinical, Consumer Health, and Visual Question Answering

Dina Demner-Fushman$^{(\boxtimes)}$ (iD)

National Library of Medicine, Bethesda, MD, USA
`ddemner@mail.nih.gov`

Abstract. This work presents an overview of three research directions that support clinical and health-related decisions: clinical question answering using both the Electronic Health Record data and the literature; consumer health question answering; and an emerging area of biomedical visual question answering (VQA).

Keywords: Clinical question answering ·
Consumer health question answering ·
Biomedical visual question answering

1 Background

People are making decisions about their health daily. Often, health-related decisions concern health problems and involve clinicians who then make clinical decisions. Both the consumers' and the clinicians' decisions are based on information, preferably in the form of answers to specific clinical and health-related questions. The National Library of Medicine (NLM), the worlds largest biomedical library, provides health-related information and resources to more than 4 million people a day. To improve the quality of its services, NLM, among other goals, focuses on understanding how searches for health-related and clinical information are initiated, how information is used, and how questions are posed and answered [1].

Questions asked by the professionals and consumers differ in their language [2], in the resources that may answer these questions, and in the types of the questions that could be asked of a resource. For example, during patient encounters, clinicians could have questions about the patient, such as the values of the latest tests, which could be answered by electronic health records (EHRs). Clinicians could also have questions about the current guidelines or new approaches to managing the patient's condition. Such questions are answered by the literature.

This work presents an overview of three research directions that support clinical and health-related decisions: clinical question answering using both the EHR data and the literature; consumer health question answering; and an emerging area of visual question answering (VQA).

© Springer Nature Switzerland AG 2019
J. A. Lossio-Ventura et al. (Eds.): SIMBig 2018, CCIS 898, pp. 1–6, 2019.
https://doi.org/10.1007/978-3-030-11680-4_1

2   Clinical Question Answering

A recent survey of information needs of clinicians revealed that attending physicians, residents, fellows, and advanced practitioners continue to encounter clinical questions at least a few times per week. Of the 292 survey participants, 79% already use mobile devices to access clinical information and are willing to use readily accessible search tools at the point of care when the need arises [3]. The types of clinical information needed and the sources that contain this information fall into several categories that include, among others, (1) information on particular patients, which could be provided by the patient, patient's record (EHR), or other clinicians and (2) medical knowledge that could be found in the textbooks, journals, and online databases [4]. We are addressing two of the information need categories: the ones that can be answered by EHRs and the literature.

2.1   Evidence-Based Answers to Clinical Questions

Developed to encourage clinicians to search the literature for the state-of-the-art in disease management, Evidence Based Medicine (EBM) framework provides clear guidelines for automating clinical question answering. EBM guidelines recommend formulating well-formed questions that describe the four aspects of information needs: (1) Patient-Problem; (2) Intervention; (3) Comparison; and (4) Outcome, together known as PICO frame. The question frame then needs to be converted to literature searches for finding the strongest evidence and the best available approach to managing a specific clinical case.

Our Clinical Question Answering system CQA 1.0 implements the framework by representing documents found by the search engines as frames in the same PICO format that is used to represent the question, and then applying fuzzy frame unification to find the best answers [5]. CQA 1.0 is the basis of the InfoBot clinical decision support system shown in Fig. 1. InfoBot formulates PICO questions and queries using clinicians' progress notes to provide personalized advice for patient care plan development [6,7].

2.2   A Natural Language Interface for EHR Questions

Translating natural language questions to structured form allows representing questions independently of the underlying structure of the EHRs. The structured questions can then be translated to EHR-specific searches to find needed information. Semantic parsing is one of the approaches to transforming questions into a structured representation. We explored a variant of first order logic (FOL) that combines FOL with lambda calculus expressions. Although difficult, automatic translation of natural language questions is feasible [8], and, in a pilot study, a combination of a rule-based and machine learning approach achieves an accuracy of 95.6% on questions manually converted to structured forms [9].

Fig. 1. A synthetic, realistic but not real patient EHR record presents the Evidence Based Dashboard at the NIH Clinical Center. The dashboard provides personalized information to support development of patient care plans. The upper-left panel displays an automatically annotated clinical note in which the disease and the treatment terms are highlighted. The system provides definitions for these terms. The terms are also used to automatically generate clinical questions and literature searches. The bottom panel displays search results and the automatically extracted bottom-line advice. The brief summary of the paper in the form of the bottom-line advice helps clinicians decide if they want to access the full paper.

3   Consumer Health Question Answering

People increasingly use online resources to find health-related information. It is important that information they find is reliable, based on the latest evidence, and provides answers at the level of the consumer's general and health literacy. National Library of Medicine receives thousands of health-related requests. The requests come in two forms: relatively long emails sent to NLM customer services and short questions typed into the search box of the NLM's consumer-oriented resource MedlinePlus [10]. Our consumer Health Information and Question Answering system (CHIQA) [11] is trained to represent both types of questions in structured form. We combine a knowledge-based approach that uses MetaMap Lite [12] to identify PICO elements in the questions, and SVM and LSTM-based approaches to identify the focus and the type of the question [13,14]. The focus is the main topic of interest, such as disease, drug or diet, whereas the type is the task for which information is needed, such as drug storage, or life-style changes.

Working with MedlinePlus and other reliable sources of information linked to it, we found that articles containing reliable answers can be found for over 65% of the questions by searching MedlinePlus for the focus and type of the question [15]. In parallel, we harness the wealth of information provided by the National Institutes of Health (NIH) resources for patients. NIH is formed by 27 institutes and centers, including NLM, each providing answers to questions frequently asked by patients. For example, the Genetic and Rare Diseases (GARD) Information Center provides answers to questions about rare diseases [16]. From these resources, we created a database of question-answer pairs that we use to train a question entailment recognition system that finds questions similar to those submitted to NLM [17]. Answers extracted from the search results and provided by the entailment module are then combined in the final answers presented to the users (see Fig. 2).

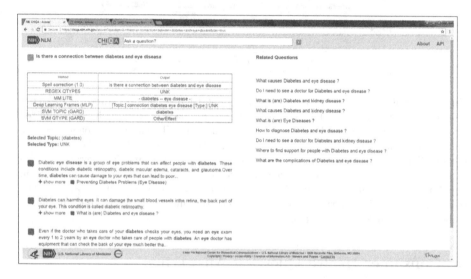

Fig. 2. The developers' view of the prototype Consumer Health Question Answering system presents analysis of the question using a knowledge-based approach, and question focus and type recognized by SVM and LSTM/CNN approaches. The answers are generated using a traditional information retrieval-based system. Related questions are found through question entailment. In the question *Is there a connection between diabetes and eye disease?*, the knowledge-based system correctly identifies both disorders. The deep learning approach captures the focus on connection between the two diseases, whereas the SVM-based approach is the only method that correctly identifies the question type, but misses one of the focuses.

4   Biomedical Visual Question Answering

Automated systems could help clinicians cope with large amounts of images by answering questions about the image contents. An emerging area, Visual Question Answering (VQA) in the medical domain, explores approaches to this form of clinical decision support. In 2018, ImageCLEF-Med released a radiology dataset and coordinated the first community-wide VQA challenge in medicine. The NLM team was one of the five participants and achieved competitive scores [18]. A closer look at the automatically generated questions and

Table 1. Official results of the ImageCLEF 2018 Medical Domain Visual Question Answering Task. The systems were evaluated using BLEU to capture similarity between the system-generated answers and the ground truth. Two metrics measured semantic similarity between the systems' answers and the ground truth: a WordNet-based WBSS (Word-based Semantic Similarity) score, and UMLS-based CBSS (Concept-based Semantic Similarity). Higher CBSS scores suggest that the systems generated more relevant clinical concepts [19]. Only the best scores for each participating team are shown. The best overall results for each score are highlighted in bold.

System	BLEU	WBSS	CBSS
Abdelmalek Essaadi University, Morocco	0.054	0.101	0.269
Jordan University of Science and Technology	0.061	0.122	0.029
National Library of Medicine	0.121	0.174	**0.338**
Tokushima University, Japan	0.135	0.174	0.334
University of Massachusetts Medical School	**0.162**	**0.186**	0.023

answers in the ImageCLEF-Med 2018 dataset and at the evaluation results (see Table 1) indicates that this potentially invaluable research area needs more manually curated datasets, innovative evaluation metrics, and wider participation to contribute to clinical decision support.

Summary

Question Answering is a long-standing area of research in artificial intelligence in which success ebbs and flows with the availability of resources, development of computational approaches, and general interest of researchers. The explosion of the deep learning approaches to image and text processing and the wider availability of resources stimulated rapid developments in clinical, consumer-health, and visual question answering, with visible contributions to how people search and find useful information to support their clinical decisions.

References

1. A Platform for Biomedical Discovery and Data-Powered Health. National Library of Medicine Strategic Plan 2017–2027. https://www.nlm.nih.gov/pubs/plan/lrp17/NLM_StrategicReport2017_2027.html. Accessed 20 Sept 2018
2. Roberts, K., Demner-Fushman, D.: Interactive use of online health resources: a comparison of consumer and professional questions. J. Am. Med. Inform. Assoc. **23**(4), 802–811 (2016)
3. Brassil, E., Gunn, B., Shenoy, A.M., Blanchard, R.: Unanswered clinical questions: a survey of specialists and primary care providers. J. Med. Libr. Assoc. **105**(1), 4–11 (2017)
4. Smith, R.: What clinical information do doctors need? BMJ **313**(7064), 1062–1068 (1996)

5. Demner-Fushman, D., Lin, J.: Answering clinical questions with knowledge-based and statistical techniques. Comput. Linguist. **33**(1), 63–103 (2007)
6. Demner-Fushman, D., Seckman, C., Fisher, C., Hauser, SE., Clayton, J., Thoma, G.R.: A prototype system to support evidence-based practice. In: Proceedings of the Annual Symposium of the American Medical Information Association, pp. 151–155. AMIA, Washington, DC (2008)
7. Demner-Fushman, D., Fisher, C., Seckman, C., Thoma, G.: Continual development of a personalized decision support system. Stud. Health Technol. Inform. **192**, 175–179 (2013)
8. Roberts, K., Demner-Fushman D.: Toward a natural language interface for EHR questions. In: Proceedings of the AMIA Joint Summits on Translational Science, pp. 157–161. AMIA, San Francisco (2015)
9. Roberts, K., Patra, BG.: A semantic parsing method for mapping clinical questions to logical forms. In: Proceedings of the Annual Symposium of the American Medical Information Association, pp. 1478–1487. AMIA, Washington, DC (2017)
10. MedlinePlus. https://medlineplus.gov/. Accessed 30 Sept 2018
11. CHIQA. https://chiqa.nlm.nih.gov/. Accessed 30 Sept 2018
12. Demner-Fushman, D., Rogers, W.J., Aronson, A.R.: MetaMap lite: an evaluation of a new Java implementation of MetaMap. J. Am. Med. Inform. Assoc. **24**(4), 841–844 (2017)
13. Roberts, K., Kilicoglu, H., Fiszman, M., Demner-Fushman, D.: Automatically classifying question types for consumer health questions. In: Proceedings of the Annual Symposium of the American Medical Information Association, pp. 15–19. AMIA, Washington, DC (2014)
14. Mrabet, Y., Kilicoglu, H., Roberts, K., Demner-Fushman, D.: Combining open-domain and biomedical knowledge for topic recognition in consumer health questions. In: Proceedings of the Annual Symposium of the American Medical Information Association, pp. 914–923. AMIA, Chicago (2016)
15. Deardorff, A., Masterton, K., Roberts, K., Kilicoglu, H., Demner-Fushman, D.: A protocol-driven approach to automatically finding authoritative answers to consumer health questions in online resources. J. Assoc. Inf. Sci. Technol. **68**(7), 1724–1736 (2017)
16. FAQs About Rare Diseases. https://rarediseases.info.nih.gov/diseases/pages/31/faqs-about-rare-diseases. Accessed 30 Sept 2018
17. Ben Abacha, A., Demner-Fushman, D.: Recognizing question entailment for medical question answering. In: Proceedings of the Annual Symposium of the American Medical Information Association, pp. 310–318. AMIA, Chicago (2016)
18. Ben Abacha, A., Gayen, S., Lau, JJ., Rajaraman, S., Demner-Fushman, D.: NLM at ImageCLEF 2018 Visual Question Answering in the Medical Domain. http://ceur-ws.org/Vol-2125/paper_165.pdf. Accessed 30 Sept 2018
19. Hasan, SA., Ling, Y., Farri, O., Liu, J., Müller H., Lungren, M.: Overview of ImageCLEF 2018 Medical Domain Visual Question Answering Task. http://ceur-ws.org/Vol-2125/paper_212.pdf. Accessed 30 Sept 2018

Which Is the Tallest Building in Europe? Representing and Reasoning About Knowledge

Ian Horrocks[(✉)]

University of Oxford, Oxford, UK
ianh@cs.ox.ac.uk

Abstract. The need for representing knowledge is ubiquitous in applications; for example, Google needs to represent knowledge about the location and height of building in order to answer questions such as "which is the tallest building in Europe". Google uses a graph to represent such knowledge, and so-called knowledge graphs are becoming increasingly popular as a knowledge representation formalism. Adding some form of rules greatly increases the power and utility of knowledge graphs, but can also lead to theoretical and/or practical tractability problems. In this papers we will briefly survey the relevant issues and possible solutions, and show that rule enhanced knowledge graphs are extremely powerful, can be given a formal logic-based semantics, and are highly scalable in practice.

1 Introduction

The need for representing knowledge is ubiquitous in applications. To mention just a few examples: travel applications need to represent knowledge about transportation and accommodation; music applications need to represent knowledge about artists and albums; (social) networking applications need to represent knowledge about user profiles and preferences; and search engines like Google need to represent a wide range of general knowledge, including, e.g., knowledge about tall buildings. These applications can then use the represented knowledge to answer questions about, e.g., seat availability on a given flight, the albums released by a given artist, the privacy settings of a given user or the height of a given building. Using this kind of knowledge enables Google, for example, to answer questions such as "what is the height of the Shard?"; if we type this question into the Google search box, we don't simply get a ranked list of relevant web pages, but a direct answer to the question, namely "306 m, 310 m to tip".

2 Knowledge Representation and Query Answering

What kind of data model should we use for storing and querying such knowledge? The use of graph-based data models has become very popular, with the

© Springer Nature Switzerland AG 2019
J. A. Lossio-Ventura et al. (Eds.): SIMBig 2018, CCIS 898, pp. 7–12, 2019.
https://doi.org/10.1007/978-3-030-11680-4_2

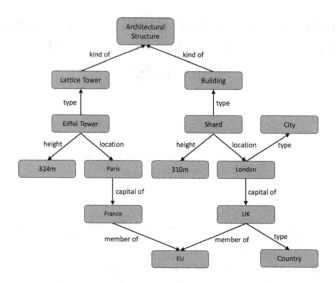

Fig. 1. A graph fragment

Google Knowledge Graph being one of the most prominent examples. Such formalisms/systems have been variously called knowledge graphs, graph databases and (RDF) triple stores, and although there are some differences, they all share the basic idea of using graph nodes to represent entities (both concrete and abstract) and labelled directed edges to represent relationships between them. Figure 1 illustrates a small fragment of such a graph in which the fact that the Shard is located in London is represented by the edge labelled "location" that connects the "Shard" entity to the "London" entity. In RDF such an edge is called a triple, as it is made up of three elements: the "subject" entity (the Shard), the edge label or "predicate" (location), and the "object" entity (London). A triple is often written ⟨subject, predicate, object⟩, e.g.:

$$\langle \text{Shard}, \text{location}, \text{London} \rangle$$

It is important to note that such information could equally well be stored in other ways, for example in a database. However, the graph data model has several advantages: it is simple yet flexible, and the simple data model, which can be thought of as a single 3-column table of edges, can be exhaustively indexed using only 6 indices (subject, predicate, object, subject-predicate, subject-object and predicate-object) [6,8]. This allows for a wide range of queries to be answered without specialised tuning of the kind that is often necessary in order to achieve good performance with relational databases.

The graph also makes for a very intuitive style of navigational query in which the basic component of a query is a fragment of a graph (called a "graph pattern" in SPARQL, a graph query language developed for use with RDF) in which some of the edges and nodes are variables. Variables act as wild-cards, and can match any edge label or entity in the graph. Figure 2, for example, illustrates

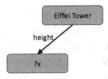

Fig. 2. A query graph pattern for the height of the Eiffel Tower

a graph pattern consisting of a single triple with "Eiffel Tower" as the subject, "height" as the predicate, and a variable "?x" as the object (a question mark is often used as the first character of a variable name). Answering the query amounts to finding fragments of the graph that match the graph pattern, with the answer being given by the entity and edge labels that match the variables in the graph pattern. The graph pattern in Fig. 2, for example, matches the triple ⟨Eiffel Tower, height, 324 m⟩, and returns the value 324 m that matches the variable ?x.

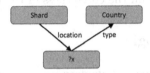

Fig. 3. A query graph pattern for the country in which the Shard is located

3　Augmenting Knowledge Graphs with Rules

Although very intuitive, the simple query model described above soon runs into problems. For example, if we want to query for the country in which the Shard is located, then we might reasonably try the graph pattern illustrated in Fig. 3. However, this will fail as the graph uses a more complex representation in which the location of the Shard is given as London, which is itself said to be the capital of the UK. Of course we could go through the graph and add location edges linking each relevant entity with the country in which it is located, but this would be very time consuming in a large graph: Google's knowledge graph, for example, is said to contain over 70 billion facts.[1]

This problem can be addressed by augmenting knowledge graphs with rules that can be used to systematically augment the facts stored in the graph. For example, a rule of the form:

$$\langle ?x, location, ?y\rangle \wedge \langle ?y, capital\ of, ?z\rangle \rightarrow \langle ?x, location, ?z\rangle \qquad (1)$$

---

[1] wikipedia.org/wiki/Knowledge_Graph.

can be read as stating that if an entity ?x is connected to an entity ?z via a path consisting of a "location" edge followed by a "capital of" edge, then ?x should also be connected to ?z by a "location" edge. Rules can also be used to capture complex knowledge about, e.g., what constitutes a building; the Eiffel Tower, for example, is not considered to be a building, but rather a "lattice tower".[2]

An additional advantage of rules is that they make it much easier to curate and maintain the correctness of the knowledge graph. For example, a query for tall buildings located in the EU would be expected to return the Shard as one of its answers; but if the UK leaves the EU, then the Shard should no longer be in the answer to this query. If the graph explicitly represents for all buildings (or even all objects) located in the UK that they are also located in the EU, then the UK leaving the EU would require the deletion of a very large number of edges, a process which would be costly and error prone. If, on the other hand, the graph explicitly represents only the cities in which objects are located, as well as the facts that the relevant cities are located in the UK and that the UK is part of the EU, and uses rules to capture the implicit location of these objects in the EU, then the UK leaving the EU would require the deletion of only a single fact in order to maintain the correctness of the knowledge graph.

4   Semantics and Logic

Knowledge graphs can be seen as a modern incarnation of semantic networks, a knowledge representation formalism dating back to the 1950s [3]. A well known problem with Semantic Networks is that their semantics is not precisely specified and so open to multiple interpretations. For example, the triple:

$$\langle \text{cat, has-colour, black} \rangle$$

could be taken to mean that some cats are black, all cats are black, or perhaps even that all black things are cats.

In order to address this issue, knowledge graphs can be given a precise semantics by systematically translating them into some suitable logical formalism, typically First Order Predicate Calculus (FOPC).[3] Edges in the graph that represent basic facts can be translated into variable-free sentences often called "ground facts" in FOPC. For example, the triple ⟨Shard, location, London⟩ can be translated into the ground fact location(Shard, London). A triple such as ⟨cat, has-colour, black⟩ is more complicated, as it represents knowledge about a class of objects (cats) rather than about some specific cat. To translate this into logic we must first decide what exactly is the intended meaning. If, for example, we intend to say that all cats are black, then we could translate this into logic as the sentence:

$$\forall x.\text{cat}(x) \rightarrow \text{black}(x.)$$

---

[2] wikipedia.org/wiki/Eiffel_Tower.

[3] Recall the use of the term "predicate" in RDF triples.

This illustrates an important point, i.e., that precision does not imply correctness: the meaning of the above sentence is very precise, but it clearly isn't true that all cats are black.

As well as making the meaning of the graph precise, a rule such as (1) above can be very naturally represented as a FOPC sentence:

$$\forall x, y, z. \text{location}(x, y) \wedge \text{capital of}(y, z) \rightarrow \text{location}(x, z) \tag{2}$$

Finally, the semantics of query answering can very naturally be translated into the semantics of logical entailment; e.g., our query for the location of the shard from Fig. 3 would return London if the sentence

$$\text{location}(\text{Shard}, \text{UK}) \wedge \text{type}(\text{UK}, \text{Country})$$

is logically entailed by the combination of (the logical translation of) the knowledge graph and rules.

5   Reasoning

The above considerations suggest that, by augmenting our knowledge graph with rules and translating into FOPC we will have a knowledge representation system that is intuitive, powerful and has a precise semantics. Moreover, we can answer queries simply by computing logical entailment. We can see the OWL language from W3C as just such a system: basic facts are represented using RDF triples, complex rules can be captured by OWL axioms, and the semantics is given by a translation into logic.

Unfortunately, although this all sounds very attractive, there are some serious problems in practice, notably the well known undecidability of entailment in FOPC. Even if we restrict ourselves to some decidable subset of FOPC, such as the two variable fragment [1], entailment computation may still be highly intractable. OWL, for example, is restricted to the $\mathcal{SROIQ}$ description logic, a well known fragment of FOPC, but although entailment for this logic is known to be decidable, it is also known to have very high worst case complexity [2].

In order to address this issue, the OWL 2 standard specifies several "profiles": subsets of the language for which reasoning is known to be tractable [7]. The OWL 2 RL profile, for example, supports rules such as (2) above, while having polynomial-time data complexity for query answering. Moreover, it has been shown that, with careful engineering, it is possible to build a system that can support complex rules and large scale graphs in real applications; RDFox, for example, is a system developed at the University of Oxford that uses in-memory storage and multi-core parallelisation to provide fast processing of rules and fast query answering over very large graphs (in the order of billions of triples) [5].

RDFox *materialises* triples that are implied by rules, i.e., it adds such triples to the graph. This is a common method for supporting rules while still providing fast query answering: materialisation can be performed in a pre-processing phase, and query answering is then independent of any rules. One possible disadvantage of materialisation-based systems is that retracting even a single fact may require the whole materialisation phase to be repeated, which is slow, and could even be infeasible if the graph is regularly changing. In RDFox, however, a novel view maintenance algorithm is used to allow for fast incremental addition and retraction of triples; for example, deleting 5,000 triples and updating the materialisation takes less than 1 s [4].

6   Discussion

The need to for representing knowledge is ubiquitous in applications, and knowledge graphs are becoming increasingly popular as a knowledge representation formalism. Augmenting basic knowledge graphs with rules greatly increases their power, and aids both query answering and knowledge curation. With careful design, such systems can be translated directly into a logical language with a precise semantics, and for which query answering is still tractable. Moreover, systems such as RDFox have demonstrated that it is possible to develop practical systems that support such a formalism, and that can handle complex rules and billions of triples while still offering fast query answering.

References

1. Grädel, E., Kolaitis, P.G., Vardi, M.Y.: On the decision problem for two-variable first-order logic. B. Symb. Log. **3**(1), 53–69 (1997)
2. Horrocks, I., Kutz, O., Sattler, U.: The even more irresistible $\mathcal{SROIQ}$. In: Proceedings of the 10th International Conference on Principles of Knowledge Representation and Reasoning (KR 2006), pp. 57–67. AAAI Press (2006). download/2006/HoKS06a.pdf
3. Lehmann, F. (ed.): Semantic Networks in Artificial Intelligence. Pergamon Press, Oxford (1992)
4. Motik, B., Nenov, Y., Piro, R., Horrocks, I.: Incremental update of datalog materialisation: the backward/forward algorithm. In: Proceedings of the 29th National Conference on Artificial Intelligence (AAAI 2015), pp. 1560–1568. AAAI Press (2015). download/2015/MNPH15b.pdf
5. Motik, B., Nenov, Y., Piro, R., Horrocks, I., Olteanu, D.: Parallel materialisation of datalog programs in centralised, main-memory RDF systems. In: Proceedings of the 28th National Conference on Artificial Intelligence (AAAI 2014), pp. 129–137. AAAI Press (2014). download/2014/MNPHO14a.pdf
6. Neumann, T., Weikum, G.: The RDF-3X engine for scalable management of RDF data. VLDB J. **19**(1), 91–113 (2010). https://doi.org/10.1007/s00778-009-0165-y
7. OWL 2 Web Ontology Language Profiles, 2nd edn. W3C Recommendation, 11 December 2012. http://www.w3.org/TR/owl2-profiles/
8. Weiss, C., Karras, P., Bernstein, A.: Hexastore: sextuple indexing for semantic web data management. PVLDB **1**(1), 1008–1019 (2008). http://www.vldb.org/pvldb/1/1453965.pdf

Data-Driven Requirements Engineering. The SUPERSEDE Way

Anna Perini[(✉)] [iD]

Fondazione Bruno Kessler (FBK), Via Sommarive 18, Trento, Italy
`perini@fbk.eu`

Abstract. This keynote addresses the challenges and opportunities for today requirements engineering, which are introduced by the ever growing amount of data generated by software at use. Data analytics techniques, which exploit artificial intelligence algorithms can be used to build tools to support requirements engineers to take faster and better quality decisions.

A concrete example is the SUPERSEDE tool-suite that supports planning new software releases on the basis of the analysis of user feedback and usage data. Main open research challenges are pointed out.

Keywords: Software requirements · Software analytics ·
Data-driven requirements engineering · Software evolution

1 Introduction

Requirements Engineering (RE) is concerned with eliciting, analyzing, prioritizing, managing changes and evolution of software requirements that capture the needs of their users. Systematic methods and techniques have been defined to support those tasks, considering the peculiarities of the different software application domains, as well as fitting different software development process models.

Today trend in software development to move faster and faster, adopting a continuous deployment approach or performing Agile development, is challenging RE [14]. For instance, it becomes key to be able to continuously assess the effect of a new software release, collecting usage data and feedback related to the user's experience. The analysis of these collected data should support project managers to take informed decisions about which new features to implement first so that customers can realize the most business value. That is, being effective in prioritizing requirements becomes extremely important.

On the other side software applications, as for example in the context of the Internet of Things, exploit and generate an ever-growing amount of data. Users of such software applications are facilitated to express their opinion about their usage experience through social media as well as dedicated feedback gathering channels.

© Springer Nature Switzerland AG 2019
J. A. Lossio-Ventura et al. (Eds.): SIMBig 2018, CCIS 898, pp. 13–18, 2019.
https://doi.org/10.1007/978-3-030-11680-4_3

These trends motivated the definition of the so-called Crowd Requirements Engineering (CrowdRE) paradigm [6], which goes beyond market-driven RE, by envisioning a continuous involvement of large, heterogeneous groups of users, for eliciting, validating and prioritizing requirements, and of the more general data-driven RE paradigm [2,4,8].

In this talk we take the perspective of data-driven RE, focusing on the case of data generated by the user masses, and recall its key ingredients, namely data, Artificial intelligence (AI) and data processing techniques, and decision-making needs. We revisit recent results of the SUPERSEDE H2020 project[1], which enables a data-driven software engineering process, and conclude pointing out main open research challenges.

2   Data-Driven RE

In data-driven RE, data generated by a software system at execution time are automatically collected and analyzed with the purpose of eliciting software system requirements and of supporting RE decisions [2,4,8]. For instance, in the context of automated driving systems development, huge sets of data are generated in field tests, e.g. through telemetry. A suitable data-driven RE framework should allow to automatically collect and organize them, and enable automatic discovery of requirements for driving behavior [4].

In the context of software applications delivered through app stores an ever increasing amount of feedback is generated by user masses, which is expressed in form of natural language textual messages, emoticons or star ratings. This data are called *explicit user feedback*, to differentiate it from usage data and interaction history that are collected through monitoring mechanisms, which is called *implicit user feedback* [12].

A considerable amount of research work has been developed in the last years on automating the analysis of explicit user feedback. Natural Language Processing (NLP) techniques are exploited to filter out irrelevant parts in feedback's textual messages, and a variety of different techniques are applied to discover information expressed by users, which are relevant for the software development teams. They include linguistic techniques, which leverage rules to identify users' intentions expressed in speech-acts [11], topic modelling, sentiment analysis and Machine Learning (ML) techniques to classify user comments into bug reports, feature requests, polarity of sentiments [8].

Understanding information needs of project managers and development team towards taking informed decisions is key to drive the development of effective feedback analytics tools [2]. Examples considered so far include understanding which software application's features are perceived positively and which one negatively, or understand which the most used features are, towards taking decisions about how to evolve the application. Decide which among a set of pending issues

---

[1] SUpporting evolution and adaptation of PERsonalized Software by Exploiting contextual Data and End-user feedback, H2020 EU funded project, http://www.supersede.eu.

(including bug fixing and feature enhancements) need to be addressed first and plan for the next software release.

2.1   The SUPERSEDE Approach

In the context of the SUPERSEDE project we have developed a set of tools that can be flexibly combined to enable an iterative and incremental development process for evolving software applications, which is driven by the analysis of data collected during software usage.

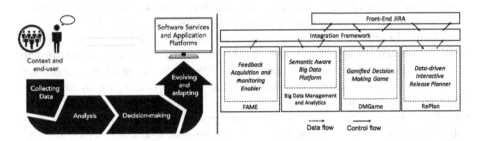

Fig. 1. Left side: the SUPERSEDE feedback loop that enable data-driven software evolution. Right side: the SUPERSEDE tool-chain components, and relative data and control flow.

Figure 1 (left side), depicts the data-driven software evolution process of SUPERSEDE where the following four main steps are performed in a loop:

- Collect: runtime and context data, together with explicit user's feedback that the user might deliver upon having used the software, are collected. The SUPERSEDE tool integrates multi-modal feedback gathering techniques, which allow users to express feedback as a combination of textual comments, emoticons, rating and pictures, with flexible and configurable monitoring components that collect data from the context and system usage [15]. The collected data are stored in a big data storage, which includes a semantic model of the software application domain at support of data analysis [13].
- Analyze: the collected data are analyzed with the purpose of supporting software evolution decisions. Different analysis techniques are provided by the SUPERSEDE tool-suite, including NLP, sentiment analysis, speech-act based analysis, and ML classification techniques, which can be combined for extracting feature requests and bug issues from user textual comments [10]. Tweet mining techniques can be used to understand quality of experience as perceived by users [7]. The tool-suite supports developers also to perform combined analyses of end-user feedback and contextual data [15].
- Decide: data analysis supports project managers and the development team to take informed decisions about how to evolve the software applications. For instance identifying new requirements and prioritizing them with respect to multiple criteria [3].

– Act: the decision taken are operationalized. Prioritized requirements and issues are included in a release plan that takes into account available resources, deadlines and organization's priorities [1].

The SUPERSEDE tool-suite components are shown in Fig. 1 (right), which depicts how they have been integrated with the JIRA issue tracking system[2]. The tool-suite components are available in Github[3]. A methodology is provided for tailoring the SUPERSEDE tool-supported software evolution method to a specific project situation [5]. This methodology builds on Situational Method Engineering, an engineering approach where a method can be described as a composition of reusable components called method chunks, which can be selected to fit specific project factors, such as project size, scope, privacy issues.

Among the different scenarios considered for the validation of the SUPER-SEDE approach, those described in [9,15] are worth to be mentioned. In the first scenario, the development team would like to get support in prioritizing existing requirements taking into account end-user's opinions, which have been expressed in a large amount of user messages. The problem to be solved can be formulated as a multi-criteria requirements prioritization problem, where user-value is computed by automatic analysis of the user feedbacks, which refer to the requirements to be prioritized. Topic modelling techniques are used to identify those feedback messages that refer to specific requirements from the large set of messages. Speech-act based analysis and sentiment analysis are applied to derive the user's preference for a given requirement. Preferences on requirements ranking associated to other criteria, such as business value and technical complexity, are elicited from human stakeholders in a collaborative process which can exploits genetic algorithms to compute optimal requirements prioritization solutions [9].

In the second scenario, the development team would like to get support in estimating how many users might be affected by a feature request or problem that they identified when performing manual analysis of about thirty user feedback messages, which were collected over a period of four months. Automated analysis of usage data from more than five thousands of users, which were collected in the same time period via the SUPERSEDE monitoring tool, can provide evidences to the development team that convince them to change their initial decision about which features to improve, or to consolidate it [15]. For instance, the analysis of monitored data helps reveal features, among those for which some users requested enhancements, which were indeed poorly used, and thus deciding to improve them will have limited impact.

3   Conclusion and Open Challenges

Data-driven RE has been proposed to fit with nowadays fast and continuous software deployment model. It builds on three key ingredients, namely data

---

[2] https://www.atlassian.com/software/jira.

[3] https://github.com/supersede-project.

generated by executing software, AI and data management techniques and RE decision-making problems. The SUPERSEDE approach provides a concrete solution for data-driven requirements prioritization and release planning in a context of continuous software evolution.

In spite of the great attention data-driven RE is receiving from the research community, relevant aspects need to be further investigated. For instance, focusing on data, how to automatically integrate data from different sources is still a challenge, as well as motivating users to provide feedback, taking into account their diversity, and how to combine push and pull mechanisms to improve the feeback collection process. More generally, data-driven RE should provide requirements engineers an easy way to answer the question "what data to collect and how much?" depending on the nature and size of a given project. For the case of data generated by users, both data collection and analysis methods that are privacy-aware and that enable transparency, should be defined. Finally, considering the ultimate objective of data-driven RE, that is to enable informed decision-making, research should adopt a decision-oriented, rather than an data analytics oriented perspective in defining new methods and tool.

Acknowledgement. This keynote leverages on results from the SUPERSEDE project, funded by the H2020 EU Framework Programme under agreement number 644018. I'd like to thank the SimBIG 2018 program co-chairs for their invitation to give this keynote, and Universidad del Pacífico for supporting my participation to the conference.

References

1. Ameller, D., Farré, C., Franch, X., Cassarino, A., Valerio, D., Elvassore, V.: Replan: a release planning tool. In: 2017 IEEE 24th International Conference on Software Analysis, Evolution and Reengineering (SANER), pp. 516–520. IEEE (2017)
2. Buse, R.P., Zimmermann, T.: Information needs for software development analytics. In: Proceedings of the 34th International Conference on Software Engineering, pp. 987–996. IEEE Press (2012)
3. Busetta, P., Kifetew, F.M., Munante, D., Perini, A., Siena, A., Susi, A.: Tool-supported collaborative requirements prioritisation. In: 2017 IEEE 41st Annual Computer Software and Applications Conference (COMPSAC), vol. 1, pp. 180–189. IEEE (2017)
4. Czarnecki, K.: Requirements engineering in the age of societal-scale cyber-physical systems: the case of automated driving. In: IEEE 26th International RE Conference, Banff, Alberta, Canada, 20–24 August 2018, pp. 3–4 (2018)
5. Franch, X., et al.: A situational approach for the definition and tailoring of a data-driven software evolution method. In: 30th International Conference on Advanced Information Systems Engineering, CAiSE 2018, Proceedings, Tallinn, Estonia, 11–15 June 2018, pp. 603–618 (2018). https://doi.org/10.1007/978-3-319-91563-0_37
6. Groen, E.C., et al.: The crowd in requirements engineering: the landscape and challenges. IEEE Softw. **34**(2), 44–52 (2017). https://doi.org/10.1109/MS.2017.33
7. Guzman, E., Alkadhi, R., Seyff, N.: An exploratory study of Twitter messages about software applications. Requirements Eng. **22**(3), 387–412 (2017)

8. Maalej, W., Nayebi, M., Johann, T., Ruhe, G.: Toward data-driven requirements engineering. IEEE Softw. **33**(1), 48–54 (2016). https://doi.org/10.1109/MS.2015. 153

9. Morales-Ramirez, I., Munante, D., Kifetew, F., Perini, A., Susi, A., Siena, A.: Exploiting user feedback in tool-supported multi-criteria requirements prioritization. In: 2017 IEEE 25th International Requirements Engineering Conference (RE), pp. 424–429, September 2017. https://doi.org/10.1109/RE.2017.41

10. Morales-Ramirez, I., Kifetew, F.M., Perini, A.: Analysis of online discussions in support of requirements discovery. In: Dubois, E., Pohl, K. (eds.) CAiSE 2017. LNCS, vol. 10253, pp. 159–174. Springer, Cham (2017). https://doi.org/10.1007/ 978-3-319-59536-8_11

11. Morales-Ramirez, I., Kifetew, F.M., Perini, A.: Speech-acts based analysis for requirements discovery from online discussions. Inf. Syst. (2018). https://doi.org/10.1016/j.is.2018.08.003, http://www.sciencedirect.com/science/ article/pii/S0306437917306087

12. Morales-Ramirez, I., Perini, A., Guizzardi, R.S.S.: An ontology of online user feedback in software engineering. Appl. Ontol. **10**(3–4), 297–330 (2015). https://doi. org/10.3233/AO-150150

13. Nadal, S., et al.: A software reference architecture for semantic-aware big data systems. Inf. Softw. Technol. **90**, 75–92 (2017)

14. Niu, N., Brinkkemper, S., Franch, X., Partanen, J., Savolainen, J.: Requirements engineering and continuous deployment. IEEE Softw. **35**(2), 86–90 (2018). https:// doi.org/10.1109/MS.2018.1661332

15. Oriol, M., et al.: FAME: supporting continuous requirements elicitation by combining user feedback and monitoring. In: IEEE 26th International RE Conference, Banff, Alberta, Canada, 20–24 August 2018, pp. 217–227 (2018)

Word Embeddings and Deep Learning for Spanish Twitter Sentiment Analysis

José Ochoa-Luna[1]([⊠])(iD) and Disraeli Ari[2]

[1] Department of Computer Science,
Universidad Católica San Pablo, Arequipa, Peru
jeochoa@ucsp.edu.pe
[2] Universidad Nacional de San Agustín, Arequipa, Peru
darim@unsa.edu.pe

Abstract. Spanish is the third language most used on the internet. However, Natural Language Processing research in this language is still far below the level of other languages like English. The aim of this paper is to fill this gap in the literature and to provide a comprehensive assessment of Deep Learning applied to Spanish sentiment analysis. We focus on the polarity detection task which, in the context of Spanish Twitter messages, remains as a challenging task. To do so, we explore the combination of several Word representations (Word2Vec, Glove, Fastext) and Deep Neural Networks models. Unlike poor performance obtained by previous related work using Deep Learning for Spanish sentiment analysis, we show promising results. Our best setting combines three word embeddings representations, Convolutional Neural Networks and Recurrent Neural Networks. This setup allows us to obtain state-of-the-art results on the TASS/SEPLN 2017 Spanish Twitter benchmark dataset, in terms of accuracy and macro F1-measure.

Keywords: Spanish sentiment analysis · Deep learning · Word embeddings

1 Introduction

Online reviews are ubiquitous. On one hand, we have a wide variety of products and services being created every day. On the other hand, customers who have either purchased products or contracted services and, ultimately, left comments in social media. With the rapid growth of Twitter, Facebook, and online review sites, sentiment analysis draws growing attention from both research and industry communities [19].

Sentiment analysis, in its basic task called polarity detection, allows us to perform an automated analysis of millions of reviews and determine whether a given opinion is positive, negative or neutral. This area has been widely researched since 2002 [17]. In fact, it is one of the most active research areas in natural language processing, data mining and social media analytics [29].

© Springer Nature Switzerland AG 2019
J. A. Lossio-Ventura et al. (Eds.): SIMBig 2018, CCIS 898, pp. 19–31, 2019.
https://doi.org/10.1007/978-3-030-11680-4_4

Polarity detection has been addressed as a text classification problem thus, can be approached by supervised and unsupervised learning methods [31]. In the unsupervised approach, a vocabulary of positive and negative words is constructed so as to polarity is inferred according the similarity between vocabulary and opinionated words. The second approach is based on machine learning. Training data and labelled reviews are used to define a classifier [17]. This last approach relies heavily on feature engineering. However, recent learning representation paradigms perform this tasks automatically [16]. In this context, Machine Learning has recently become the dominant approach for sentiment analysis, due to availability of data, better models and hardware resources [30].

In this paper we adopt a Deep Learning approach for sentiment analysis. In particular, we aim at performing automated classification of short texts in sentences and Twitter messages for the Spanish language. This is challenging because of the limited contextual information that they normally contain.

In our proposed approach, sentence words are mapped to word representations. Three kinds of word representations (Word2vec [20], Glove [26], Fastext [4]) are used in our setting. This combination, which is novel for Spanish sentiment analysis, can be useful in several domains. Overall, our goal is to provide a general setup that can be applied with less effort in several contexts.

The Deep Learning architecture proposed is composed by a Convolutional Neural Network [15], a Recurrent Neural Network [13] and a final dense layer. In order to avoid overfitting, besides traditional dropout schemes, we propose a novel data augmentation approach. Data augmentation is useful for low resources languages such as Spanish—Specially for Spanish sentiment analysis.

Those design choices allow us to obtain state-of-the-art results, in terms of accuracy and macro F1 measure, on the InterTASS 2017 dataset. This dataset was proposed in the TASS workshop at SEPLN. In the last six years, this workshop has been the main source for Spanish sentiment analysis datasets and proposals [18].

The remainder of the paper is organized as follows. Related work is presented in Sect. 2. Background concepts are described in Sect. 3. Our proposal is presented in Sect. 4. Results are described in Sect. 5. Finally, Sect. 6 concludes the paper.

2   Related Work

There is a plethora of related works for sentiment analysis but, we are only interested in contributions for the Spanish language. Arguably, one of the most complete Spanish sentiment analysis systems was proposed by Brooke et al. [6], which had a linguistic approach. That approach integrated linguistic resources in a model to decide about polarity opinions [31]. Recent successful approaches for Spanish polarity classification have been mostly based on machine learning [10].

In the last six years, the TASS at SEPLN Workshop has been the main source for Spanish sentiment analysis datasets and proposals [11,18]. Benchmarks for both the polarity detection task and aspect-based sentiment analysis task have been proposed in several editions of this Workshop. Spanish Tweets have been emphasized.

Nowadays deep learning approaches emerge as powerful computational models that discover intricate semantic representations of texts automatically from data without feature engineering. These approaches have improved the state-of-the-art in many sentiment analysis tasks including sentiment classification of sentences/documents, sentiment extraction and sentiment lexicon learning [29]. However, these results have been mostly obtained for English Language. Since our proposal is based on Deep Learning, the related work that follows emphasizes these kinds of algorithms.

Arguably, the first approach using Deep Learning techniques for Spanish Sentiment Analysis was proposed in the TASS at SEPLN workshop in 2015 [32]. The authors presented one architecture that was composed by a RNN layer (LSTMs cells), a dense layer and a Sigmoid function as output. The performance over the general dataset was poor, 0.60 in terms of accuracy (the best result was 0.69 in TASS 2015).

The first Convolutional Neural Network approach for Spanish Sentiment Analysis was described by Segura-Bedmar et al. [28]. The CNN model proposed for sentiment analysis was mostly based on Kim's work [15]. It was comprised by only a single convolutional layer, followed by a max-pooling layer and a Softmax classifier as the final layer. Word embeddings were used in three ways: a learned word embedding from scratch and two pre-trained word2vec models. In terms of accuracy they obtained 0.64, which was far from the best result (0.72 was the best result in TASS 2016 [11]).

Another CNN approach for Spanish Sentiment Analysis was presented by Paredes et al. [25]. First, a preprocessing step (tokenization and normalization) was performed which was followed by a Word2vec embedding. Then, this model was comprised of a 2D convolutional layer, a max pooling and a final Softmax layer, i.e., it was also similar to Kim's work [15]. It was reported a F-measure of 0.887 over a non-public Twitter corpus of 10000 tweets.

Most of the Deep Learning approaches for Spanish sentiment analysis have been presented in TASS 2017 [18]. For instance, Rosa et al. [27] used word embeddings within two approaches: SVM (with manually crafted features) and Convolutional Neural Networks. Pre-trained Word2vec, Glove and Fastext embeddings were used. Unlike our approach, these embeddings were used separately. In fact, the best results of this paper were obtained using Word2vec. When CNN was employed, unidimensional convolutions were performed. While several convolutional layers were tested, the best model had three convolutional layers, using 2, 3 and 4 word filters. However, their best results were obtained when combined SVM with CNN. They simply used a decision rule based on both probability results. Interesting results, in terms of accuracy, were obtained. It was reported a 0.596 value for the InterTASS dataset (the best accuracy result was 0.608 for TASS 2017 [18]).

Garcia-Vega et al. [12] used word embeddings with shallow classifiers. In addition, they also tested recurrent neural networks with LSTM nodes and a dense layer. Two kinds of experiments were performed using word embeddings and TFIDF values as inputs. Both experiments obtained poor results (0.333 and 0.404 in terms of accuracy for the 2017 InterTASS dataset).

Araque et al. [1] explored recurrent neural networks in two ways (i) a set of LSTM cells whose input were word embeddings, (ii) a combination of input word vector and polarity values obtained from a sentiment lexicon. As usual, a last dense layer with a Softmax function was used as final output. While interesting, experimental results showed that the best performance was obtained by the second model, i.e., LSTM + Lexicon + dense layer. They obtained 0.562 for 2017 IntertTASS dataset, in terms of accuracy. This value was far from the top results.

In the last years, the best results were obtained for the ELiRF group [14]. In TASS 2017, they obtained the second best result for the InterTSS task, 0.607, in terms of accuracy (The first place presented an ensemble approach [7]). It is worth noting that ELiRF best results were obtained using a Multilayer perceptron (MLP) with word embeddings as inputs. This MLP had two layers with ReLu activation functions. A Second approach used a stack of CNN and LSTM models with pre-trained word embeddings. The architecture was composed by one convolutional layer, 64 LSTM cells and a fully connected MLP layer. This last architecture had a poor performance (0.436 in terms of Accuracy).

3 Background

3.1 Sentiment Analysis

Sentiment analysis (also known as opinion mining) is an active research area in natural language processing [30]. Sentiment classification is a fundamental and extensively studied area in sentiment analysis. It targets at determining the sentiment polarity (positive or negative) of a sentence (or a document) based on its textual content [29]. Polarity classification tasks have usually based on two main approaches [5]: a supervised approach, which applies machine learning algorithms in order to train a polarity classifier using a labelled corpus; an unsupervised approach, semantic lexicon-based, which integrates linguistic resources in a model in order to identify the polarity of the opinions.

Since the performance of a machine learner heavily depends on the choices of data representation, many studies devote to building powerful feature extractor with domain expert and careful engineering [22].

As stated by Liu [17], sentiment analysis has been researched at three levels:

- Document level: The task at this level is to classify whether a whole opinion document expresses a positive or negative sentiment [24]
- Sentence level: The task at this level goes to the sentences and determines whether each sentence expressed a positive, negative, or neutral opinion. Neutral usually means no opinion.

– Entity and Aspect level [3]: Both the document level and the sentence level analyses do not discover what exactly people liked and did not like. Aspect level performs finer-grained analysis.

3.2 Deep Neural Networks

Several deep neural network approaches have been successfully applied to sentiment analysis in the last years [33]. However, these results have been mostly obtained for English Language [18]. In this section we only focus on word representations, Convolutional Neural Networks (CNNs) and Recurrent Neural Networks (RNNs). They are the main building blocks of our proposal.

Word Representations (Word2vec, Glove, Fastext). Nowadays, word representations are paramount for sentiment analysis [33]. In order to model text words as features within a machine learning framework, a common approach is to encode words as discrete atomic symbols. These encodings are arbitrary and provide no useful information to the system regarding the relationships that may exist between the individual symbols [30].

The discrete representation has some problems such as missing new words. This representation also requires human labor to create and adapt. It is also hard to compute accurate word similarity and is quite subjective. To cope with these problems, the distributional similarity based representations propose to represent a word by means of its neighbors, its context [29].

Word2vec [20] is a particularly computationally-efficient predictive model for learning word embeddings from raw text. It takes a vector with several hundred dimensions where each word is represented by a distribution of weights across those elements [2,8]. Thus, instead of a one-to-one mapping between an element in the vector and a word, the representation of a word is spread across all the elements in the vector. In addition, each element in the vector contributes to the definition of many words. Such a vector comes to represent in some abstract way the "meaning" of a word. And simply by examining a large corpus it is possible to learn word vectors that are able to capture the relationships between words in a surprisingly expressive way.

Unlike Word2vec, Glove [26] seeks to make explicit what Word2vec does implicitly: encoding meaning as vector offsets in an embedding space. In Glove, it is stated that the ratio of the co-occurrence probabilities of two words (rather than their co-occurrence probabilities themselves) is what contains information and so look to encode this information as vector differences.

Instead of directly learning a vector representation for a word, Fastext [4] learns a representation for each character n-gram. In this sense, each word is represented as a bag of characters n-grams. Thus, the overall word embedding is a sum of these characters n-grams. The advantage of Fastext is that generates better embeddings for rare and out-of-corpus words. By using different n-grams, Fastext explores key structural components of words.

Convolutional Neural Networks. While Convolutional Neural Networks (CNN) have been primarily applied to image processing, they have also been used for NLP tasks [15].

In the image context [16], given a raw input (2D arrays of pixel intensities), several *convolutional* layers allow us to capture features images at several abstraction levels. In this context, a discrete convolution takes a filter matrix and multiply its values element-wise with the original matrix, then sum them up. To get the full convolution we do this for each element by sliding the filter over the whole matrix.

The convolved map feature denotes a level of abstraction obtained after the convolution operations (there are also ReLU activation, Pooling and Softmax layers). CNN exploits the property that many natural signals are compositional hierarchies: higher-level features are obtained by composing lower-level ones. In images, local combinations of edges form motifs, motifs assemble into parts, and parts from objects [16]. All this learning representation is performed in an unsupervised manner. The amount of filters and convolutional layers denote how rich features and abstraction levels we wish to obtain from images.

Conversely, if we wish to apply CNNs in natural language tasks several changes are needed [15]. Text data is tokenized and must be encoded as numbers (input numerical variables are usual in neural networks algorithms). In the last five years, word embeddings representations (but also character and paragraph) have been preferred. This is due to semantical/syntactical similarity is better expressed in a distributed manner [20].

A sentence can be represented as a matrix. The sentence length denotes the number of rows and the word embedding dimension denotes the number of columns. This allows us to perform discrete convolutions as in the image case (2D input matrix). However, one must be careful when defining filter sizes which usually have the same width as word embeddings [15].

Instead of working with 2D representation, we may also work with 1D representation, i.e., to concatenate several word embeddings in a long vector and then apply several convolutional layers.

Recurrent Neural Networks. Recurrent Neural Networks (RNN) [9] are a kind of neural network that makes it possible to model long-distance dependencies among variables. Therefore, RNNs are best suited for tasks that involve sequential inputs, such as speech and language [16]. RNNs process an input sequence one element at a time, maintaining in their hidden units a state vector that implicitly contains information about the history of all the past elements of the sequence. To do so, a connection is added that references the previously hidden states h_{t-1} when computing hidden state h, formally [23]:

$$h_t = tanh(W_{xh}x_t + W_{hh}h_{t-1} + b_h)$$

$h_t = 0$ when the initial step is $t = 0$. The only difference from the hidden layer in a standard neural network is the addition of the connection $W_{hh}h_{t-1}$ from the hidden state at time step $t - 1$ connecting to that at time step t. Since this is a recursive equation that uses h_{t-1} from the previous time step.

In the context of Sentiment Analysis, an opinionated sentence is a sequence of words. Thus, RNNs are suitable for modeling this input sequence [13]. Similar to CNNs, the input is given as words (character) embeddings which can be learned during training or may also be pre-trained (Glove, Word2vec, Fastext).

Each word is mapped to a word embedding which is the input at every time step of the RNN. The maximum sequence length denotes the length of the recurrent neural network. Each hidden state models the dependence among a current word and all the precedent words. Usually the final hidden state, which ideally denotes all the encoded sentence, is connected to a dense layer so as to perform sentiment classification [13].

RNNs are very powerful dynamic systems, but training them has proved to be problematic. The backpropagated gradients either grow or shrink at each time step. Thus, over many time steps they typical explode or vanish. A sequence of words comprise a sequence of RNNs cells. These cells can have some gate mechanism in order to avoid gradient vanishing longer sequences. In this setting Long Short Term Memory Cells (LSTM) or Gated Recurrent Units (GRU) are common choices [23].

4 Proposal

The aim of this paper is to explore several Deep Learning algorithms possibilities in order to perform sentiment analysis. The focus is to tackle the polarity detection task for Spanish Tweets. In this sense, some models were tested. Details of these experiments are given in Sect. 5.

In this section, we present our best pipeline for Spanish sentiment analysis given short texts. Basically, it is composed by Word embeddings, CNN and RNN models. The pipeline is showed in Figure 1. A concise description is given as follows.

- Basic pre-processing is performed as the focus is given to data augmentation.
- The input is a sequence of words—a short opinionated sentence. These words are mapped to three pre-trained Spanish word embeddings (Word2vec, Glove, Fastext).
- The three channels are the input to a 3D Convolutional Neural Network. After several convolutional and max pooling layers we obtain a feature vector of a given length.
- The feature vector obtained from the CNN is mapped to a sequence and passed to a RNN. It is a simple RNN model, with LSTM cells.
- The final hidden state of the RNN is completely connected to a dense layer to train a classifier.

Further details about these design choices are given as follows.

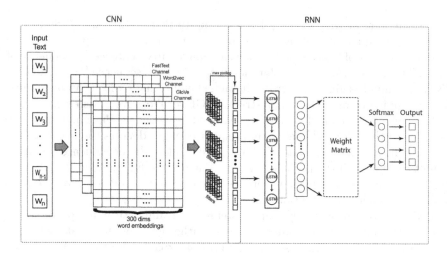

Fig. 1. Pipeline of our proposal: Word Embeddings+CNN+RNN.

4.1 Data Augmentation

In general a few pre-processing steps are performed over raw data. Since we have few training examples in Spanish and Deep Learning techniques are susceptible to overfitting, we would rather focus on data augmentation. We propose a novel approach for data augmentation. Basically, we identify nouns, adjectives and verbs on sentences by performing Part-Of-Speech tagging[1]. By doing so, we emphasize tokens that are prone to be opinionated words. Then, more examples are created by combining bigrams and trigrams from the former tokens. In addition, we augment data based on word synonyms [33]. Opinionated words are replaced by synonyms. Overall, this process allowed us to obtain better generalization results.

4.2 Word Embeddings Choice

One of the main contributions of this paper was to find the best word embedding setting. We have trained Word2vec and Glove embeddings on Spanish corpus. Moreover, we have used a pre-trained Fastext embedding. At the end, empirical tests allowed us to decide for using these three mappings as channels in our CNN building block. In the context of Spanish sentiment analysis, none of the previous works had used three embedding channels in CNNs before. We argue that by using Word2vec, Glove and Fastext at the same time, we were able to take advantage of several "word meanings". These word meanings are accordingly linked to word context, co-occurrence probabilities and rare words representations.

---

[1] The following tool was used to perform POS tagging: http://www.cis.uni-muenchen. de/~schmid/tools/TreeTagger/.

4.3   CNN Architecture

Our CNN architecture is based on Kim's work [15]. Since three word embeddings are used, the first convolutional layer receives a 3D input. Each word embedding has 300 dimensions. Filters have the same width as embeddings dimension and we perform convolutions from 1 to 5 words. The number of filters is 300. The pooling layer allows us to control the desired feature vector obtained.

Some other hyper paremeters are given as follows:

- batch size: 32
- dropout keep probability: 0.5
- filter sizes: 3,4,5
- hidden unit: 300
- maximum pooling size: 4
- number of epochs: 10

4.4   RNN Architecture

The RNN receives a CNN vector as input and LSTMs cells are defined accordingly. There is one layer of LSTM cells, these cells have 300 hidden units. The last hidden state is fully connected to a dense layer which allows us to define a classifier [13].

5   Experiments

Experiments were performed using Deep Learning algorithms. CNNs and RNNs were tested separately. Our best result was obtained by composing word embeddings, CNNs and RNNs. We first describe the benchmark dataset used. Then, accuracy results are showed.

5.1   Dataset

The dataset used to perform comparisons was InterTASS which is a collection of Tweets written in Spanish. It is composed of more than 3,000 tweets annotated at four opinion intensity level (positive, neutral, negative and none). We only focus on task 1. This task aims to evaluate polarity classification at tweet level. This shared task has been proposed at the TASS SEPLN workshop in 2017 [18]. We have used this dataset since it is the most recent benchmark that allows us to compare among Deep Learning approaches for Spanish sentiment analysis. The dataset is further detailed in Table 1.

Table 1. InterTASS dataset (TASS 2017)

Corpus	Tweets
Training	1,008
Development	506
Test	1,899
Total	3,413

5.2   Results

We have implemented several deep neural networks models, and the dataset InterTASS 2017 was used for training. For this implementation we use Tensorflow[2]. In order to find the best hyper parameters, we have used a ten-fold cross validation process. The test set has only been used to report results. In Table 2 we report results in terms of accuracy.

A first attempt was to test several RNNs models (many-to-one architecture , single layer, multilayer, bidirectional). The reported model, RNN in Table 2, has a many-to-one architecture. The input is a sequence of words and the output is the resulting polarity. There is only a hidden layer, and the input is a pre-trained sequence of Word2vec embeddings.

A second attempt was to test several CNN models, i.e., 1D CNN, 2D CNN and 3D CNNs, until 4 convolutional/pooling layers. The reported model, CNN in Table 2, is a 3D CNN. Thus, the input received three channels of pre-trained word embeddings. It had only three layers: a convolutional, a pooling and a dense layer.

It is worth noting that our best result was obtained by the model described in Sect. 3 (CNN+RNN in Table 2). This is a combination of a 3D CNN and a many-to-one RNN. A 3D CNN architecture whose outputs where mapped to a sequence of LSTM cells. Our data augmentation scheme was also used in order to avoid overfitting.

Table 2. Deep learning approaches results on InterTASS dataset (TASS 2017)

Our DL attempts	Accuracy
CNN+RNN	**0.609**
CNN	0.5552
RNN	0.4972

In Table 3, we compare, in terms of accuracy and Macro F1 measure, our best model with the state-of-the-art InterTASS 2017 results. It is worth noting that

---

[2] https://www.tensorflow.org/.

Table 3. State-of-the-art results on InterTASS dataset (TASS 2017)

System	Accuracy	Macro F1
CNN+RNN (our approach)	**0.609**	**0.551**
jacerong-run1 [7]	0.608	0.459
ELiRF-UPV-run1 [14]	0.607	0.493
RETUYT-svm cnn [27]	0.596	0.471
tecnolenguasent [21]	0.595	0.456

our approach outperforms all the other approaches. In addition, our proposal is the only top result using a Deep Learning approach.

6 Conclusion

Despite being one of the three most used languages at Internet, Spanish has had few resources developed for natural language processing tasks. Unlike English sentiment analysis, Deep Learning approaches were unable to obtain state-of-the-art results on Spanish benchmark datasets in the past. The aim of this work was to provide empirical evidence on the performance of Deep Learning algorithms for Spanish Twitter sentiment analysis. Thus, we have showed that a combination of data augmentation, at least three kinds of word embeddings, a 3D Convolutional Neural Network, followed by a Recurrent Neural Network allows us to obtain state-of-the-art results, in terms of accuracy and macro F1 measure on the InterTASS 2017 benchmark. In addition, this setup can be easily adapted to other domains.

References

1. Araque, O., Barbado, R., Sanchez-Rada, J.F., Iglesias, C.A.: Applying recurrent neural networks to sentiment analysis of Spanish tweets. In: Proceedings of TASS 2017: Workshop on Sentiment Analysis at SEPLN, pp. 71–76 (2017)
2. Bengio, Y., Ducharme, R., Vincent, P., Janvin, C.: A neural probabilistic language model. J. Mach. Learn. Res. **3**, 1137–1155 (2003). http://dl.acm.org/citation.cfm?id=944919.944966
3. Blair-goldensohn, S., Neylon, T., Hannan, K., Reis, G.A., Mcdonald, R., Reynar, J.: Building a sentiment summarizer for local service reviews. In: NLP in the Information Explosion Era (2008)
4. Bojanowski, P., Grave, E., Joulin, A., Mikolov, T.: Enriching word vectors with subword information. Trans. Assoc. Comput. Linguist. **5**, 135–146 (2017)
5. Brody, S., Elhadad, N.: An unsupervised aspect-sentiment model for online reviews. In: Human Language Technologies: The 2010 Annual Conference of the North American Chapter of the Association for Computational Linguistics, HLT 2010, pp. 804–812. Association for Computational Linguistics, Stroudsburg (2010). http://dl.acm.org/citation.cfm?id=1857999.1858121

6. Brooke, J., Tofiloski, M., Taboada, M.: Cross-linguistic sentiment analysis: from English to Spanish. In: Proceedings of RANLP 2009, pp. 50–54 (2009)

7. Ceron-Guzman, J.A.: Classier ensembles that push the state-of-the-art in sentiment analysis of Spanish tweets. In: Proceedings of TASS 2017: Workshop on Sentiment Analysis at SEPLN, pp. 59–64 (2017)

8. Collobert, R., Weston, J., Bottou, L., Karlen, M., Kavukcuoglu, K., Kuksa, P.P.: Natural language processing (almost) from scratch. CoRR abs/1103.0398 (2011). http://arxiv.org/abs/1103.0398

9. Elman, J.L.: Finding structure in time. Cogn. Sci. **14**(2), 179–211 (1990)

10. Garcia, M., Martinez, E., Villena, J., Garcia, J.: TASS 2015 - the evolution of the spanish opinion mining systems. Procesamiento de Lenguaje Natural **56**, 33–40 (2016)

11. Garcia-Cumbreras, M.A., Villena-Roman, J., Martinez-Camara, E., Diaz-Galiano, M., Martin-Valdivia, T., Ureña Lopez, A.: Overview of TASS 2016. In: Proceedings of TASS 2016: Workshop on Sentiment Analysis at SEPLN, pp. 13–21 (2016)

12. Garcia-Vega, M., Montejo-Raez, A., Diaz-Galiano, M.C., Jimenez-Zafra, S.M.: Sinai in TASS 2017: tweet polarity classification integrating user information. In: Proceedings of TASS 2017: Workshop on Sentiment Analysis at SEPLN, pp. 91–96 (2017)

13. Graves, A.: Supervised Sequence Labelling with Recurrent Neural Networks. Studies in Computational Intelligence, vol. 385. Springer, Berlin (2012). https://doi.org/10.1007/978-3-642-24797-2. https://cds.cern.ch/record/1503877

14. Hurtado, L.F., Pla, F., Gonzalez, J.A.: ELiRF-UPV at TASS 2017: sentiment analysis in twitter based on deep learning. In: Proceedings of TASS 2017: Workshop on Sentiment Analysis at SEPLN, pp. 29–34 (2017)

15. Kim, Y.: Convolutional neural networks for sentence classification. In: Proceedings of the 2014 Conference on Empirical Methods in Natural Language Processing, EMNLP 2014, Doha, Qatar, 25–29 October 2014, A meeting of SIGDAT, a Special Interest Group of the ACL, pp. 1746–1751 (2014). http://aclweb.org/anthology/D/D14/D14-1181.pdf

16. LeCun, Y., Bengio, Y., Hinton, G.: Deep learning. Nature **521**(7553), 436–444 (2015)

17. Liu, B.: Sentiment Analysis and Opinion Mining. Morgan and Claypool Publishers (2012)

18. Martinez-Camara, E., Diaz-Galiano, M., Garcia-Cumbreras, M.A., Garcia-Vega, M., Villena-Roman, J.: Overview of TASS 2017. In: Proceedings of TASS 2017: Workshop on Sentiment Analysis at SEPLN, pp. 13–21 (2017)

19. McGlohon, M., Glance, N., Reiter, Z.: Star quality: Aggregating reviews to rank products and merchants. In: Proceedings of Fourth International Conference on Weblogs and Social Media (ICWSM) (2010)

20. Mikolov, T., Sutskever, I., Chen, K., Corrado, G.S., Dean, J.: Distributed representations of words and phrases and their compositionality. In: Burges, C.J.C., Bottou, L., Welling, M., Ghahramani, Z., Weinberger, K.Q. (eds.) Advances in Neural Information Processing Systems 26, pp. 3111–3119. Curran Associates, Inc. (2013). http://papers.nips.cc/paper/5021-distributed-representations-of-words-and-phrases-and-their-compositionality.pdf

21. Moreno-Ortiz, A., Perez-Hernendez, C.: Tecnolengua Lingmotif at TASS 2017: Spanish twitter dataset classification combining wide-coverage lexical resources and text features. In: Proceedings of TASS 2017: Workshop on Sentiment Analysis at SEPLN, pp. 35–42 (2017)

22. Narayanan, V., Arora, I., Bhatia, A.: Fast and accurate sentiment classification using an enhanced naive bayes model. In: Yin, H., et al. (eds.) IDEAL 2013. LNCS, vol. 8206, pp. 194–201. Springer, Heidelberg (2013). https://doi.org/10.1007/978-3-642-41278-3_24

23. Neubig, G.: Neural machine translation and sequence-to-sequence models: A tutorial. CoRR abs/1703.01619 (2017). http://arxiv.org/abs/1703.01619

24. Pang, B., Lee, L.: Opinion mining and sentiment analysis. Found. Trends Inf. Retr. 2(1–2), 1–135 (2008). https://doi.org/10.1561/1500000011

25. Paredes-Valverde, M.A., Colomo-Palacios, R., Salas-Zarate, M.D.P., Valencia-Garcia, R.: Sentiment analysis in Spanish for improvement of products and services: a deep learning approach. Sci. Program. 6, 1–6 (2017)

26. Pennington, J., Socher, R., Manning, C.D.: Glove: global vectors for word representation. In: Empirical Methods in Natural Language Processing (EMNLP), pp. 1532–1543 (2014). http://www.aclweb.org/anthology/D14-1162

27. Rosa, A., Chiruzzo, L., Etcheverry, M., Castro, S.: RETUYT in TASS 2017: sentiment analysis for Spanish tweets using SVM and CNN. In: Proceedings of TASS 2017: Workshop on Sentiment Analysis at SEPLN, pp. 77–83 (2017)

28. Segura-Bedmar, I., Quiros, A., Martínez, P.: Exploring convolutional neural networks for sentiment analysis of Spanish tweets. In: Proceedings of the 15th Conference of the European Chapter of the Association for Computational Linguistics: vol. 1, Long Papers, pp. 1014–1022. Association for Computational Linguistics (2017). http://aclweb.org/anthology/E17-1095

29. Tang, D., Wei, F., Qin, B., Yang, N., Liu, T., Zhou, M.: Sentiment embeddings with applications to sentiment analysis. IEEE Trans. Knowl. Data Eng. 28(2), 496–509 (2016)

30. Tang, D., Qin, B., Liu, T.: Deep learning for sentiment analysis: successful approaches and future challenges. Wiley Interdisc. Rev.: Data Mining Knowl. Discov. 5(6), 292–303 (2015)

31. Turney, P.D.: Thumbs up or thumbs down?: semantic orientation applied to unsupervised classification of reviews. In: Proceedings of the 40th Annual Meeting on Association for Computational Linguistics, ACL 2002, pp. 417–424. Association for Computational Linguistics, Stroudsburg (2002). https://doi.org/10.3115/1073083.1073153

32. Vilares, D., Doval, Y., Alonso, M.A., Gomez-Rodriguez, C.: LyS at TASS 2015: deep learning experiments for sentiment analysis on Spanish tweets. In: Proceedings of TASS 2015: Workshop on Sentiment Analysis at SEPLN, pp. 47–52 (2015)

33. Zhang, L., Wang, S., Liu, B.: Deep learning for sentiment analysis: a survey. CoRR abs/1801.07883 (2018). http://arxiv.org/abs/1801.07883

Twitter Event Detection in a City

Martín Steglich$^{(\boxtimes)}$, Raúl Speroni, and Juan José Prada

Facultad de Ingeniería, Universidad de la República, Montevideo, Uruguay
msteglichc@gmail.com, raulsperoni@gmail.com, prada@fing.edu.uy

Abstract. Large cities and metropolitan areas are complex systems with connections between their environments and individuals. Citizens express themselves daily about events related to the city on the Internet. This information has great value due to its freshness, diversity of points of view and impact on public opinion.

Information technologies allow us to imagine other types of interfaces for communication between people and institutions. Interfaces capable of extracting useful information even if it is not directed to the corresponding institutions.

In this work a framework that combines different techniques for the events extraction in a city from social networks is built. Using the city of Montevideo as a case study and its waste management as a domain, it was possible to correctly identify 94% of the events reported with only 4% false positives.

Keywords: Smart city · Event extraction · Event detection · Social networks · Twitter · Natural language processing · Machine learning

1 Introduction

1.1 Motivation

We are currently facing two social phenomena relevant to the history of humanity: the acceleration of urbanization and the digital revolution. Particularly in Uruguay, access to the Internet from mobile devices and their use to consume and disseminate information by citizens has grown explosively [4,15].

As the population grows, the challenges grow. Increasingly, large cities and metropolitan areas are seen as complex systems with connections between their environments and individuals. Services such as traffic, public transport, waste collection and public safety, among others, require more planning and dynamic decision-making mechanisms that take into account the inclusion of citizen participation processes [3].

Everyday the perspective of the citizens about events related to the city is expressed in blogs and social networks. The information given by the citizens, more and more frequently with the use of mobile devices, does not always go

© Springer Nature Switzerland AG 2019
J. A. Lossio-Ventura et al. (Eds.): SIMBig 2018, CCIS 898, pp. 32–45, 2019.
https://doi.org/10.1007/978-3-030-11680-4_5

through the formal channels provided by the different organizations and nevertheless has great value due to its freshness, diversity of points of view and impact on the public opinion.

The state of the art in disciplines of information technologies such as Natural Language Processing, allow us to imagine other types of interfaces for communication between citizens and institutions. Interfaces capable of extracting useful information for the management of the city even if this information was not generated with the intention of being transmitted formally or was not directed to the corresponding institution. It is possible to imagine that simple complaints or comments on the Internet can become relevant for the management of a city. Therefore it is the object of this work to build event sensing mechanisms able to transform comments on social networks into a valuable resource for the city.

1.2   Objective

The main goal of this work is to develop an information extraction platform from text published on social networks. This solution could be instantiated to obtain events in any city and domain for which quality data exist. The social network Twitter will be used, since it is often the choice of people that wants to make a complaint in real time and because allows the extraction of information easily via API. The solution will seek to identify those tweets in Spanish, which are claims or complaints related to waste management in Montevideo, using techniques of Machine Learning and Natural Language Processing.

The city of Montevideo was chosen for this work because of the quantity and quality of the open data it offers and the waste management of the city is the choice as the domain of the problem. Waste management is one of the most sensitive aspects for the population, so there is a significant amount of complaints in social networks that can serve as a corpus of data for this work.

Fig. 1. Tweets about waste in Montevideo

Although there are centralized mechanisms for the reception of complaints by the Municipality of Montevideo (IM, for its acronym in Spanish), they are not always used by citizens. In many cases, as shown in the Fig. 1, these people prefer the immediacy of Twitter to report a problem in the city.

The solution and techniques proposed in this paper are part of a broader academic work [16].

2   State of the Art

For the purposes of this paper, we can define event as a real-world activity that occurs during a certain period of time in a certain geographic space.

There is abundant previous research on events extraction from written texts. These jobs can be categorized according to the types of events, data sources and methods used [2].

The task of detecting trends in written media generally seeks to identify new issues or topics that have growing importance within the corpus [13]. Following the same line of thought, different techniques of event burst detection in traditional written media has been investigated [6,8–10,12,20].

An example can be seen in [19], Snowsill et al. present an online approach to detect events in news streams based on tests of statistical significance over n-gram frequencies within a time frame. The direct application of these techniques on a large volume of information with noise like that coming from social networks does not seem feasible especially since not all bursts are of interest.

A Twitter-based news processing system called TwitterStand is proposed by Jagan Sankaranarayana et al. in [18]. In that work, a Naive Bayes Classifier is used to determine if a tweet corresponds to a news item or irrelevant information and then a clustering algorithm based on term vectors is used, using Tf-Idf similarity to group the news.

Identifying controversial events that were the origin of public discussions on Twitter is the object of the framework developed by Popescu and Pennacchiotti in [14]. The framework is based on Twitter snapshots. They distinguish between snapshots about events and those irrelevant using supervised decision trees trained on a manually annotated corpus [5].

Another good example of use is seen in [17], where Sakaki et al. used tweets to detect specific types of events such as earthquakes and typhoons. They formulated the event detection problem as a classification problem by training an SVM over manually tagged data to separate positive tweets from the negative ones.

Alqhtani et al. combine in [1] the extraction of Twitter events with the use of Image Mining, a technique that uses Computer Vision and Image Processing concepts so that the images attached to a tweet are considered in the classification.

3   Proposed Solution

It is estimated that an average of 6,000 tweets per second are published in the world [21], the solution must be able to identify those in Spanish, which are claims related to waste management in Montevideo. We will call these tweets "Useful Tweets".

From the moment a tweet is generated by a citizen until it becomes an event to be analyzed by the users of the platform, four stages are covered. The Fig. 2 illustrates how information is enriched in each stage.

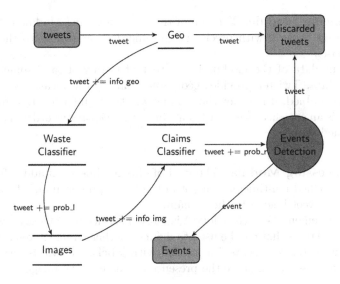

Fig. 2. Module interaction

3.1   Stage 1: Information Retrieval

First of all the platform must recover and store all those tweets that are candidates to be "Useful Tweets".

Twitter Streaming API is used to collect the tweets. The chosen search criteria was created from the combination of two word lists assembled using observation methods:

– Montevideo related keywords (Montevideo, im, imm, @montevideoim, etc.).
– Waste management related keywords (contenedor, recolector, papelera, basura, etc.).

Each tweet to be recovered, is stored maintaining a common structure with can be used for other data sources that may exist. The stored information keeps the original text, the location, if it exists, and metadata to be used later by the platform.

3.2   Stage 2: Information Enrichment

In this stage, several modules that make up the platform, act. These modules task is to enrich the information obtained in Stage 1. Each module responds to a different strategy to determine if a tweet belongs to the set of "Useful Tweets". Modules provide new features that enrich the original information and help determine in Stage 3 if a tweet is an event or not.

Georeferencing Module. The aim of the Georeferencing Module is to enrich a tweet during Stage 2 with precise geographic data according to the domain and city chosen.

The input data of the module is the text of the tweet, and some optional geographic data. Twitter provides geographic data if the location of origin of the tweet in enabled, if it is not the module will try to infer location from text. The output is an enriched tweet with a list of possible locations ordered according to an assigned score.

Image Processing Module. The goal of this module is to add information about the attached multimedia content in the tweets, determining if the images attached to a tweet belong to the domain.

For its implementation, the Cloud Vision API service from Google[1] was used, which returns labels that can be used to define whether an image refers to waste management. After some experimentation with labels, the words "waste" and "litter" were chosen to indicate the presence of waste in the image.

Claims Classifier Module. The objective of this module is to determine if a tweet belongs to the set of tweets in Spanish that can be considered claims or complaints.

The training corpus used for the implementation was assembled using a subset of the collection of complaints of the Unique Response System (SUR, for its acronym in Spanish) of the IM. The corpus consisted of a total of 120 thousand claims. These claims are in the catalog of open data of the Uruguayan State and are classified by date and category[2].

Additionally, the Tass General 2015 corpus was used[3], which has approximately 60,000 tweets in Spanish written by different influential personalities from different areas and countries. The themes of the tweets covers politics, football, literature and entertainment [7] and allows the assumption of absence of complaints in its content.

8 thousand tweets that were previously discarded by the georeferencing module were also added because they were originated in other countries. These tweets are of special interest because most of them are not written in Spanish.

The resulting corpus (Claim corpus) was built balancing classes (Claim and Non-Claim) and was divided in train and tests sets for later use.

A classifier was built and an estimator was trained based on the Support Vector Machines (SVM) algorithm using the Scikit-Learn's implementation of LinearSVC[4].

---

[1] https://cloud.google.com/vision.
[2] https://catalogodatos.gub.uy/dataset/reclamos-registrados-en-el-sistema-unico-de-reclamos-sur-de-la-intendencia-de-montevideo.
[3] http://www.sepln.org/workshops/tass/2015/tass2015.php#corpus.
[4] http://scikit-learn.org/stable/modules/generated/sklearn.svm.LinearSVC.html.

In addition, for this classifier the Scikit-Learn's transformers Count Vectorizer (converts the documents into a matrix of tokens count) and TfIdfTransformer (converts a matrix of tokens count into a normalized matrix using the Tf-Idf representation) were used.

The classifier is then a composition of several steps between the transformers and the LinearSVC estimator.

Each transformer as well as the estimators have a set of parameters whose variation can result in a model better adjusted to the particular problem.

To look for the best performance of the classifier it is necessary to train and evaluate the complete model with each combination of parameters.

According to the Scikit-Learn's documentation, it was decided to explore the parameters ngram_range of Count Vectorizer (specifies if the count will be done on unigram, bigrams, trigrams or a combination them) and C of LinearSVC (determine to what extent to favor the search for a hyperplane whose minimum distance to the examples is as large as possible or the search for a hyperplane that separates the examples of different classes as best as possible).

Waste Classifier Module. The objective of this module is to determine during Stage 2 if a tweet belongs to the set of tweets that deal with waste management.

In this module, the SUR corpus was used. By being categorized by domain it is possible to separate the claims that deal with waste from those that do not. Using the domain to which each claim belongs, a corpus (Waste corpus) with approximately 50% of the claims belonging to the Waste class and 50% claims of the Non-Waste class was obtained.

For the implementation of this module a regression model was used. This model is also a composition of the Count Vectorizer and TfIdfTransformer transformers together with an estimator called SVC (another implementation of SVM that allows the prediction of probabilities).

3.3   Stage 3: Information Classification and Events Generation

In Stage 3, the platform must decide whether a tweet is an event or not. Each piece of enriched information incorporated in Stage 2 will be inputs for this decision.

Events Detection Module. The objective of this module is to determine from the results of the Stage 2 modules whether a tweet is an event.

The corpus used for training and testing this module consists of the tweets obtained in the information retrieval stage between September 2016 and January 2017 (Tweet corpus). Each one of those tweets was manually annotated with one of four classes:

– **"VERY_USEFUL"**: Refers to the waste management in Montevideo and contains a specific location.

- **"USEFUL"**: Refers to the waste management in Montevideo and contains some location reference.
- **"BIT_USEFUL"**: Refers to waste management in Montevideo, without location reference.
- **"NO_USEFUL"**: It does not refer to waste management in Montevideo.

For the purposes of this module, a tweet labeled with "VERY_USEFUL" will be considered as an event and others as no events.

For the implementation Random Forests [11] was used, which is a method of assembling Decision Trees. Random Forests starts from the construction of several independent trees during the training and predicts the class of a new example taking the class predicted by the majority of the trees.

The built model is a composition of four transformers, one per module of Stage 2, and a RandomForestClassifier estimator of the Scikit-Learn library. Each transformer takes a tweet and returns a value corresponding to the output of each module. The four values make up the feature vector that will be classified by the model as an event or no event.

3.4   Stage 4: Events Visualization

In the final stage of the system, the events detected in Stage 3 are recovered and displayed to the user. To this end, a webpage was built, showing the map of Montevideo together with the detected events marked according to their location. The location is the best of the ones obtained by the Georeferencing Module of the Stage 2.

4   Results

In this section we present the results obtained by the different modules of the platform.

4.1   Information Retrieval Stage

From the date 27-09-2016 until the date 10-01-2017 the platform retrieved 15,528 tweets, which after being filtered by the georeferencing module, and being categorized and annotated manually, are divided as observed in the Table 1.

4.2   Georeferencing Module

Of the total of imported tweets, the georeferencing module ruled out 7,857 because they are georeferenced tweets located outside the city of Montevideo.

Of the 238 tweets annotated with the label VERY_USEFUL the georeferencing module found correct solutions for 213 of them. Of the remaining 25, 12 had no solution and 13 had wrong solutions.

Table 1. Imported tweets (Tweet corpus)

	Quantity
Discarded by Georeferencing	7,857
Tagged as VERY_USEFUL	238
Tagged as USEFUL	133
Tagged as BIT_USEFUL	2,046
Tagged as NO_USEFUL	5,254
Total	15,528

4.3   Image Processing Module

As explained in Sect. 3.2 for the construction of this module, the Cloud Vision API provided by Google was used. Despite having a small corpus of information pieces with attached images, the results obtained by this module shown in the Table 2 are favorable.

Table 2. Results about tweets with images

	Precision	Recall	F1-score	Examples	% of the corpus
Without-Garbage	0.86	1.00	0.93	840	66.5%
With-Garbage	1.00	0.68	0.81	423	33.5%
Total	0.91	0.89	0.89	1263	100%

To obtain the results, the collected tweets that had an attached image were taken into account. These tweets were inspected manually to determine how many were incorrectly classified by the module. The resulting confusion matrix is shown in Table 3.

Table 3. Confusion matrix over tweets with images

		Estimated Class		
		Without-Garbage	With-Garbage	
Real Class	Without-Garbage	840	0	*False Positives*
	With-Garbage	135	288	
		False Negatives		

In the final results, and as can be seen in the models generated in the Subsect. 4.6 to classify a tweet as an event, it appears that the influence of this module in the definitive classification is low; In general, the images illustrate a reality expressed in the text, causing some redundancy between modules. This redundancy is most noticeable when configuring tweet extraction to work with keywords related to the domain. In a different situation, for example making an extraction of all the tweets in Spanish or all those related to Montevideo, the Image Processing Module would take more relevance in the final result.

4.4   Claims Classifier Module

The evaluation of this module on the test set (33% of the Claim corpus, 42,664 examples) gives a precision and a recall of 1.0, for both categories (Claim and Non-Claim).

The results on the test set are exceptionally good, nevertheless it is good to remember that the model will be used on tweets and not on claims of the SUR system. In Table 4 the indicators can be seen for the Tweet corpus and in Table 5 the confusion matrix can be found. In this case, although the recall is an acceptable 84%, the accuracy is only 14% due to the large number of false positives.

Table 4. Results on the Tweet corpus

	Precision	Recall	F1-score	Examples	% of the corpus
Non-claim	0.99	0.83	0.91	7433	97.53%
Claim	0.14	0.84	0.24	238	3.12%
Total	0.97	0.83	0.88	7621	100%

Table 5. Confusion matrix for the Tweet corpus

		Estimated Class		
		Non-Claim	Claim	
Real Class	Non-Claim	6183	1250	*False Positives*
	Claim	38	200	

False Negatives

4.5   Waste Classifier Module

For the implementation of this module a regression model that assigns probabilities of belonging to the Waste class for each tweet was chosen. This type of models allows to vary the threshold above which a document will be considered within a given class.

As an example, taking the threshold $= 0.5$ and evaluating this module on the test set (33% of the total corpus, 42,076 examples) we obtain an accuracy of 0.97 and a recall of 0.98 for both categories (Waste and Non-Waste).

As in the Subsect. 4.4 the results on the test set of the Waste corpus are very good, but it is important to consider that the model will be used to classify tweets and not claims originated in the SUR system. It can be seen in the Table 6 the indicators when evaluating the model on the Tweet corpus and in the Table 7 the confusion matrix for the same set can be found. In a similar way to the previous section a good recall is obtained for the class sought but a very poor precision, this is caused by the high number of false positives.

Table 6. Results on the Tweet corpus, threshold $= 0.5$

	Precision	Recall	F1-score	Examples	% of the corpus
Non-waste	0.99	0.42	0.59	7433	97.53%
Waste	0.05	0.90	0.09	233	3.12%
Total	0.96	0.44	0.58	7671	100%

Table 7. Confusion matrix for the Tweet corpus, threshold $= 0.5$

		Estimated Class		
		Non-Waste	Waste	
Real Class	Non-Waste	3128	4305	*False Positives*
	Waste	23	215	

False Negatives

In the Subsect. 4.6, the importance of both domain classifiers in the final classification of a tweet as an event is measured. Even with good individual recall the classifiers seem to have little participation in the final classification, especially the waste classifier. This behavior is due in part to the fact that tweets are retrieved by keywords related to waste, a scenario in which the module has little influence on the final result.

4.6   Events Detection Results

The parametric adjustment was made comparing several alternatives favoring the recall by means of a cross validation evaluation with five partitions.

The outputs of the Stage 2 modules acted as attributes for the Random-Forests model used. Scikit-Learn allows to measure the importance of each feature in the resulting model. The attributes of the model trees were evaluated and the values are seen in the Table 8.

Results for the test set of the Tweet corpus can be seen in the Table 9 and the results and the associated confusion matrix for the complete Tweet corpus are found in the Table 10 and the Table 11. According to this data, the probability

Table 8. Parameters

Feature	Importance
Georeferencing module	0.740
Claims classifier module	0.175
Classifier waste module	0.065
Image processing module	0.018

Table 9. Events detection results for the test set of the Tweet corpus

	Precision	Recall	F1-score	Examples	% of the corpus
No event	1.00	0.96	0.98	2459	97%
Event	0.40	0.95	0.56	73	3%
Total	0.98	0.96	0.97	2532	100%

Table 10. Events detection results for the complete Tweet corpus

	Precision	Recall	F1-score	Examples	% of the corpus
No event	1.00	0.96	0.98	7433	96.9 %
Event	0.44	0.94	0.59	238	3.1 %
Total	0.98	0.96	0.97	7671	100 %

Table 11. Confusion matrix for the complete Tweet corpus

		Estimated Class		
		No Event	Event	
Real Class	No Event	7144	289	*False Positives*
	Event	15	223	

False Negatives

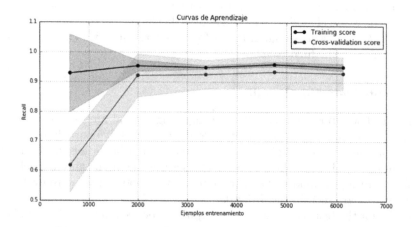

Fig. 3. Learning curves

of classifying an event correctly is 94% (recall) and the probability of incorrectly classifying an event, that is, obtaining a false positive is 4% (fall out).

It is clear that the number of positive instances for the "event" label is low and therefore the results should be taken with caution. However, as seen in Fig. 3 the recall difference between the validation and training corpus starting from certain number of examples seems stabilize, which indicates the ability of the model to generalize predictions.

It is interesting to note that both the classifiers of Stage 2 and the event classifier always sought to favor the recall over the precision in the training stage. The decision follows from the concept that it is better to have fewer false negatives regardless of the number of false positives, that is: from the perspective of city management it matters more to find as many events as possible, even if this means finding more instances that are not events.

5   Conclusions

Everyday more and more citizens interact with each other and with different organizations through technology and social media. In this new era of information exchange, it can be observed that the perception that can be obtained about what happens in a city is more diverse, integral and exact than it used to be.

Artificial Intelligence has taken a great leap in terms of its practical applications thanks to the abundant amount of data generated by Internet users. New tools and algorithms that are applicable to areas so far unrelated to these technologies are released on a daily basis. Industry and public services are examples where diverse techniques of Artificial Intelligence such as Natural Language Processing and different methods of automatic learning can have a great impact.

The generation of data is the consequence of the massive use of the Internet and the main condition for the development of AI applications that are really useful. The availability of quality open data on different aspects of a city is what makes possible research like the one presented in this work. The fact that public institutions have good policies about open data, as in the case of IM in Uruguay, is what allows to move forward in the investigation and analysis of this data for the improvement of processes and services that could benefit citizens.

The platform built was fed with 15,528 tweets during a period of 105 days, these tweets were retrieved by keywords associated with the city of Montevideo and the chosen domain: Waste management. Of the 234 tweets manually marked as possible events, the platform identified 94% correctly while getting only 4% false positives (289 tweets).

The results obtained show that it is possible to identify city events associated to the domain for most of the tweets recovered that complied with the conditions. For these events, it is possible to have a precise geographic location associated to elements of the urban furniture that are the object of the complaint or claim.

The identification of events, located in time and space, from the diverse perspective of citizens and with aggregate information such as photographs or videos allow a level of analysis that would make possible, for example, the detection of problematic geographic areas, the identification of patterns, or sentiment analysis for any given public service.

Going one step further, if each event identified by the platform could be assigned directly to a team for its resolution, and that team could report the result through images directly to the platform, it would close the communication circle with the citizen.

It is in this context of technological advances, and after analyzing the results of this work that we can conclude that it is possible to build sensing mechanisms of social networks that can become a new type of interface between citizens and public organizations. These interfaces would allow a greater participation of citizens in a familiar terrain: social networks, and could offer a more transparent resolution of some claims. These interfaces could become a relevant aspect of the transformation of a city into a smart city.

References

1. Alqhtani, S.M., Luo, S., Regan, B.: Fusing text and image for event detection in twitter. arXiv preprint arXiv:1503.03920 (2015)
2. Atefeh, F., Khreich, W.: A survey of techniques for event detection in twitter. Comput. Intell. **31**(1), 132–164 (2015)
3. Bouskela, M.: La ruta hacia las smart cities, January 2017. http://goo.gl/3Y0tWK. Accessed 03 June 2018
4. ICT Facts: Figures-the world in 2015. The International Telecommunication Union (ITU), Geneva (2015)
5. Friedman, J.H.: Greedy function approximation: a gradient boosting machine. Ann. Stat. **29**, 1189–1232 (2001)
6. Fung, G.P.C., Yu, J.X., Yu, P.S., Lu, H.: Parameter free bursty events detection in text streams. In: Proceedings of the 31st International Conference on Very Large Data Bases, pp. 181–192. VLDB Endowment (2005)
7. García Cumbreras, M.Á., Martínez Cámara, E., Villena Román, J., García Morera, J.: Tass 2015-the evolution of the spanish opinion mining systems (2016)
8. Goorha, S., Ungar, L.: Discovery of significant emerging trends. In: Proceedings of the 16th ACM SIGKDD International Conference on Knowledge Discovery and Data Mining, pp. 57–64. ACM (2010)
9. He, Q., Chang, K., Lim, E.P.: Analyzing feature trajectories for event detection. In: Proceedings of the 30th Annual International ACM SIGIR Conference on Research and Development in Information Retrieval, pp. 207–214. ACM (2007)
10. He, Q., Chang, K., Lim, E.P., Zhang, J.: Bursty feature representation for clustering text streams. In: SDM, pp. 491–496. SIAM (2007)
11. Ho, T.K.: Random decision forests. In: Proceedings of the Third International Conference on Document Analysis and Recognition, vol. 1, pp. 278–282. IEEE (1995)
12. Kleinberg, J.: Bursty and hierarchical structure in streams. Data Min. Knowl. Discov. **7**(4), 373–397 (2003)
13. Kontostathis, A., Galitsky, L.M., Pottenger, W.M., Roy, S., Phelps, D.J.: A survey of emerging trend detection in textual data mining. In: Berry, M.W. (ed.) Survey of text mining, pp. 185–224. Springer, New York (2004). https://doi.org/10.1007/978-1-4757-4305-0_9

14. Popescu, A.M., Pennacchiotti, M.: Detecting controversial events from twitter. In: Proceedings of the 19th ACM International Conference on Information and Knowledge Management, pp. 1873–1876. ACM (2010)

15. Radar, G.: Perfil internauta uruguayo 2016 - resumen ejecutivo, January 2017. http://www.gruporadar.com.uy/01/wp-content/uploads/2016/11/El-Perfil-del-Internauta-Uruguayo-2016-Resumen-Ejecutivo.pdf. Accessed 03 June 2018

16. Raúl Speroni, M.S.: Extracción de eventos en una ciudad a partir de redes sociales. https://www.fing.edu.uy/inco/grupos/pln/prygrado/Informe_Speroni_Steglich.pdf. Accessed 19 July 2018

17. Sakaki, T., Okazaki, M., Matsuo, Y.: Earthquake shakes twitter users: real-time event detection by social sensors. In: Proceedings of the 19th International Conference on World Wide Web, pp. 851–860. ACM (2010)

18. Sankaranarayanan, J., Samet, H., Teitler, B.E., Lieberman, M.D., Sperling, J.: Twitterstand: news in tweets. In: Proceedings of the 17th ACM SIGSPATIAL International Conference on Advances in Geographic Information Systems, pp. 42–51. ACM (2009)

19. Snowsill, T., Nicart, F., Stefani, M., De Bie, T., Cristianini, N.: Finding surprising patterns in textual data streams. In: 2010 2nd International Workshop on Cognitive Information Processing, pp. 405–410. IEEE (2010)

20. Wang, X., Zhai, C., Hu, X., Sproat, R.: Mining correlated bursty topic patterns from coordinated text streams. In: Proceedings of the 13th ACM SIGKDD International Conference on Knowledge Discovery and Data Mining, pp. 784–793. ACM (2007)

21. www.internetlivestats.com/: Twitter usage statistics. http://www.internetlivestats.com/twitter-statistics. Accessed 1 June 2018

ANEW for Spanish Twitter Sentiment Analysis Using Instance-Based Multi-label Learning Algorithms

Rodrigo Palomino[1]([✉]), Carlos Meléndez[1], David Mauricio[1], and Jorge Valverde-Rebaza[2]

[1] Universidad Peruana de Ciencias Aplicadas, Lima, Peru
rodrigopalominosilva@gmail.com, dms_research@yahoo.com
[2] Department of Scientific Research, Visibilia, São Carlos, SP 13560-647, Brazil
jvalverr@visibilia.net.br

Abstract. In the last years, different efforts have been made to extract information that users express through online social networking services, *e.g.* Twitter. Despite the progress achieved, there are still open gaps to be addressed. Related to the sentiment analysis issue, we stand out the following gaps: (a) low accuracy in sentiment classification task for short texts; and, (b) lack of tools for sentiment analysis in several languages. Aiming to fill these gaps, in this paper we apply the Spanish adaptation of ANEW (Affective Norms for English Words) as resource to improve the Twitter sentiment analysis by applying a variety of multi-label classifiers in a corpus of Spanish tweets collected by us. To the best of our knowledge, this is the first work using a Spanish adaptation of ANEW for sentiment analysis.

Keywords: Classification · Sentiment analysis ·
Multi-label classification · Twitter · ANEW · Affective word lists

1 Introduction

In the last years, the volume of generated content through online social networking services, *e.g.* Twitter, has significantly increased. Until April 2018, Twitter has reached 330 million of active users; who share their opinions about diverse topics, like the products or services they consume. Hence that, the tasks related to mining and analysis of this data have become hot topics.

Sentiment analysis is a task aiming to determine, automatically, whether a text document received either a positive, negative or neutral opinion [12]. Clearly, this problem could be addressed as a traditional classification problem. However, given the variety of emotions that could be expressed in the same text, short documents as tweets commonly have associated more than one category. This fact makes the task more complex being necessary deal with it as a multi-label classification problem.

© Springer Nature Switzerland AG 2019
J. A. Lossio-Ventura et al. (Eds.): SIMBig 2018, CCIS 898, pp. 46–53, 2019.
https://doi.org/10.1007/978-3-030-11680-4_6

To perform sentiment analysis in a multi-label context, one of the most important steps is the right identification of sentiments existing in the text corpus. From this identification, is possible selecting the labels that will categorize the instances. To aid in this task, in the literature exist several word lists labeled with emotional valence. Some word list for sentiment analysis extensively studied by the research community are: ANEW, General Inquirer, OpinionFinder, SentiWordNet, and WordNet-Affect [7].

ANEW is one of the most accurate affective word lists since the scores for words offered by it have been validated by several scientific psycholinguistic studies. Considering its good performance, adaptations to Spanish [9], Portuguese [10] and Italian [6], have been developed.

In this work, the Spanish adaptation of ANEW is used to determine the labels and the attributes for the categorization of Spanish tweets. We use a group of instance-based multi-label classifiers to perform our experimental evaluations. The contributions of this paper are twofold, we: (i) make available a new dataset of Spanish tweets in attribute-value table form built based on ANEW, and (ii) carried out a comprehensive evaluation to quantify the classification abilities of well-known instance-based multi-label algorithms on short-text Spanish corpus.

The remainder of this paper is organized as follows. In Sect. 2 we discuss about the previous works in sentiment analysis. In Sect. 3, we briefly describe the algorithms and evaluation metrics for multi-label classification. In Sect. 4, we briefly describe the construction of the dataset based on the Spanish adaptation of ANEW. In Sect. 5, we show our experimental results. Section 6 closes with our conclusions and future work.

2   Sentiment Analysis

Several studies have focused on sentiment analysis task. Most of them address the problem as a classification of sentiments found in a portion of text. In this context, a sentiment is commonly considered as *positive, negative* or *neutral*, according to its nature. Nevertheless, a comment/post in a social networking service can involve multiple sentiments at the same time. Therefore, the main dilemma lies in identifying which sentiments to classify or how to detect them within a comment/post.

Raja and Swamynathan [8] analyzed the different ways in which sentiments in comments/posts can impact the accuracy of any recommendation task. Bobicev [1] faced the problem of sentiment detection in health-related forum posts, proposing a set of new labels: *confusion, encouragement, gratitude*, and *factual*. Zhao et al. [14] developed the *Social Sentiment Sensor* (SSS) system, which aims the hot topic detection and topic-oriented sentiment analysis from social network data. Tellez et al. [12] built a tool which, automatically, identify the necessary preprocessing techniques to improve the performance of some classifiers.

Despite the variety of techniques and tools existing in the literature, there is a lack of them directed to Spanish language [5]. This fact is probably due to the low amount of linguistic resources available to support the development of solutions for sentiment analysis in the cited language. Therefore, this work shows that is possible obtain accurate multi-label classification results in the context of sentiment analysis of short-texts written in Spanish by using an adaptation of ANEW.

3   Instance-Based Multi-label Classifiers

In the literature exists a variety of multi-label classifiers. Some of the most commonly used, by its simplicity, are the called instance-based (or lazy), such as: MLkNN, BRkNN, MLMUT and MLnotMUT.

Zhang and Zhou [13] proposed *MLkNN*, which uses the MAP principle to determine the labels of each test instance according to the number of its k-nearest neighbors belonging to each class. Spyromitros et al. [11] proposed *BRkNN*, which performs independent predictions for each label of each test instance following a single search of its k-nearest neighbors. Two extensions are added to this algorithm: *a* and *b*. Both extensions are based on the confidence score calculation of each label, which is obtained by considering the percentage of the k-nearest neighbors that include it. *BRkNN-a* considers a confidence score greater than half of the number of k-nearest neighbors; if no label satisfies this condition, it outputs the label with the greatest confidence score. Conversely, *BRkNN-b* classifies the test instance with a group of x labels, which have the greatest confidence score, where x is nearest integer of the average size of the multi-labels of the k-nearest neighbors of the instance.

Cherman et al. [3] developed the *MLMUT* and *MLnotMUT* algorithms, which use mutuality strategies and a voting system to improve the BRkNN algorithm. Since both algorithms are based on BRkNN, the corresponding extensions can be applied. Therefore, the algorithms *MLMUT-a* and *MLnotMUT-b* are obtained.

Metz et al. [4] proposed *GeneralB*, an algorithm specifically developed to be used as baseline. GeneralB ranks the single labels in the set of labels according to their individual relative frequencies on the multi-labels in the dataset, and the x most frequent single-labels are included in the predicted label set. As the predicted set of labels should have a reasonable number of single-labels such that it is not too strict (including too few single-labels) or too flexible (including too many single-labels), x is defined as the closest integer value of the label cardinality of dataset.

It is important to comment that here exists a variety of metrics to evaluate the performance of classifiers mentioned. However, we focus on Hamming loss due to be considered as better than precision in multi-label context since it considers the fraction of the wrong labels, instead of right, to the total number of labels. Therefore, hamming loss is defined as: *i.e. $HammingLoss(H, D) =$* $\frac{1}{N} \sum_{t=1}^{N} \frac{|Y_i \Delta Z_i|}{|L|}$, where Δ denotes the symmetric difference between the true labels Y_i and the predicted labels Z_i, and L represents the set of labels. The smaller hamming loss value, the better the multi-label classifier performance.

4   Dataset Description

In this section, we present the characteristics of dataset collected as well as the preprocessing applied over it to be used in our experiments.

4.1   Spanish Adaptation of ANEW

Redondo et al. [9] provided a set of 1034 emotional words in Spanish from the original proposal of Bradley and Lang [2]. The sentiments represented in ANEW are categorized into three dimensions: *valence*, which ranges from unpleasant to pleasant; *arousal*, which ranges from calm to excited; and *dominance*, which ranges from out-of-control to in-control. The categorization is done by scoring each word in a scale from 1 to 9 on each of the three dimensions.

To transform this categorization to multi-label, we use the minimum and maximum values from each dimension to determine the labels: unpleasant, pleasant, calm, excited, out-of-control and in-control. We assign a value of 0 or 1 to each label of each word according to its score, *i.e.* if the score in a specific dimension is lower than 5, then we assign 1 to the label that represents the minimum value of the dimension, and 0 to the label that represents the maximum value of the dimension; but if the score is greater or equal to 5, then we assign 1 to the label that represents the maximum value of the dimension and 0 to the label that represents the minimum value of the dimension.

We use the *Meaning Cloud*[1] service to lemmatization process of list of words. Considering the fact that repeated words can be outputted, the support of an expert in psychology is requested to determine the final list of words as well as its corresponding categorization. As a result, a list of 1020 words is obtained.

4.2   Spanish Tweets Data Collection and Preprocessing

We collected 3000 tweets posted on April 19th, with the support of *Twitter Archiver*[2] tool by using the following query: *"academia OR alumno OR aprendizaje OR biblioteca OR catedra OR ciencia OR colegio OR conocimiento OR diploma OR docente OR doctorado OR enseñanza OR escuela OR facultad OR guarderia OR instituto OR instructor OR intelecto OR inteligencia OR licenciatura OR liceo OR maestro OR magisterio OR mentor OR museo OR pedagogia OR pedagogo OR profesor OR titulo OR tutor OR universidad"*.

A lemmatization process supported also by the Meaning Cloud service was applied on the collected tweets. After, the tweets are categorized in different labels by checking the presence of any of the words from ANEW within its content. If there is a match, we assigned the corresponding labels of each encountered word on the tweet. When matching the words, we discard those instances which do not present any ANEW word in its content. Therefore, we keep a total of 1723 tweets. Figure 1 shows the general process implemented for tagging a collected tweet and put it into the dataset.

---

[1] https://www.meaningcloud.com/es/.
[2] https://goo.gl/DkbRbs.

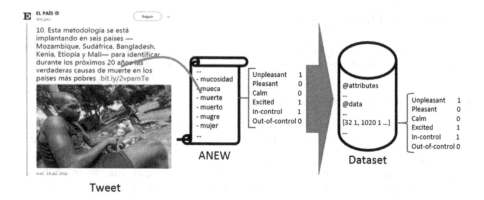

Fig. 1. General idea of the process for tagging a tweet and put it into the dataset.

Subsequently, an attribute selection task is performed by using Mulan[3] library. Specifically, the IG-BR method with Rankeras base algorithm was used by setting the threshold to 0.001 as the minimum correlation value. This configuration reduces the number of attributes to 142.

5 Experimental Setup and Results

The objective of our experiments is to evaluate the performance of instance-based multi-label classifiers in the context of short-texts in Spanish language, based on the sentiment dimensions found in ANEW. Therefore, our evaluation considers the Mulan implementations of MLkNN, BRkNN, BRkNN-a, BRkNN-b, MLMUT, MLMUT-a, MLMUT-b, MLnotMUT, MLnotMUT-a and MLnotMUT-b. Moreover, we use GeneralB as baseline algorithm. Since all the instance-based multi-label classifiers are based on k-nearest neighbors principle, we perform our experiments considering variations of k parameter. Moreover, we also consider three distance functions: Euclidean, Chebyshev and Cosine.

Table 1 shows the results of multi-label sentiment classification obtained for different values of nearest neighbors (k). Each value in Table 1 represents the average hamming loss obtained by 10-fold cross validation process. Highlighted values represent the best results for each k value and for each distance function.

From Table 1 we observe that for Euclidean distance, classifiers based on mutual strategies (MLMUT, MLnotMUT, and their variations) performed better. Specifically, MLMUT is the best classifier and reached its best performance at $k = 9$. A similar behavior is observed for results obtained for Chebyshev distance, i.e. classifiers based on mutual strategies performed better, specifically the MLMUT with $k = 9$. For both distances, all evaluated classifiers outperformed GeneralB independently of k value. For Cosine distance, none of the algorithms can performed better than baseline; however, some classifiers such

---

[3] http://mulan.sourceforge.

Table 1. Experimental results of instance-based algorithms evaluated. Different values of k and Euclidean (Eu), Chebyshev (Ch) and Cosine (Co) distances have been used. Best results for each k value are highlighted in bold.

Algorithm		$k=1$	$k=3$	$k=5$	$k=7$	$k=9$	$k=11$	$k=13$	$k=15$
GeneralB	Eu. 0.3192	0.3192	0.3192	0.3192	0.3192	0.3192	0.3192	0.3192
	Ch. 0.3192	0.3192	0.3192	0.3192	0.3192	0.3192	0.3192	0.3192
	Co. **0.3192**	**0.3192**	**0.3192**	**0.3192**	**0.3192**	**0.3192**	0.3192	0.3192
MLkNN	Eu. 0.1056	**0.0760**	0.0937	0.0936	0.0932	0.0995	0.0994	0.0999
	Ch. 0.1438	0.0872	0.0818	0.0877	0.0944	0.0934	0.0991	0.0952
	Co. 0.5624	0.4507	0.3732	0.3480	0.3686	0.3802	0.4002	0.3964
BRkNN	Eu. 0.0684	0.0996	0.1220	0.1465	0.1550	0.1618	0.1821	0.1877
	Ch. 0.0810	0.1138	0.1416	0.1678	0.1750	0.1815	0.2021	0.2065
	Co. **0.3192**	**0.3192**	**0.3192**	**0.3192**	**0.3192**	**0.3192**	**0.3192**	**0.3192**
BRkNN-a	Eu. 0.0684	0.0996	0.1220	0.1465	0.1550	0.1618	0.1821	0.1877
	Ch. 0.0810	0.1138	0.1416	0.1678	0.1750	0.1815	0.2021	0.2065
	Co. **0.3192**	**0.3192**	**0.3192**	**0.3192**	**0.3192**	**0.3192**	**0.3192**	**0.3192**
BRkNN-b	Eu. 0.0701	0.1021	0.1238	0.1496	0.1566	0.1626	0.1823	0.1880
	Ch. **0.0754**	0.1056	0.1291	0.1537	0.1622	0.1712	0.1915	0.1972
	Co. 0.3288	0.3288	0.3288	0.3288	0.3288	0.3288	0.3288	0.3288
MLMUT	Eu. 0.0959	0.0810	**0.0760**	**0.0718**	**0.0706**	**0.0709**	**0.0716**	0.0706
	Ch. 0.0998	0.0883	0.0817	0.0818	0.0817	0.0817	0.0810	0.0810
	Co. **0.3192**	**0.3192**	**0.3192**	**0.3192**	**0.3192**	**0.3192**	**0.3192**	**0.3192**
MLMUT-a	Eu. 0.0959	0.0810	**0.0760**	**0.0718**	00.0706	**0.0709**	**0.0716**	**0.0705**
	Ch. 0.0998	0.0883	0.0817	0.0818	0.0817	0.0817	0.0810	0.0810
	Co. **0.3192**	**0.3192**	**0.3192**	**0.3192**	**0.3192**	**0.3192**	**0.3192**	**0.3192**
MLMUT-b	Eu. 0.0959	0.0906	0.0821	0.0774	0.0742	0.0748	0.0736	0.0733
	Ch. 0.0925	**0.0874**	**0.0807**	**0.0787**	**0.0772**	**0.0766**	**0.0754**	0.0754
	Co. 0.3288	0.3288	0.3288	0.3288	0.3288	0.3288	0.3288	0.3288
MLnotMUT	Eu. **0.0678**	0.0769	0.0872	0.0961	0.1021	0.1069	0.1161	0.1238
	Ch. 0.1025	0.1129	0.1166	0.1033	0.1162	0.1241	0.1309	0.1355
	Co. 0.5624	0.5025	0.4733	**0.3192**	**0.3192**	**0.3192**	0.3239	0.3231
MLnotMUT-a	Eu. 0.0679	0.0770	0.0872	0.0961	0.1021	0.1069	0.1161	0.1238
	Ch. 0.1025	0.1129	0.1166	0.1033	0.1162	0.1241	0.1309	0.1355
	Co. 0.5624	0.5025	0.4733	**0.3192**	**0.3192**	**0.3192**	0.3239	0.3231
MLnotMUT-b	Eu. 0.0903	0.1048	0.1204	0.1322	0.1376	0.1433	0.1534	0.1585
	Ch. 0.1044	0.1191	0.1244	0.1124	0.1219	0.1313	0.1374	0.1435
	Co. 0.5624	0.5152	0.4757	0.3321	0.3304	0.3307	**0.3192**	0.3239

as BRkNN, BRkNN-a, MLMUT and MLMUT-a which at most can reach the value of GeneralB. We can also observe that MLnotMut-b can reach this value at $k=13$.

To examine the difference in performance between the use of the distance functions, Fig. 2 plots the results for the best five algorithms along the distances used. We can observe that algorithms MLMUT, MLMUT-a and MLMUT-b performed better than the rest and also present less dispersion. This leads us to think that the algorithms that involve mutuality strategies are more consistent than others that do not include them.

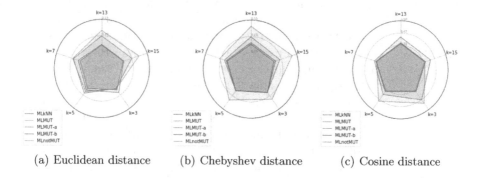

(a) Euclidean distance   (b) Chebyshev distance   (c) Cosine distance

Fig. 2. Results of the best five algorithms on the three distances

The source code, dataset collected and other resources related to our experimental setup and results are publicly available[4].

6   Conclusion

In this work we used the Spanish adaptation of ANEW to determine the labels and attributes for the automatic categorization of Spanish tweets. After the preprocessing tasks, we obtained 1723 Tweets and 142 attributes with a confidence of at least 0.001. The categories used were pleasant, unpleasant, calm, excited, in-control and out-of-control.

In other hand, the experiments show that algorithms with mutuality strategies are more stable despite the variations of k value, and also outperform the other lazy algorithms when using the Euclidean and Chebyshev distances. However, when using the Cosine distance, none of the algorithms could outperform the baseline. The best value obtained using Cosine distance was equal to the baseline, which was reached by BRkNN, BRkNN-a, MLMUT, and MLMUT-a. We also observed that, for the three distances, the "a" configuration of the BRkNN implementation performs slightly better than the "b" configuration.

The results reported in this paper could be complemented with other experiments using other preprocessing techniques and different comparison metrics. In that sense, as future work, we plan to complement this work focusing on the preprocessing phase, experimenting with diverse techniques in order to reduce the sparsity of the dataset and also present results with more evaluation metrics.

Acknowledgements. We would like to thank psychologist María Soledad Silva for her collaboration.

---

[4] https://github.com/RpalominoSil/ANEW-for-Sentiment-Analysis.

References

1. Bobicev, V.: Text classification: the case of multiple labels. In: 2016 International Conference on Communications (COMM), pp. 39–42. IEEE (2016)
2. Bradley, M.M., Lang, P.J.: Affective norms for English words (ANEW): Instruction manual and affective ratings. Technical report. Citeseer (1999)
3. Cherman, E.A., Spolaôr, N., Valverde-Rebaza, J., Monard, M.C.: Lazy multi-label learning algorithms based on mutuality strategies. J. Intell. Robot. Syst. **80**(1), 261–276 (2015)
4. Metz, J., de Abreu, L.F.D., Cherman, E.A., Monard, M.C.: On the estimation of predictive evaluation measure baselines for multi-label learning. In: Pavón, J., Duque-Méndez, N.D., Fuentes-Fernández, R. (eds.) IBERAMIA 2012. LNCS (LNAI), vol. 7637, pp. 189–198. Springer, Heidelberg (2012). https://doi.org/10.1007/978-3-642-34654-5_20
5. Miranda, C.H., Guzman, J.: A review of sentiment analysis in Spanish. Tecciencia **12**(22), 35–48 (2017)
6. Montefinese, M., Ambrosini, E., Fairfield, B., Mammarella, N.: The adaptation of the affective norms for English words (ANEW) for Italian. Behav. Res. Methods **46**(3), 887–903 (2014)
7. Nielsen, F.Å.: A new ANEW: evaluation of a word list for sentiment analysis in microblogs. arXiv preprint arXiv:1103.2903 (2011)
8. Raja, M., Swamynathan, S.: Tweet sentiment analyzer: sentiment score estimation method for assessing the value of opinions in tweets. In: Proceedings of the International Conference on Advances in Information Communication Technology & Computing, p. 83. ACM (2016)
9. Redondo, J., Fraga, I., Padrón, I., Comesaña, M.: The Spanish adaptation of ANEW (affective norms for English words). Behav. Res. Methods **39**(3), 600–605 (2007)
10. Soares, A.P., Comesaña, M., Pinheiro, A.P., Simões, A., Frade, C.S.: The adaptation of the affective norms for English words (ANEW) for European Portuguese. Behav. Res. Methods **44**(1), 256–269 (2012)
11. Spyromitros, E., Tsoumakas, G., Vlahavas, I.: An empirical study of lazy multilabel classification algorithms. In: Darzentas, J., Vouros, G.A., Vosinakis, S., Arnellos, A. (eds.) SETN 2008. LNCS (LNAI), vol. 5138, pp. 401–406. Springer, Heidelberg (2008). https://doi.org/10.1007/978-3-540-87881-0_40
12. Tellez, E.S., Miranda-Jiménez, S., Graff, M., Moctezuma, D., Siordia, O.S., Villaseñor, E.A.: A case study of Spanish text transformations for twitter sentiment analysis. Expert Syst. Appl. **81**, 457–471 (2017)
13. Zhang, M.L., Zhou, Z.H.: A k-nearest neighbor based algorithm for multi-label classification. In: 2005 IEEE International Conference on Granular Computing, vol. 2, pp. 718–721. IEEE (2005)
14. Zhao, Y., Qin, B., Liu, T., Tang, D.: Social sentiment sensor: a visualization system for topic detection and topic sentiment analysis on microblog. Multimedia Tools Appl. **75**(15), 8843–8860 (2016)

An Operational Deep Learning Pipeline for Classifying Life Events from Individual Tweets

Xinsong Du[1], Jiang Bian[1], and Mattia Prosperi[2]($\boxtimes$)

[1] Department of Health Outcomes and Biomedical Informatics,
University of Florida, Gainesville, FL 32611, USA
{xinsongdu, bianjiang}@ufl.edu
[2] Department of Epidemiology, University of Florida,
Gainesville, FL 32611, USA
m.prosperi@ufl.edu

Abstract. We here present an operational deep learning pipeline for classifying life events from individual tweets, using job loss as a use case and Twitter data collected between 2010 and 2013 (historic sample from the public stream). The pipeline includes identification of keywords through snowball sampling, multiple rater manual annotation, supervised deep learning, text processing (word embedding, bag of words) and architecture selection (convolutional, shallow-and-wide convolutional, and long-short-term memory) with parameter optimization, external validation and feedback learning. After model optimization, a shallow-and-wide network with a pre-trained 200-dimensional word2vec achieved a precision of 78% (over an average single keyword precision of 50%) and an area under receiver operating characteristic of 86%. Precision and recall also increased by 5% using bag of words. When tested on tweets with ambiguous annotations (i.e. tweets that were hard for human annotators to classify), the network achieved 65% precision. Finally, on a random set of tweets that did not contain any of the snowballed keywords, 30% were classified as job loss events; this putatively false positive set can be used to reinforce the learner's training. In conclusion, the pipeline streamlines both the manual and automated process, providing feedback reinforcement (snowballing and external tweets), and shows good performance on classifying individual tweets on the use case, potentially saving human resources needed to collate such data for research studies.

Keywords: Deep learning · Job loss · Twitter · Classification

1 Introduction

Twitter is a popular resource for data mining and information retrieval due to its ease of access, large sample size, and diverse information content. Natural language processing (NLP) and machine learning methods have been employed for automated annotation of tweets, which is a necessary precursor for subsequent works, e.g. behavioral assessments, epidemiology surveillance, and demoscopic studies. For instance, in relation to

J. A. Lossio-Ventura et al. (Eds.): SIMBig 2018, CCIS 898, pp. 54–66, 2019.
https://doi.org/10.1007/978-3-030-11680-4_7

individual tweet classification, Reece et al. [1] used random forests to forecast the onset and cause of mental illness, in which Linguistic Inquiry and Word Count (LIWC) [2], tweet sentiment, average word count per tweet and tweet frequency of the user were used as features; Alsaedi et al. [3] employed naïve Bayes to detect disruptive events. Bag of words, text sentiment, etc. were used as features; Sumner et al. [4] applied LIWC to extract features, using support vector machines (SVM), decision trees, random forests, and naïve Bayes to predict the 'dark triad', which is the anti-social triad of psychopathy, narcissism and Machiavellianism; Makazhanov et al. [5] employed naïve Bayes and social network information (e.g. number of times a user has retweeted party candidates) to identify political preferences; Conover et al. [6] employed term frequency-inverse document frequency (TF-IDF) [7] and latent semantic analysis [8] to extract features from tweets, and used SVM also to predict political alignment among Twitter users. Deep learning is often combined with NLP due to its flexibility in architecture, text embedding, and ability to approximate complex target functions. For tweet classification, Won et al. [9] applied a deep residual network to detect images related to protest activities in tweets; Zhang et al. [10] used restricted boltzman machine and long-short-term memory (LSTM) networks to detect traffic accidents. They did stemming for tweets and used binary features; Founta et al. [11] Badjatiya et al. [12] and Pitsilis et al. [13] identified abusive behavior, such as offensive language, in social media with deep learning models. In these three papers, Founta et al. used word embedding to extract features; Badjatiya et al. applied TF-IDF, n-gram, and word embedding [14, 15] to different models; and Pitsilis et al. used term frequency as feature.

Given the widespread use of NLP/deep learning in Twitter analytics, it is valuable to develop operational procedures apt to facilitate manual data annotation, streamline automated classification, and validate generalizability. Therefore, we developed a multi-step pipeline to classify life events from individual tweets; as a use case, we chose detection job loss. The pipeline included identification of keywords through snowball sampling, multiple-rater manual annotation, supervised deep learning, architecture selection – convolutional neural network (CNN), shallow-and-wide convolutional neural network (SWCNN), and LSTM – with parameter optimization, external validation and feedback 'active' learning.

2   Methods

The pipeline is applied upon the definition of a research outcome, i.e. the life event of interest. In our case, it is job loss. The following steps were executed (with internal feedbacks as needed).

   i. **Initial Keyword Identification:** The study team agreed on a list of seed keywords related to the outcome of interest.
   ii. **Data Collection and Preprocessing:** We used a Twitter crawler tool, tweetf0rm [16], leveraging the Twitter API to collect random public tweets. Xapian [26] was used for index and search due to the large volume of data that we have collected. Non-English tweets were removed and the remaining ones were standardized. Tweets were mapped to vectors of real numbers (to be used as input to the deep

learners) using the GloVe word embedding tool. In addition to GloVe, the following text tweaks were made: (1) Replaced all links (e.g. http://4ms.me/bcIVe0) in tweets with "<url>"; (2) All user mentions (e.g. @FoxNews) were replaced with "<user>"; (3) Hashtags (e.g. #jobloss) were replaced by "<hashtag> PHRASE" (e.g. "<hashtag> jobloss"). We further padded all tokenized tweets to 37 tokens, which is the length of the longest one, to achieve the same vector dimension. From the standardized set, the tweets containing the keywords were separated from those not containing the keywords.

iii. **Manual Annotation:** We did annotation on tweet level. A subset of the tweets (from hundreds to a few thousands) containing the selected keywords was manually categorized by 3 raters as related to the outcome or not (i.e. job loss event vs. not) with possibility to flag ambiguous instances. The initial set was calibrated to the number of available raters. Ties were discussed and if not resolved, ambiguous tweets were kept apart. For each keyword, precision, recall and raters' kappa agreement were calculated [17].

iv. **Snowball Sampling:** From the initial annotated set, and on the basis of precision (the number of job loss events divided by total number of tweets generated by the particular keyword), the raters agreed on a set of additional keywords to look for, and steps i to iv were repeated until the results were satisfactory. This technique is known as snowball sampling [18] (see detailed section below). For instance, when annotating tweets with the keyword *"got fired"*, a rater may find the tweet *"Since I started working for my dad I've been fired 4 times and kicked out of his shop at least 15+. I need a job"*, in which *"kicked out"* could be added to the keyword set.

v. **Model Training and Selection:** The non-ambiguous set of manually annotated tweets after snowball sampling was used to train, optimize learning parameters (details given in the next paragraphs), cross-validate, and select the best deep learning model among the list of choice (here, CNN, SWCNN, LSTM using word embedding, and SVM, LASSO logistic regression with bag of words) based on a complexity/performance tradeoff. Performance indices included precision, recall and area under the curve (AUC) of the receiver operating characteristic. Two levels of cross-validation (CV) were used: the first one was a plain n-fold CV that optimizes the parameters (e.g. filter, learning rate, word embedding dimensions); the second one was a repeated n-fold CV (i.e. $r \cdot n$-fold CV) that drew distributions on the performance indices and – since repeated CV usually yields Gaussian – compared them using a t-test corrected for sample overlap [19]. The models were also cross-validated by using increasing data subsamples to estimate if the manually annotated sample size is large enough to guarantee flattening of performance, i.e. increasing the number of manually annotated tweets does not significantly increase learner's performance.

vi. **Robustness Assessment and Model Reinforcement:** To assess the robustness, the best model was tested on the ambiguous tweets and on the random sample of tweets that do not contain the keywords. Since our models were trained only on tweets containing the keywords, it is necessary to make it more general by reinforcing it with random tweets that do not have the keywords. We randomly collected tweets without the keywords, did prediction using trained model. Putatively false positive tweets were added to the training set to repeat steps v to vi.

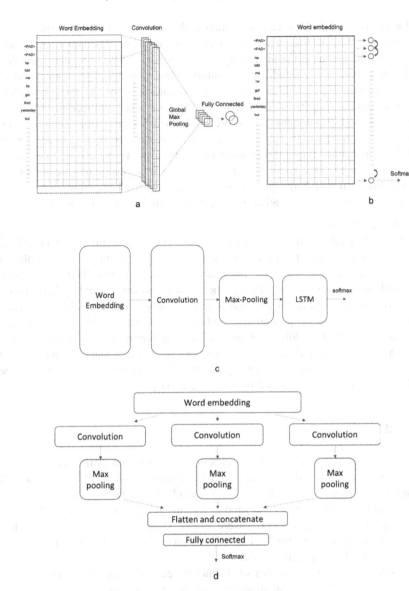

Fig. 1. Neural network architectures

Details on Deep Learning Architectures and Parameter Settings

We compared different deep learning methods: (i) CNN [20] (Fig. 1a); (ii) SWCNN (Fig. 1d) with word embedding [21]; (iii) LSTM [22] (Fig. 1c); and (iv) the combination of CNN and LSTM (Fig. 1d). CNN has been widely used in computer vision and NLP, and the convolutional layer allows for a direct multi-dimensional input and neighborhood/context manifolds, i.e. it is appropriate for input encodings such as word embedding, and the output of global max-pooling for the convolutional layer is the vector for the tweet. LSTM is also a popular architecture, apt to capture time-series information such the word sequence information in a sentence. The tweet vector was represented by the output of the hidden layer in the last state. SWCNN is an extension of CNN, which has only one hidden convolutional layer, but the layer contains three branches each with different filter length. The concatenation of the three branches' global max-pooling outputs is the tweet vector. For word embedding we used 100-dimensions and 200-dimensions GloVe pre-trained Twitter word vectors, developed on 2 billion tweets with 1.2 million words vocabulary. In brief, a tweet can be converted into an embedding n-dimensional matrix by mapping each word to a certain vector, i.e. each word in a tweet is transformed into a corresponding vector in the matrix. The programming language Python 3.6.5 [27] and the Tensorflow library [28], including the Adam routine for adaptive parameter optimization, were used.

Details on Snowballing

Snowball sampling (also known as chain sampling, referral sampling, or chain-referral sampling) is a non-probability and non-random sampling procedure widely employed in social science research, where current study participants recruit future ones using their network and acquaintances [18]. In other words, snowball sampling is based on referrals from initial subjects to generate additional subjects. It is useful used when properties peculiar to the samples are rare and difficult to find. The snowball sampling can be: linear, where the addition of new subjects is limited to one suggestion per subject at a time; exponential, when any subject can suggest new subjects and they can suggest others with no limitations; or exponential discriminative, where some of the referred subjects are excluded based on certain criteria [29].

3   Results

We run the API crawler on historical tweets collected between 2010 and 2013, randomly sampling a month of the year (with the exception of 2012 which was not available). On average, 47 million tweets were available for each month.

The initial keyword set was: "got fired", "lost my job", and "unemployed". Upon snowballing, the following keywords were added: "been fired", "laid off", "kicked out", "was fired" and "were fired". For high-frequency keywords associated to low precision, subsampling was carried out.

After sampling, the total number of distinct tweets including one or more of the keywords was 3,017. The manual annotation was done by two independent raters plus a third for tie breaking, and took about one person's month time. Out of the total, raters' labelling agreed on 2,845 tweets (94% agreement, with a kappa score of 85%), whilst the remaining 172 tweets were flagged as ambiguous and sent to the third rater for further annotation.

Of the 2,845 unambiguous tweets, 1,498 (52.6%) were labelled as a job loss event, whereas 99 of the 172 ambiguous ones (57.6%) were classified as job loss by two out of three raters. Table 1 shows the summary of the snowball keyword sampling and of the labeling process.

Using the unambiguously labelled data set, we passed on to the model optimization and selection phase, using CNN, LSTM, SWCNN, and CNN+LSTM architectures with either 100- or 200-dimensions for the word2vec embedding. For the bag of words input encoding we fit a LASSO logistic regression and a linear SVM optimizing the penalty parameters. Table 2 shows the parameter ranges used for optimization as well as performance obtained from the repeated cross-validation. In terms of parameters, all models reached optimal performance with a 0.001 learning rate and 32 filters for CNN, one 256 dimensional hidden layer for the LSTM, while there were variations in the optimal dropout rate and filter length. Table 3 shows the odds ratios and p-values of the keywords from the logistic regression.

The best model under repeated cross-validation (10×5 runs) using word embedding was a SWCNN with 200-dimensional word embedding, which yielded top performance in all three indices (78% precision, 78% recall and 86% AUC). When using the bag of words, the best model was logistic regression, with LASSO feature selection (84% precision, 84% recall and 91% AUC). When comparing the error distributions across models fitted with the word embedding, using the t-test, we found that performance of CNN, LSTM and CNN+LSTM were, respectively: 75% precision, 74% recall, 82% AUC; 77% precision, 77% recall, 84% AUC; and 77% precision, 77% recall, 84% AUC. The difference in average performance between the SWCNN and the other models were deemed significant at the 5% level. The difference in performance between the best bag of words model and the best word embedding model was yielded a p-value of $1.9 * 10^{-10}$. We also fit a base-minimal model made by a logistic regression combining the keywords presence/absence altogether, which yielded worse precision (68%), recall (68%) and AUC (72%) as compared to any of the deep learners significantly below the 1% alpha-level.

In order to verify if the absolute performance was dependent on the training set size, we retrained the SWCNN on increasing subsamples of the data using repeated cross-validation, and compared the AUC in relation to the sample size. Figure 2 shows that the SWCNN, with both 100- and 200-dimensional word embedding, increases steeply in AUC with sample size below 1,000, and then slowly plateau toward the total training set size, which is indicative that the training set size of choice was adequate.

Finally, we tested the SWCNN model on the ambiguous tweets and on the random Twitter sample not containing any of the keywords (n = 17,490). For the ambiguous set, the model reached a precision of 65.06%, and for the random set the precision was 70.52%.

In terms of speed of annotation, the final SWCNN model was able to label about 0.4 million tweets per minute on a standard medium-end laptop. In detail, the model labeled 17,490 random tweets in 2.5 s on an Apple Macbook with 2.2 GHz Intel Core i7 and 16 GB of memory.

Table 1. Summary of the snowball keyword sampling and of the labeling process

Keyword	Month-year	Number of tweets (total)	Number of tweets containing the keyword	Number of labelled tweets	First annotator's precision	Second annotator's precision	Percent agreement score	Kappa score
got fired	2010-08	41,595,507	582	582	78.87%	82.99%	92.78%	76.69%
lost my job			107	107	69.16%	61.68%	88.78%	75.36%
unemployed			3,159	150	28.67%	26.67%	96.67%	91.68%
laid off			639	150	56.67%	52.00%	90.00%	79.89%
kicked out			2,361	75	1.33%	0.00%	98.67%	0%
been fired			315	150	36.67%	36.67%	97.33%	94.26%
were fired			97	97	27.84%	29.90%	93.81%	84.95%
was fired			368	150	78.87%	82.99%	92.78%	76.69%
got fired	2011-10	21,067,082	240	100	74.00%	72.00%	96.00%	89.86%
lost my job			64	64	53.13%	68.75%	84.38%	68.00%
unemployed			1,285	150	27.33%	24.00%	95.33%	87.78%
laid off			194	194	64.95%	64.95%	95.88%	90.94%
been fired			86	86	33.72%	33.72%	100%	100%
lost my job	2013-01	78,999,809	88	88	71.59%	70.45%	98.86%	97.24%
unemployed			1,940	150	36.67%	37.33%	96.67%	92.85%
laid off			254	247	59.92%	59.11%	98.38%	96.64%
been fired			208	150	38.67%	39.33%	99.33%	98.60%
lost my job	2013-02	48,172,409	118	118	56.78%	56.78%	96.61%	93.09%
unemployed			1,458	150	35.33%	38.00%	96.00%	91.39%
laid off			264	150	36.00%	37.33%	98.67%	97.13%
been fired			225	214	36.92%	39.25%	97.67%	95.05%

Table 2. Models' performances and parameter tuning details

Model	Input	Filter length[a]	Learning rate[b]	Vector dimension[c]	Number of filters[d]	Dropout[e]	LSTM hidden layer size[f]	Regul. coef[g]	10 × 5-fold cross validation		
									Precision	Recall	AUC
CNN	Word2vec	4	0.001	100	32	0.7	-	-	0.75 ± 0.004	0.74 ± 0.004	0.82 ± 0.003
LSTM		-	0.001	100	-	0.5	256	-	0.77 ± 0.004	0.76 ± 0.005	0.84 ± 0.002
CNN+LSTM		3	0.001	100	32	0.7	256	-	0.75 ± 0.006	0.74 ± 0.005	0.82 ± 0.003
SWCNN		3, 4, 5	0.001	100	32	0.7	256	-	0.77 ± 0.003	0.77 ± 0.003	0.86 ± 0.003
CNN		4	0.001	200	32	0.5	-	-	0.74 ± 0.007	0.73 ± 0.006	0.81 ± 0.004
LSTM		-	0.001	200	-	0.5	256	-	0.77 ± 0.005	0.76 ± 0.005	0.84 ± 0.002
CNN+LSTM		3	0.001	200	32	0.7	256	-	0.77 ± 0.006	0.77 ± 0.006	0.84 ± 0.002
SWCNN		3, 4, 5	0.001	200	32	0.5	-	-	**0.78 ± 0.004**	**0.78 ± 0.003**	**0.86 ± 0.003**
Linear SVM	Bag of words	-	-	-	-	-	-	**2**	0.81 ± 0.006	0.81 ± 0.006	0.89 ± 0.004
Logistic regression (LASSO)		-	-	-	-	-	-	**100**	**0.84 ± 0.004**	**0.84 ± 0.004**	**0.91 ± 0.003**
Logistic regression	Keywords	-	-	-	-	-	-	-	0.67 ± 0.000	0.68 ± 0.001	0.74 ± 0.001

a[3, 4, 5]
b[0.0001, 0.001, 0.01, 0.1]
c[100, 200]
d[2, 4, 8, 16, 32, 64, 128, 256]
e[0.2, 0.5, 0.7]
f[128, 256, 512]
g$[2e-5, 2e-4, 2e-3, \ldots, 2e15]_{SVM}$ and $[1e-6, 1e-5, \ldots, 1e6]_{LASSO}$

Table 3. Keywords importance from logistic regression

Variable	Odds ratio [95% confidence interval]	P-value
Intercept	0.7372 [0.3471–1.5654]	0.4275
"lost my job"	2.4025 [1.0986–5.2537]	0.0281
"got fired"	5.8369 [2.7199–12.5261]	5.94e−6
"been fired"	0.7710 [0.3618–1.6429]	0.5004
"unemployed"	0.6429 [0.2999–1.3781]	0.2562
"laid off"	1.8900 [0.8824–4.0481]	0.1014
"kicked out"	0.0352 [0.0073–0.1701]	3.11e−5
"was fired"	2.0097 [0.8204–4.9230]	0.1267
"were fired"	0.5032 [0.2061–1.2280]	0.1314

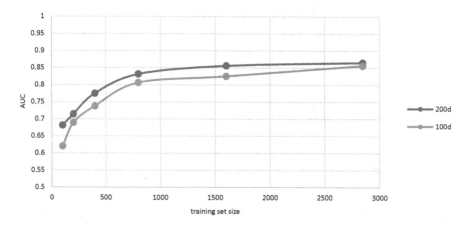

Fig. 2. SWCNN performance by increasing training set size

4   Discussion

In this work, we presented a combined man-machine modelling pipeline for characterizing life events from Twitter. Our pipeline takes advantage of a robust model selection as well as two feedback loops (in the manual and validation phases) that can improve quality of data collection and prediction performance.

In regards to the neural architecture selection for text classification, here we empirically compared different architectures with word embedding, i.e. CNN, LSTM, CNN+LSTM, and SWCNN, each of which had no extra hidden layer (CNN had one convolutional layer, LSTM had only one recurrent layer, CNN+LSTM had one convolutional hidden layer and one recurrent hidden layer, and SWCNN had only one hidden convolutional layer). In our use case, the SWCNN performed best. Le et al. [21] previously claimed that a shallow neural network, i.e. one hidden layer, is usually enough for text classification. Yin et al. [24] claimed that, while a CNN is good at extracting local and position-invariant features, recurrent architectures can capture the structure-dependent information of the whole input. They also concluded based on a qualitative analysis that RNN would outperform CNN if a sentence contains no less than 10 tokens. For our experiments, each input has 37 tokens, and LSTM, which is a variant of RNN, performed better than the CNN, but not than SWCNN. Zhang et al. [25] pointed out that 1-max pooling is the best choice, and combining several filters with region sizes close to the optimal single one would enhance the performance. For this reason we applied 1-max pooling for CNN and SWCNN, and we also found that 4 was the best region size for convolutional layer.

When comparing the word embedding + deep learning with other approaches, we found that they largely outperform a linear score based on keywords' weighting, but are inferior to logistic regression that uses selection of bag of words. One possible reason might be the small size of our dataset. Deep learning methods usually outperform traditional ones in large datasets (e.g. million-level sample size) while traditional machine learning methods performed better in data that contains only hundreds or thousands of samples [25].

The relatively small sample size is one of the limitations of our work. Another limitation of this study is that the manual annotation step requires considerable effort, in relation to: (a) keyword search and snowballing, (b) potential high number of tweets that need to be annotated in order to reach good prediction performance, and (c) because of potential ambiguities in the texts. This is also highly dependent on the outcome of choice. Using crowdsourcing reduces time and increases quantity, at a price of quality (although annotations can be weighted by rater's reliability) and potential added monetary costs.

In regards to contribution of confounders or other predictors to the outcome variance, rather than the tweet texts, the demographics (e.g. age, gender, race, and ethnicity) of Twitter users might need to be controlled. Nevertheless, there is not an easy way to identify Twitter users' demographics, as Twitter does not require its users to provide such information.

Another critical limitation is sampling bias. First, although Twitter has a rich set of tools and a relatively open policy for data collection, gathering relevant data to answer a specific scientific question is not easy. Second, even with a list of well-developed keywords, the data had many false positives, which affirms the necessity of building incremental classifiers to further narrow the search results. Finally, the users of social media tend to be younger and not representative of the real population. All these issues are likely to create sample bias and miss important information for representing a broader population. The feedback loops that we introduced are indeed a practical implementation of active learning [23] and have the purpose to reduce in part the sampling bias occurring from search terms.

In terms of extensions to the approach, the pipeline can be easily extended to include different NLP approaches in addition to word embedding and bag of words, e.g. TF-IDF, n-grams, linguistic inquiry and word count [2], as well as other machine learning techniques. The repeated cross-validation ensures robust parameter and model selection and the testing on error distributions allows the least complex model to be selected (e.g. in terms of free parameters) by maintaining optimal performance. This can be useful especially in the case where interpretable techniques are used, such as decision trees or rules, helpful if the problem requires evaluation of feature importance and interactions besides mere prediction performance, e.g. for mechanistic or causal studies, or hypothesis testing.

Another possible extension of the pipeline is in relation to study design. For instance, a longitudinal design could include data mining on tweet histories where past data is used to predict future life events, i.e. the history of a single user's tweets before a time t is input for a life event that is described in a tweet at time t or after. In this case, a time-series word embedding should be defined, and likely a recurrent deep learning architecture to capture the temporal signal.

In summary, our tweet annotation pipeline streamlines both the manual and automated process, providing feedback reinforcement (snowballing and external tweets), and shows good performance on classifying individual tweets on the use case, potentially saving human resources needed to collate such data for subsequent research studies.

5　Conclusion

Automatization, optimization and accuracy of annotation are key to research that lies on quality data foundations, such as survey and behavioral studies. Social media are a large, potentially invaluable source for such studies, but their information content is highly unstructured and noisy. Therefore, systematization of social media content for research is highly needed.

By introducing a reproducible procedure, Twitter analytics becomes as legit as any research based on traditional survey methods, like questionnaires or analysis of structured data bases. Nonetheless, the initial collaborative man-aided step is still a necessity in absence of reliable self-organized learning of concepts, and needs to be optimized to allow the machine to learn effectively from the human-labelled data.

Acknowledgements. MP, JB, and XD are in part supported by US NSF grant SES 1734134.

References

1. Reece, A.G., Reagan, A.J., Lix, K.L.M., Dodds, P.S., Danforth, C.M., Langer, E.J.: Forecasting the onset and course of mental illness with Twitter data. Sci. Rep. **7**(1), 13006 (2017). https://doi.org/10.1038/s41598-017-12961-9
2. Tausczik, Y.R., Pennebaker, J.W.: The psychological meaning of words: LIWC and computerized text analysis methods. J. Lang. Soc. Psychol. **29**(1), 24–54 (2010). https://doi.org/10.1177/0261927X09351676
3. Alsaedi, N., Burnap, P., Rana, O.: Can we predict a riot? Disruptive event detection using Twitter. ACM Trans. Internet Technol. **17**(2), 18:1–18:26 (2017). https://doi.org/10.1145/2996183
4. Sumner, C., Byers, A., Boochever, R., Park, G.J.: Predicting dark triad personality traits from Twitter usage and a linguistic analysis of tweets. In: 2012 11th International Conference on Machine Learning and Applications, vol. 2, pp. 386–393 (2012). https://doi.org/10.1109/ICMLA.2012.218

5. Makazhanov, A., Rafiei, D.: Predicting political preference of Twitter users. In: 2013 IEEE/ACM International Conference on Advances in Social Networks Analysis and Mining (ASONAM 2013), pp. 298–305 (2013). https://doi.org/10.1145/2492517.2492527

6. Conover, M., Gonçalves, B., Ratkiewicz, J., Flammini, A., Menczer, F.: Predicting the political alignment of Twitter users. In: 2011 IEEE Third International Conference on Privacy, Security, Risk and Trust and 2011 IEEE Third International Conference on Social Computing, pp. 192–199 (2011). https://doi.org/10.1109/PASSAT/SocialCom.2011.34

7. Ramos, J.P.: Using TF-IDF to determine word relevance in document queries. Presented at the First International Conference on Machine Learning, New Brunswick, NJ, USA (2003)

8. Hofmann, T.: Unsupervised learning by probabilistic latent semantic analysis. Mach. Learn. **42**(1–2), 177–196 (2001). https://doi.org/10.1023/A:1007617005950

9. Won, D., Steinert-Threlkeld, Z.C., Joo, J.: Protest activity detection and perceived violence estimation from social media images. ArXiv:1709.06204 [Cs] (2017). http://arxiv.org/abs/1709.06204

10. Zhang, Z., He, Q., Gao, J., Ni, M.: A deep learning approach for detecting traffic accidents from social media data. Transp. Res. Part C: Emerg. Technol. **86**, 580–596 (2018). https://doi.org/10.1016/j.trc.2017.11

11. Founta, A.-M., Chatzakou, D., Kourtellis, N., Blackburn, J., Vakali, A., Leontiadis, I.: A unified deep learning architecture for abuse detection. ArXiv:1802.00385 [Cs] (2018). http://arxiv.org/abs/1802.00385

12. Badjatiya, P., Gupta, S., Gupta, M., Varma, V.: Deep learning for hate speech detection in tweets. In: Proceedings of the 26th International Conference on World Wide Web Companion, pp. 759–760. International World Wide Web Conferences Steering Committee, Republic and Canton of Geneva, Switzerland (2017). https://doi.org/10.1145/3041021.3054223

13. Pitsilis, G.K., Ramampiaro, H., Langseth, H.: Detecting offensive language in tweets using deep learning. ArXiv:1801.04433 [Cs] (2018). http://arxiv.org/abs/1801.04433

14. Pennington, J., Socher, R., Manning, C.: Glove: global vectors for word representation, pp. 1532–1543. Association for Computational Linguistics (2014). https://doi.org/10.3115/v1/D14-1162

15. Mikolov, T., Sutskever, I., Chen, K., Corrado, G.S., Dean, J.: Distributed representations of words and phrases and their compositionality, p. 9 (n.d.)

16. Bian, J., et al.: Mining Twitter to assess the public perception of the "internet of things". PLoS One **11**(7), e0158450 (2016). https://doi.org/10.1371/journal.pone.0158450

17. McHugh, M.L.: Interrater reliability: the kappa statistic. Biochem. Med. **22**(3), 276–282 (2012)

18. Goodman, L.A.: Snowball sampling. Ann. Math. Stat. **32**(1), 148–170 (1961). https://doi.org/10.1214/aoms/1177705148

19. Nadeau, C., Bengio, Y.: Inference for the generalization error. In: Proceedings of the 12th International Conference on Neural Information Processing Systems, pp. 307–313. MIT Press, Cambridge (1999). http://dl.acm.org/citation.cfm?id=3009657.3009701

20. LeCun, Y., Bengio, Y.: Convolutional networks for images, speech, and time-series, p. 15 (n.d.)

21. Le, H.T., Cerisara, C., Denis, A.: Do convolutional networks need to be deep for text classification? ArXiv:1707.04108 [Cs] (2017). http://arxiv.org/abs/1707.04108

22. Hochreiter, S., Schmidhuber, J.: Long short-term memory. Neural Comput. **9**(8), 1735–1780 (1997). https://doi.org/10.1162/neco.1997.9.8.1735

23. Das, S., Wong, W.K., Dietterich, T., Fern, A., Emmott, A.: Incorporating expert feedback into active anomaly discovery. In: 2016 IEEE 16th International Conference on Data Mining (ICDM), pp. 853–858 (2016). https://doi.org/10.1109/ICDM.2016.0102

Using Behavior and Text Analysis to Detect Propagandists and Misinformers on Twitter

Michael Orlov[1]([⊠]) and Marina Litvak[2]([⊠])

[1] NoExec, Beer Sheva, Israel
orlovm@noexec.org
[2] Shamoon College of Engineering, Beer Sheva, Israel
marinal@ac.sce.ac.il

Abstract. There are organized groups that disseminate similar messages in online forums and social media; they respond to real-time events or as persistent policy, and operate with state-level or organizational funding. Identifying these groups is of vital importance for preventing distribution of sponsored propaganda and misinformation. This paper presents an unsupervised approach using behavioral and text analysis of users and messages to identify groups of users who abuse the Twitter micro-blogging service to disseminate propaganda and misinformation. Groups of users who frequently post strikingly similar content at different times are identified through repeated clustering and frequent itemset mining, with the lack of credibility of their content validated through human assessment. This paper introduces a case study into automatic identification of propagandists and misinformers in social media.

Keywords: Propaganda · Misinformation · Social networks

1 Introduction

The ever-growing popularity of social networks influences everyday life, causing us to rely on other people's opinions when making large and small decisions, from the purchase of new products online to voting for a new government. It is not surprising that by spreading disinformation and misinformation social media became a weapon of choice for manipulating public opinion. Fake content and propaganda are rampant on social media and must be detected and filtered out. The problem of information validity in social media has gained significant traction in recent years, culminating in large-scale efforts by the research community to deal with "fake news" [7], clickbait [6], "fake reviews" [2], rumors [8], and other kinds of misinformation.

We are confident that detecting and blocking users who disseminate misinformation and propaganda is a much more effective way of dealing with fake content, as it enables prevention of its massive and consistent distribution in social media. Therefore, in this paper we deal with detection of propagandists.

© Springer Nature Switzerland AG 2019
J. A. Lossio-Ventura et al. (Eds.): SIMBig 2018, CCIS 898, pp. 67–74, 2019.
https://doi.org/10.1007/978-3-030-11680-4_8

We define propagandists as groups of people who intentionally spread misinformation or biased statements, typically receiving payment for this task, similarly to the definition of "fake reviews" disseminators in [2]. An article in the Russian-language Meduza media outlet [12] describes one example of paid propagandists performing their task on a social network[1] while neglecting to delete the task description and requirements, as illustrated in Fig. 1.

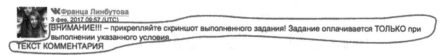

Fig. 1. A comment on VK social network that includes paid propaganda task description. The marked section translates as: *"ATTENTION!!!—attach a screenshot of the task performed! The task is paid ONLY when this condition is fulfilled. TEXT OF COMMENT."* The rest of the post promotes a municipal project.

Twitter is one of the most popular platforms for dissemination of information. We would expect Twitter to attract focused attention of propagandists—organized groups who disseminate similar messages in online forums and social media, in response to real-time events or as a persistent policy, operating with state-level or organizational funding. We explore, below, an unsupervised approach to identifying groups of users who abuse the Twitter micro-blogging service to disseminate propaganda and misinformation. This task is accomplished via behavioral analysis of users and text analysis of their content. Users who frequently post strikingly similar content at different times are identified through repeated clustering, and their groups are subsequently identified via frequent itemset mining. The lack of credibility of their content is validated manually. The most influential disseminators are detected by calculating their PageRank centrality in the social network and the results are visualized. Our purpose is to present a case study into automatic identification of propagandists in social media.

2   Related Work

The subject of credibility of information propagated on Twitter has been previously analyzed. Castillo et al. [5] observed that while most messages posted on Twitter are truthful, the service also facilitates spreading misinformation and false rumors. Dissemination of false rumors under critical circumstances was analyzed in [14], and the aggregation analysis on tweets was performed in order to

---

[1] VK is a social network popular in Russia, see https://vk.com.

differentiate between false rumors and confirmed news. Discussion about detecting rumors and misinformation in social networks remains very popular nowadays. Authors of [3] demonstrate the importance of social media for fake news suppliers by measuring the source of their web traffic. Hamidian and Diab [8] performed supervised rumors classification using the tweet latent vector feature. Large-scale datasets for rumor detection were built in [17] and [21].

However, not much attention has been paid to detection of *propagandists* in social media. Some works used the term *propaganda* in relation to spammers [13]. Metaxas [15]associated the theory of propaganda with the behavior of web spammers and applied social anti-propagandistic techniques to recognize trust graphs on the web. Lumezanu et al. [11] studied the tweeting behavior of assumed Twitter propagandists and identified tweeting patterns that characterize them as users who consistently express the same opinion or ideology. The first attempt to automatically detect propaganda on Twitter was made in [20], where linguistically-infused predictive models were built to classify news posts as suspicious or verified, and then to predict four subtypes of suspicious news, including propaganda.

In this paper, we address the problem of *automatically identifying paid propagandists*, who have an agenda, but do not necessarily spread false rumors, or even false information. This problem is principally different from what had been stated in other papers, classifying propaganda as rumor or equating it with spam, which is a much wider concept. Our approach is very intuitive and unsupervised.

3 Methodology

When using Twitter as an information source, we would like to detect tweets that contain propaganda[2], and users who disseminate it. We assume that propaganda is disseminated by professionals who are centrally managed and who have the following characteristics (partly supported by [11]): (1) They work in groups; (2) Disseminators from the same group write very similar (or even identical) posts within a short timeframe; (3) Each disseminator writes very frequently (within short intervals between posts and/or replies); (4) One disseminator may have multiple accounts; as such, a group of accounts with strikingly similar content may represent the same person; (5) We assume that propaganda posts are primarily political; (6) The content of tweets from one particular disseminator may vary according to the subject of an "assignment," and, as such, each subject is discussed in disseminator's accounts during some *temporal frame* of its relevance; (7) Propaganda carries content similar to an official governance "vision" depicted in mass media.

---

[2] Propaganda is defined as: *"posts that contain information, especially of a biased or misleading nature, that is used to promote or publicize a particular political cause or point of view"* (Oxford English Dictionary, 3rd Online Edition).

Based on the foregoing assumptions, we propose to perform the following analysis for detection of propagandists:

- Based on (1) and (2), given a time dimension, repeatedly cluster tweets posted during the same time interval (timeframe), based on their content. For each run, a group of users who posted similar posts (clustered together) can be obtained. Given N runs for N timeframes, we can obtain a group of users who consistently write similar content—these are users whose tweets were clustered together in most of the runs. The retrieved users can be considered good suspects for propaganda dissemination.
- Based on (3), the timeframes must be small, and clustering must be performed quite frequently.
- Based on (4), we do not distinguish between different individuals. Our purpose is to detect a set of accounts, where each individual (propagandist) can be represented by a single account or by a set of accounts.
- Based on (5), we can verify the final results of our analysis and see whether the posts published from the detected accounts indeed contain political content.
- Based on (6) and (7), we collect data that belongs to content that is discussed in mass media.

We outline, below, the main algorithm steps for the proposed methodology.

1. *Filtering and pre-processing tweets.* We consider only tweets in English and perform standard preprocessing using tokenization, stopword removal, and stemming. We also filter out numbers, non-textual content (like emoji symbols), and links.
2. *Split data set into timeframes.* We split the data set into N timeframes, so that each split contains tweets posted at the same period of time (between two consecutive timeframes n_i and n_{i+1}). The timeframes must be relatively short, according to assumption (3).
3. *Cluster tweets at each timeframe.* We cluster tweets at each timeframe n_i in order to find a group of users who posted similar content (clustered together). K-means has been chosen as the unsupervised clustering method, using the elbow method to determine the optimal number of clusters. The simple vector space model [18] with adapted tf-idf weights[3] was used for tweets representation. We denote the clustering results (set of clusters) for timeframe n_i by $C_i = \{c_{i_1}, c_{i_2}, \ldots, c_{i_k}\}$. The final clusters are composed of user IDs (after replacing tweet IDs by IDs of users who posted them), therefore the clusters are not disjointed.
4. *Calculate groups of users[4] frequently clustered together.* We scan the obtained clusters and, using adapted version of the AprioriTID algorithm [1,10], compute groups of users whose posts were frequently clustered together. We start from generating a list L_1 from all single users u_i appearing in at least T (the minimum threshold specified by the user) timeframes. Then, we generate a

---

[3] A tweet was considered as a document, and collection of all tweets as a corpus.
[4] By "user" we mean account and not individual, based on assumption (4).

list of pairs $L_2 = \{\langle u_l, u_m, i \rangle, u_l \in L_1, u_m \in L_1\}$ of users that are clustered together in at least T timeframes. According to the Apriori algorithm, we then join pairs from L_2 in order to obtain L_3 and so forth. This step is necessary if we want to detect organized groups of propaganda disseminators.

5. *Identifying the most influential disseminators with PageRank centrality.* We construct an undirected graph, with nodes standing for users. We add an edge between two users if they have been clustered together at least once (in one timeframe). The weights on edges are proportional[5] to the number of times they were clustered together. As an option, edges having weights below the specified threshold t can be removed from the graph. We calculate PageRank centrality on the resultant graph and keep the obtained scores for detected accounts as a disseminator's "influency" measure, as illustrated in Fig. 2. Using an eigenvalue centrality metric for measuring influence in graph structure of a social network considers its "recursive" nature. For example, in [2] HITS algorithm [9] is adapted for computing the honesty of users and goodness of products.

6. *Visualize the "dissemination" network structure and analyze results.* We visualize the graph obtained in the prior step, where the PageRank centrality for each node affects its size. We also apply topic modeling in order to visualize main topics in the content that was detected as propaganda.

Row ID	S Node	D PR
0	Col_Connaughton	2.895
32	syrializer	2.277
3	mrsn_34	2.087
26	AmericanSyrians	1.983
16	AgendaOfEvil	1.873
19	awesomeseminars	1.826
12	Shababeeksouria	1.794
28	VitalAnon	1.559
9	mooredavid1970	0.967
17	ferozwala	0.917
23	Eve_on_Syria	0.905

Fig. 2. Partial example of a list of PageRank centrality values that were computed for the disseminators graph in step 5 above.

The algorithm's flow is shown in Fig. 3.

---

[5] Edge weights are normalized to be in range of $[0, 1]$.

4 Case Study

Dataset. Military airstrikes in Syria in September 2017 attracted worldwide criticism. Reflection of these events in Twitter can be tracked using the keyword *#syria*, determined via Hamilton 68 [19] as the most popular hashtag for 600 monitored Twitter accounts that were linked to Russian influence operations. Our case study was carried out on a dataset obtained from Twitter, collected using the Twitter Stream API with the *#syria* hashtag. The dataset covers 10,848 tweets posted by 3,847 users throughout September 9–12, 2017.

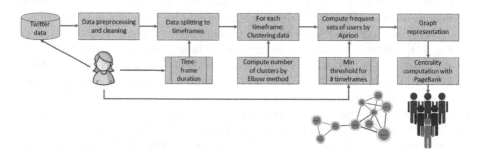

Fig. 3. Pipeline for detecting users who consistently post similar content.

Parameters/Settings. We performed clustering with 10 ($K = 10$) clusters, as an optimal clusters number according to the elbow method, 8 ($N = 8$) times (every 12 h, according to our assumption that organized propagandists work regular hours), and looked for a group of accounts that consistently (all timeframes without exceptions, with $T = 100\%$) post similar content.

Tools. We have implemented the above-described process in KNIME, a data analytics, reporting, and integration platform [4].

Results. Our algorithm detected seven suspicious accounts. The content of messages posted by these accounts confirmed our suspicions of organized propaganda dissemination. Speaking formally, we manually approved 100% of precision. However, the recall was not measured due to the absence of manual annotation for all accounts in our data.

Topic modeling[6] results confirmed that most topics in the detected posts aligned well with political propaganda vocabulary. For example, the top topic words *attack, russia, report, isis, force, bomb, military* represent Russia's military operations in Syria, and *trump, attack, chemical, false, flag, weapons* represent an insinuated American undercover involvement in the area.

Activity analysis of the detected accounts confirmed assumption (3) about propagandists posting significantly more frequently than regular Twitter users. While regular users had 12.8 h mean time between posts, propagandists featured 1.8 h mean time. This assumption has been also confirmed by empirical analysis in [20].

---

[6] Topic modeling was performed using KNIME's LDA implementation.

5   Conclusions

This paper introduces initial stages in our research related to automatic detection of propagandists, based on analysis of users' behavior and messages. We propose an intuitive unsupervised approach for detecting Twitter accounts that disseminate propaganda. We intend to continue this research in several directions: (a) Extend our experiments with respect to other (baseline) methods, commercial domains, and various (standard) IR evaluation metrics; (b) Evaluate in depth the contribution of each separate stage of our pipeline; (c) Incorporate additional (or alternative) techniques, like topic modeling, graph clustering, or analyzing web traffic of news sources, into our pipeline; (d) Adapt and apply our approach to tweets written in different languages, with focus on Russian, due to high popularity of Twitter among organized dissemination groups [16]; (e) Combine the proposed approach with authorship analysis to detect actual users that might use several accounts, according to assumption (4); (f) Perform geolocation prediction and analysis on the detected accounts to provide additional important information related to geographical distribution of organized propaganda dissemination activity; (g) Perform supervised classification of detected tweets for more accurate analysis; (h) Incorporate retweeting statistics into our network centrality analysis (step 6) to detect the most influential disseminators.

Our approach can be of great assistance in collecting a high quality dataset of propaganda and its disseminators, which then can be used for training supervised predictive models and for automatic evaluations. An automatic evaluation of our approach can be performed via verification of automatically detected accounts with accounts identified by public annotation tools, such as PropOrNot[7].

References

1. Agrawal, R., Mannila, H., Srikant, R., Toivonen, H., Verkamo, A.I., et al.: Fast discovery of association rules. Adv. Knowl. Discov. Data Min. **12**(1), 307–328 (1996)
2. Akoglu, L., Chandy, R., Faloutsos, C.: Opinion fraud detection in online reviews by network effects. In: ICWSM 2013, pp. 2–11 (2013)
3. Allcott, H., Gentzkow, M.: Social media and fake news in the 2016 election. J. Econ. Perspect. **31**(2), 211–36 (2017)
4. Berthold, M.R., et al.: KNIME: the Konstanz information miner. In: Preisach, C., Burkhardt, H., Schmidt-Thieme, L., Decker, R. (eds.) Data Analysis, Machine Learning and Applications. Studies in Classification, Data Analysis, and Knowledge Organization, pp. 319–326. Springer, Heidelberg (2008). https://doi.org/10.1007/978-3-540-78246-9_38
5. Castillo, C., Mendoza, M., Poblete, B.: Information credibility on Twitter. In: Proceedings of the 20th International Conference on World Wide Web, WWW 2011, pp. 675–684. ACM, New York (2011). https://doi.org/10.1145/1963405.1963500
6. Chen, Y., Conroy, N.J., Rubin, V.L.: Misleading online content: recognizing clickbait as false news. In: Proceedings of the 2015 ACM on Workshop on Multimodal Deception Detection, pp. 15–19. ACM (2015)

---

[7] See http://www.propornot.com.

7. Conroy, N.J., Rubin, V.L., Chen, Y.: Automatic deception detection: methods for finding fake news. Proc. Assoc. Inf. Sci. Technol. **52**(1), 1–4 (2015)
8. Hamidian, S., Diab, M.T.: Rumor identification and belief investigation on Twitter. In: WASSA@ NAACL-HLT, pp. 3–8 (2016)
9. Kleinberg, J.M., Kumar, R., Raghavan, P., Rajagopalan, S., Tomkins, A.S.: The web as a graph: measurements, models, and methods. In: Asano, T., Imai, H., Lee, D.T., Nakano, S., Tokuyama, T. (eds.) COCOON 1999. LNCS, vol. 1627, pp. 1–17. Springer, Heidelberg (1999). https://doi.org/10.1007/3-540-48686-0_1
10. Li, Z.C., He, P.L., Lei, M.: A high efficient AprioriTid algorithm for mining association rule. In: 2005 International Conference on Machine Learning and Cybernetics, vol. 3, pp. 1812–1815. IEEE, August 2005. https://doi.org/10.1109/ICMLC.2005.1527239
11. Lumezanu, C., Feamster, N., Klein, H.: #bias: measuring the tweeting behavior of propagandists. In: Sixth International AAAI Conference on Weblogs and Social Media (2012). http://www.aaai.org/ocs/index.php/ICWSM/ICWSM12/paper/view/4588
12. Meduza: Authors of paid comments in support of Moscow authorities forgot to edit assignment (2017). https://meduza.io/shapito/2017/02/03/avtory-platnyh-kommentariev-v-podderzhku-moskovskih-vlastey-zabyli-otredaktirovat-zadanie
13. Mehta, B., Hofmann, T., Fankhauser, P.: Lies and propaganda: detecting spam users in collaborative filtering. In: Proceedings of the 12th International Conference on Intelligent User Interfaces, pp. 14–21. ACM (2007). https://doi.org/10.1145/1216295.1216307
14. Mendoza, M., Poblete, B., Castillo, C.: Twitter under crisis: can we trust what we RT? In: Proceedings of the First Workshop on Social Media Analytics, SOMA 2010, pp. 71–79. ACM, New York (2010). https://doi.org/10.1145/1964858.1964869
15. Metaxas, P.: Using propagation of distrust to find untrustworthy web neighborhoods. In: 2009 Fourth International Conference on Internet and Web Applications and Services, ICIW 2009, pp. 516–521. IEEE (2009). https://doi.org/10.1109/ICIW.2009.83
16. Paul, C., Matthews, M.: The Russian "Firehose of Falsehood" Propaganda Model. RAND Corporation, Santa Monica (2016)
17. Qazvinian, V., Rosengren, E., Radev, D.R., Mei, Q.: Rumor has it: identifying misinformation in microblogs. In: Proceedings of the Conference on Empirical Methods in Natural Language Processing, EMNLP 2011, pp. 1589–1599. Association for Computational Linguistics, Stroudsburg (2011). https://www.aclweb.org/anthology/D11-1147
18. Salton, G., Wong, A., Yang, C.S.: A vector space model for automatic indexing. Commun. ACM **18**(11), 613–620 (1975)
19. The Alliance for Securing Democracy: Hamilton 68 (2017). https://dashboard.securingdemocracy.org
20. Volkova, S., Shaffer, K., Jang, J.Y., Hodas, N.: Separating facts from fiction: linguistic models to classify suspicious and trusted news posts on Twitter. In: Proceedings of the 55th Annual Meeting of the Association for Computational Linguistics, Short Papers, vol. 2, pp. 647–653 (2017)
21. Zubiaga, A., Liakata, M., Procter, R., Bontcheva, K., Tolmie, P.: Towards detecting rumours in social media. In: AAAI Workshop: AI for Cities (2015)

Analyzing the Retweeting Behavior of Influencers to Predict Popular Tweets, with and Without Considering their Content

Matías Gastón Silva[✉], Martín Ariel Domínguez[✉],
and Pablo Gabriel Celayes[✉]

FaMAF, Universidad Nacional de Cordoba, Córdoba, Argentina
{mgs0113,mdoming,celayes}@famaf.unc.edu.ar

Abstract. Twitter and social networks in general, participate more and more in everyday life. This is why they have become a fundamental source of information that reflects the ideas and opinions of their users. This paper shows how the most influential users, called influencers, can be decisive in defining whether a publication becomes popular or not, regardless of its content. To achieve this, we build a dataset of Spanish-writing users sampled from Twitter, along with the content generated and shared by them within a year. In a first phase, we use different algorithms to detect users who are "influencers". In a second phase, we train a binary classifier to predict if a given tweet will be a trending publication, based on information about the activity of the influencers on the given tweet. We obtain a model with an F_1-score close to 79%, based on the retweeting behavior of a 10% of the users dataset considered as influencers. Finally, we add two Natural Language Processing (NLP) techniques to analyze the content: Twitter-LDA topic modeling, and FastText word embeddings. While both models alone have an F_1 of less than 50% for trending prediction, FastText combined with the social model reaches an 86.7% score. We conclude that while analyzing the content can help to predict the popularity of a tweet, the influence of a user's environment in the retweeting decision is surprisingly high.

Keywords: Retweet prediction · Social Network Analysis · Machine learning · LDA · FastText · Word embeddings

1 Introduction

The evolution of technology and the constant growth of its infrastructure allow us to be connected to our social networks, anytime, anywhere. Because of this state of permanent communication, social networks today are a vast reservoir of valuable information. One example of how this data is used to the advantage of businesses is the marketing field, where this kind of information is used to learn about the tastes and needs of the population to promote brands. In this sense,

© Springer Nature Switzerland AG 2019
J. A. Lossio-Ventura et al. (Eds.): SIMBig 2018, CCIS 898, pp. 75–90, 2019.
https://doi.org/10.1007/978-3-030-11680-4_9

influencers have been acknowledged as message replicators and, as so, they are also used as marketing tools. Political campaigns are another instance of the use of social network data. Campaigners develop massive communication strategies that direct specific messages, even fake news, based on profiled users. Therefore, the analysis of these data becomes essential to understand social phenomena and its impact on how a piece of content can be massively spread.

This work attempts to contribute to the understanding of how publications in social networks become popular. In particular, we concentrate on trying to quantify the importance of the behavior of central users in the propagation of information. More specifically, this work is done on Twitter, an online real-time social network, where users can post, read and share information in multiple formats, mostly in the form of short text messages (originally 140 characters, extended to 280 characters in late 2017). In this case, we only analyze written content. Twitter tags each post with a unique timestamp and places the publication on the timeline of its emitter. The users and their timelines are mostly public and can be downloaded through the public API provided by Twitter. On this social network, users have a front page where they can find posts from the people they follow. If someone thinks a message is of interest or likes the content, she can republish it over her timeline. This action is called *retweeting* and represents, at least for us, acceptance of the tweet[1]. The repetition of the retweeting action by multiple users on a given post is the way in which a publication becomes "popular" in Twitter. Consequently, the subject of the tweet becomes a trending topic.

To address the issue of how a tweet becomes a trending topic, in a first phase, we evaluate different algorithms to effectively detect influencers, which will allow us to rank them by importance. In a second stage, we separate a part of the most influential users and use their retweeting activity to train a binary classifier over tweets. The set of selected features refers to whether a portion of these central users has shared the tweet or not. The target binary variable is whether each publication is popular or not. A tweet is defined as popular if it has been retweeted more than a certain number of times, which we will establish opportunely. The model obtained is evaluated on a set of unseen tweets, reaching an F_1 score of 79.2% in predicting which tweets are popular. Note that these predictions were made without taking into account the content of publications. Subsequently, we add two NLP techniques to analyze the content: word embeddings, with the FastText [10] algorithm, and a Twitter-specific adaptation of the Latent Dirichlet Allocation (LDA)[30] topic modeling technique. The result of combining the model based on central users behavior with FastText, reaches a performance of 86.7%, taking 10% of the users ranked as influencers.

---

[1] We assume that acceptance is the most usual way to use a retweet. However, it is true that not always a retweet represents acceptance, in some cases a retweet could be used to be ironic about a publication, or also, to make visible some topic with which we disagree.

Summarizing, the present work was carried out in the following phases:

- Construction of datasets: a set of Twitter users, the network of follower relations among them and a set of tweets produced or shared by them.
- Selection of an influencer detection algorithm.
- Study of the network of selected users and detection of most relevant ones in terms of activity and network position, splitting users in two groups: a set of ranked influencers and a set of regular users.
- Comparison of models to learn and predict general retweeting preferences on a dataset of tweets, based on information about the influencers set.
- Study of possible improvements to social prediction models, introducing NLP techniques such as topic modeling and sentence embeddings.

The rest of this paper is structured as follows: In Sect. 2, we analyze related works in the area, comparing them to our work. In Sect. 3, we describe how we build the datasets from Twitter for our experiments. Next, in Sect. 4, we describe the details of the construction of our social model for prediction of popular tweets. We also include information on how we add content-based features using the Twitter LDA topic modeling and FastText word embeddings. Finally, Sect. 5 contains the analysis of the results obtained and in Sect. 6, we present our conclusions and possible lines of future research.

2   Related Work

Along with the evolution of social networks, the academic studies based on them have increased in quantity and quality, with many works studying the problem of predicting popular or viral content.

A recurring topic among these works is the analysis of the content of the publications as in [11,22,28]. In particular, in [11], a genetic algorithm is proposed to optimize the composition of the message to increase its outreach. In this case, the authors take a different approach from ours, generating a simulation over an artificial network similar to Twitter, where nodes decide in a deterministic way whether or not to retweet a given message. Here, the focus is on the generation of content, without considering social features. Among these purely content-based works, [17] is more closely related to our study. They develop purely content-based models for predicting the likelihood of a given tweet being retweeted by general users. The performance of their models is reported only through ROC curves, without providing any overall performance score to establish a precise quantitative comparison to our model. However, a visual comparison between their ROC curves and the ones produced by our final models indicates a higher AUC score in our results. This study also provides a feature importance analysis, which produces very revealing insights about what makes a tweet popular.

Another point of view, more similar to ours, is the focus on the social environment of users rather than the content being spread, which can be found in [26,29]. In [29], the authors work with different mechanisms to infer when people are likely to initiate a new activity. After the experiments, the conclusion

was that the testimonial comments of neighbors were more relevant than promotion messages showing the advantages of such activity. In addition to the
increase in registration, permanence was also improved more by peers influence
than by typical promotion. As expected, without any promotion, the inscription
and permanence rates were much lower than the ones in the scenarios described
above. This case is a practical experiment that only shows the conclusions, but
no models are provided at the end of the investigation. In [6], the author predicts retweets from a given user based mostly on the retweeting behavior in her
second-degree social neighborhood with an average F_1-score of 87.9%. Our work
tries to expand this idea to a more general model, focusing on a community
instead of a single user.

Finally, the work in [18,27] conducts trendy research that analyzes the flow
of fake information. Here the authors evaluate the propagation of fake news over
Twitter and find out that this kind of news is more viralized than real ones.
Another revealing insight was that the propagation was faster for publications
with fake information. Once again, this work gives more importance to the content, but it also captures the idea of influencing users by a synthetic environment
with fake content or users.

3   Dataset

In this section, we describe the dataset used in this work for all experiments.
The base dataset (social graph and tweets) is taken from the previous work [6].
We extend this base with more content (almost double), keeping the same social
graph of users. We explain the construction of our dataset in two steps: first
building the social graph of users and then getting content shared by them.

3.1   Social Graph

To perform the experiments of this paper, we reuse a dataset created for the
previous work [6], which contains Twitter users and the who-follows-whom relations between them. Back then, the idea was to create a minimal representative
dataset of Twitter where all users would have a similar amount of social information about their neighborhood of connected users. The decision was to build
a homogeneous network where each user has the same number of followed users.

To this end, a two-step process was performed. Initially, a large enough *universe graph* was built, which was subsequently filtered to obtain a smaller but
more homogeneous subgraph.

The *universe graph* was built starting with a singleton graph containing just
one Twitter user account $\mathcal{U}_0 = \{u_0\}$ and performing 3 iterations of the following
procedure: (1) Fetch all users followed by users in $\mathcal{U}_i$; (2) From that group, filter
only those having at least 40 followers and following at least 40 accounts; (3)
Add filtered users and their edges to get an extended $\mathcal{U}_{i+1}$ graph.

This process generated a *universe graph* $\mathcal{U} := \mathcal{U}_3$ with 2, 926, 181 vertices and
10, 144, 158 edges.

For the second step, in order to get a homogeneous network (note that many users added in the last step might have no outgoing edges), a subgraph was taken following this procedure:

- We started off with a small sample of seed users S, consisting of users in $\mathcal{U}$ having out-degree 50, this is, users following exactly 50 other users.
- For each of those, we added their 50 most socially affine followed users. The affinity between two users was measured as the ratio between the number of users followed by both and number of users followed by at least one of them.
- We repeated the last step for each newly added user until there were no more new users to add.

This procedure returns the final graph $\mathcal{G}$ with $5,180$ vertices and $229,553$ edges, called the homogeneous K-degree closure ($K = 50$ in this case) of S in the universe graph $\mathcal{U}$.

3.2   Content

The content dataset is composed of $1,636,480$ tweets inherited from previous work extended with a set of $2,237,287$ new tweets. These tweets result from extracting the content written in Spanish from user's timelines in $\mathcal{G}$ for dates between March 2016 and February 2017. This does not mean that we have all the tweets of every user in this period of time. Due to the limitations of the API (30 days at the moment of collecting the data) it is impossible to fetch old tweets.

4   Experimental Setup

In this work, we aim to build models capable of accurately predicting the acceptance that a tweet t could have over the general audience of users ($U_G \subset \mathcal{G}$), based only on the reaction of influencers ($U_I \subset \mathcal{G}$) to the publication. This section describes how we set up models for this purpose over a selection of users and tweets from the ($\mathcal{G}, \mathcal{T}$) dataset defined before.

First, we start with the predictive model based only on social features. Then we move on to explain how additional content-based features were incorporated to improve predictions, giving details about NLP techniques, namely an adaptation of LDA topic modeling to Twitter and sentence embeddings based on the FastText algorithm.

4.1   Social Prediction

The primary focus of this work is to predict if a tweet t will have enough retweets from general users to consider it as *trending tweet* based on information on which of the influencers from U_I has shared it.

Even though the dataset is homogeneous enough considering connections, there are still inactive users in the network. Users that only use the social network in passive mode without engaging in any tweeting or retweeting activity are omitted. As regards the content dataset, as expected, most of the tweets are shared only by its author. This behavior causes an imbalance in the classification that affects the performance. It can be fixed filtering out those irrelevant tweets.

Therefore, we begin this section with an explanation of our filtering processes to select relevant users and tweets. After that, we detail how we proceed to get the influencers U_I from $\mathcal{G}$ and which algorithms we use to that purpose. Finally, we explain the feature extraction and dataset splitting for training and testing the models without any data overlap between those tasks.

User Selection. As mentioned before, the inactive users are omitted in this experiment because they are unpredictable by nature. We consider that a user in our dataset is passive if she has less than ten retweets in her timeline. Filtering those out leaves us with a set of only 3626 active users in $\mathcal{G}$. We restrict the analysis to those users, also removing content shared only by inactive users from $\mathcal{T}$.

Trending Tweets. We call a tweet *trending* if we consider it popular enough to possibly become a trending topic. This consideration is related to the number of retweets it earns over the general public U_G. To get the *golden value* of retweets considered enough to consider a given tweet as popular, we analyzed and built a histogram of how many retweets each tweet in $\mathcal{T}$ receives.

Initially, we wanted to use the value in the 90th percentile as our golden value, but given the fact that most tweets are shared only by their author, this value turned out to equal 1. So we decided to discard all the tweets with less than 3 retweets, which caused this percentile to increase to 13, allowing us to implement more accurate models. Therefore, we consider a tweet *trending* if it was retweeted at least 13 times.

On the other hand, it is important to remark that the experiments carried out make sense only within the context of U_G users, keeping in mind that the goal of this work is to analyze the influence of the U_I group over general users. That is why we are interested only in those tweets from $\mathcal{T}$ that showed up on the *timeline* of at least one user in U_G, defining $T' := \left(\bigcup_{x \in U_G} timeline(x) \right)$.

Influencers Detection. Much effort has been made by the research community in influencers detection [1,7,19,25]. However, most of the works are based on supervised methods, which are not applicable in our case, since we do not have a labeled corpus of influencers.

We decided to use the ideas included in [1], which proposes a combination of three types of features: network centrality, activity level and profile features. Since we didn't have any extended profile information in our dataset, we focused on centrality and activity. This has the advantage of making the results more generalizable to other social networks without depending on specific information that might be available only in Twitter, and for certain users.

To measure the *centrality* of a user we apply an average of metrics computed by the following algorithms: PageRank [20], Betweenness [9], closeness [23], Eigenvector centrality [3] and Eccentricity [4] included in `igraph` Python package [8]. The *activity* level of a user is computed simply as the average of the number of tweets and the number of retweets posted by users.

To decide the best option to rank users as influencers, we compared different weighted combinations of centrality and activity measures, $\alpha * Centrality + (1 - \alpha) * Activity$, where α controls the importance given to *centrality*. In Fig. 1, we can see that the best results were obtained for a simple mean of both metrics ($\alpha = 0.5$). To compare the performance of these options a subset of 500 random tweets from $\mathcal{T}$ was set aside. This sample called T_{SI} is removed from $\mathcal{T}$ to avoid considering them as part of the test set, where trending prediction models will be evaluated later.

In Fig. 1, we show the results for the different alternatives. Each curve is plotted using the selected ranking and running the purely social prediction over T_{SI}, splitting $75\% - 25\%$ for training and test. The y-axis details F_1 score for prediction, while the x-axis reflects the number of influencers, chosen with the evaluated ranking, used for social feature extraction in the models, detailed later in this Section.

Figure 1 reveals that a very central user would be useless for this study if she has a low level of activity and, similarly, a very active user has no value as an influencer if she is not sufficiently well connected. The comparison of these results indicates that the best choice for measuring the influence level of users is the average of centrality and activity.

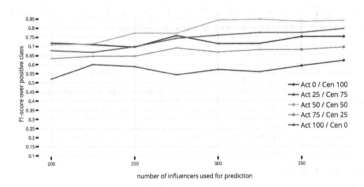

Fig. 1. Comparison of alternatives of influence detection where *Act* involves features related with Activity and *Cen* those related with Centrality. The curves correspond with the pure social model performance prediction over T_{SI}.

Now that we have selected our metric, we apply it to $\mathcal{G}$ without these 500 tweets from T_{SI}, to get a ranking of all users by level of influence. We take the top 25% as our set of *influencers* and call it U_I, the rest of the users are considered the general audience and called U_G. The goal of the social models described later is to predict the level of acceptance of tweets among the general audience U_G, based on knowledge about the activity of the influencers U_I on them. The idea for the experiments described in the following sections, is to vary the number of influencers taken from U_I to predict the popularity of tweets.

Social Features. As mentioned earlier, we need to train a classifier model to make predictions. For that purpose, it is necessary to define the feature vector and the target vector. For the feature vector, in the social based model, we only consider the retweeting behaviour the selected influencers have over tweets from the training set. For each tweet t, we can define a binary vector $T_t :=$ $\begin{bmatrix} i_{t1} & i_{t2} & \dots & i_{tn} \end{bmatrix}$, where n is the number of influencers, and each i_{tj} is 1 if the tweet t was retweeted by the influencer j, and 0 otherwise. More formally, let the function $TM(j)$ return the set of tweets in the timeline for influencer j. Grouping in a matrix all the vectors associated with the m tweets, the input for the model becomes:

$$features := \begin{bmatrix} i_{11} & i_{12} & \dots & i_{1n} \\ \dots & \dots & \dots & \dots \\ i_{t1} & i_{t2} & \dots & i_{tn} \\ \dots & \dots & \dots & \dots \\ i_{m1} & i_{m2} & \dots & i_{mn} \end{bmatrix} \quad \text{where } i_{tj} = \begin{cases} 1 \text{ if } t \in TM(j) \\ 0 \quad \text{otherwise} \end{cases}$$

Note that the content of tweet t is not considered, we only include the information about which of the users in U_I retweeted t. Now, as part of the supervised method, we use the following objective vector, calculated over the training set of tweets. Let $RT(t)$ be a function that returns the number of retweets in U_G for the tweet t; we define the target vector as follows:

$$classification = \begin{bmatrix} r_1 \\ \dots \\ r_t \\ \dots \\ r_m \end{bmatrix} \quad \text{where } r_t = \begin{cases} 1 \ RT(s) >= golden\ value \\ 0 \qquad\qquad otherwise \end{cases}$$

4.2 Splitting the Dataset

To evaluate the performance of our models, we divide our dataset of tweets into two parts, one for training and another for evaluation. As usual, these datasets are not overlapping. In other words, the evaluation data is not seen by the training algorithms.

Regardless of the chosen number of influencers for prediction, we want the training and evaluation datasets to remain disjoint. In this sense, as we explained previously in this section, the left diagram in Fig. 2 shows how we split the set $\mathcal{G}$ in two disjoint parts, U_I (influencers users) and U_G (common users). For the all other experiments of this paper, U_I is defined as the 25% best-ranked users from $\mathcal{G}$, using the average of centrality and activity to detect influencers (Fig. 1).

To determine well-formed training and test sets for tweets, we drop from the $\mathcal{T}$ dataset the tweets posted by users in U_I named T_I. In addition, it is also necessary to cut from $\mathcal{T}$ the set T_{SI} used previously in this section to detect influencers. The remaining tweets, i.e. $T_G = \mathcal{T}' - T_I - T_{SI}$ are split again. To do so, T_G is randomly split in training (75%) and test (25%) datasets to evaluate prediction models. For clarification, the right diagram in Fig. 2 describes these splits.

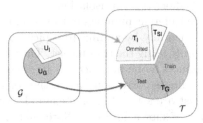

Fig. 2. The left chart distinguishes general users (set U_G) from influencers (set U_I). The right chart shows how to obtain training and test datasets

4.3   Adding Content-Based Features

To achieve an increase in the quality of trending tweet prediction, we apply NLP techniques to to extend the purely social model with content-based features. Representing text content with vocabulary-based representations such as TF-IDF introduces problems of efficiency and overfitting due to large dimensionality. That is why it is convenient to use more compact vector representations that somehow manage to encode semantic similarity between texts. Trying the most popular algorithms for this task, we found that Twitter-LDA as a topic extractor and FastText as a sentence embedder were the options that best fit in our experiments. Both are described later in this section.

Preprocessing. To begin with, we enumerate the sequential transformations performed to turn a tweet into a vector of numeric features describing its content.

- **Normalization.** In the first step, we remove the following for normalizing purposes: URLs, accents, unusual characters, numbers and stopwords.
- **Tokenization.** Next, we convert the text to lowercase, split it into tokens and apply lemmatization for Spanish language to all words. We use the spaCy package [12] for this stage. The resulting representation as a sequence of normalized tokens is the basis for both Twitter-LDA and FastText representations.

Twitter-LDA. Twitter-LDA [30] is a variant of the classic LDA topic modelling algorithm used in [6], specially tuned for short text documents like tweets. The LDA model enables us to discover a given number of underlying topics within a given corpus, generating a representation of each topic as a probability distribution over the words. Additionally, it reduces dimensionality by representing the each text with a topic-based distribution. The Twitter-LDA adaptation modifies the assumptions of LDA by restricting each tweet to just one relevant foreground topic, and adding an extra "phantom" topic of background words used to model uninformative vocabulary in each tweet. Moreover, tweets are grouped by user during the training phase, allowing the model to pick up more topical patterns than it would by treating short texts in isolation.

We experimented with different numbers of topics on the training dataset: 5, 10, 15, 20 and then incrementally by adding 10 topics up to 80. In all cases, we validate the experiments only using the training set. The best results are obtained using 10 topics. This produces a one-hot encoded 10-dimensional representation of tweets, where the coordinate corresponding to the topic assigned to a tweet is set to 1, and all the rest are set to 0. Some examples of the resulting topics and their top-five words are shown in Fig. 3. Note that words that represent a topic bear a semantic relation between them. The first topic in the Figure groups "legales" (legal), "acreedores" (creditors) and "pagarles" (to pay) which belong to the same semantic field.

futuro	construir	atencion	metropolitanas	paso	pais	escuelas	medios	progreso	local
legales	presidente	integral	areas	legal	argentinos	seminariocti	responsabilidad	inmigrantes	argentina
acreedores	gobierno	alimentaria	descentralizados	gobernador	paises	innovacion	economicos	contribucion	global
pagarles	cfkargentina	infantil	ciudades	escrutinio	orgullo	seminario	voceros	multicultural	brasil
trabajaremos	chaco	mortalidad	habana	transparencia	tecnopolisarg	aulas	hegemonicos	espectaculo	mercosur

Fig. 3. An example of top words in 10 Twitter-LDA topics from Twitter dataset

FastText. *Word embeddings* refers to a family of different techniques that associate vector representations to input words. Conceptually, the idea is to map a discrete large-dimensional bag-of-words representation of a corpus into a continuous space of fewer dimensions. The resulting representations have the property that words with similar meanings correspond to nearby vectors as we can see in the left plot of Fig. 4. As a consequence, this kind of representation improves efficiency and reduces overfitting without loss of information.

In this work, we use the FastText implementation [10] of word embeddings, which is presented as an alternative to the traditional Word2Vec model [15]. One of its most prominent features is the possibility of assigning vectors to words not seen during the model training, looking for matches on character n-grams to vectorize those out-of-vocabulary words. This makes it more robust for handling misspelled words that are commonly found in social media text. We use a pre-trained model of 100 dimensions, included in the FastText library from [10]. Although word embeddings models provide vector representations for single words only, convolution functions can be applied to obtain vectors of the same dimensionality that represents whole sentences or paragraphs. In the case of FastText library, a given text is represented as the average of all the vectors of its component words.

The left plot in Fig. 4 shows some examples of Spanish words with similar meaning which are plotted in the same color, and are close to each other. For example, the words "jajaja" and "jejej" are different ways to indicate laughing. The right side plot of Fig. 4 shows the distance of FastText vectors for the tweets at the bottom of the Figure. We plotted with the same color tweets with similar meanings: tweets 1 and 2 are very close to each other (in English: *"it can't be possible, lol"* and *"no way, lol"*, respectively). For the 2D visualization of the 100 dimension FastText vectors we used the Multi-Dimensional Scaling algorithm included in `scikit-learn.manifold` package. As expected, tweets representations are close if their content is similar.

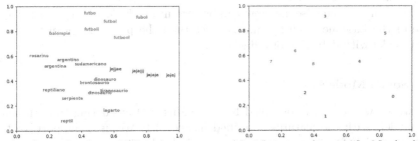

Numbered examples of tweets: 0."jajajaj que risa que me da" - **1.**"esto no puede ser jejeje" - **2.**"no hay forma de ser jajjja" - **3.**"Restos de tiranosaurios fueron encontrados en Neuquén." - **4.**"Este es otro tweet que habla de dinosaurios y cocodrilos" - **5.**"restos dinosaurios dar lugar lagartos y lagartijas" - **6.**"Hablar de fútbol en argentina es muy difícil" - **7.**"Uruguay dejo un buen papel ante argentina y empato el partido para lograr su clasificacion." - **8.**"sigue complicado el inicio del torneo argentino de futbol la afa cada vez peor"

Fig. 4. Two-dimensional visualization of FastText vectors for selected examples of words (left) and tweets (right).

5   Results

Now we describe how we build our predictive models and the results obtained with and without content analysis. We will compare our models to a baseline built from a purely social model where users considered influencers are selected randomly instead of using an influencer detection algorithm. With this we want to show the utility of using an algorithm to detect influencers, and the relevant information those provide for learning about the behavior of general users.

5.1   Baseline

As a baseline, we use a model that is sufficiently demanding to be compared with our proposals. We decided to use the same kind of features as in the pure social version, but randomly selecting a set of 25% of the users from $\mathcal{G}$ as the set of influencers U_I .

To make a fair comparison with our models we do a new split from the dataset $\mathcal{T}$ to T_I and T_G with the content of users in the random selection of U_I and U_G respectively. In turn, a $75\% - 25\%$ train-test split is performed on T_G for the training and evaluation of the baseline models under the same conditions as in the social alternative. We keep the datasets disjoint and evaluate over general users with influencer behavior data as input.

Following the same pattern as in the other social models, we then proceed to evaluate the social baseline over increasingly large numbers of users from U_I taken as the source of social features. In this case we do not have a ranking of users to draw the top ones from, so we make these selections randomly as well. In order to calculate the baseline performance, for each value of the number of source influencers (let us call this k), we randomly select k users from U_I, and train and test a model using the train-test split of T_G. To avoid lucky and potentially misleading results, we repeat this process five times for each value of k, reporting the average F1-score.

The results of the baseline score can be seen in Fig. 5. As expected, the yield curve of the baseline is always much lower than the performance of the pure social model with detection of influencers.

5.2   Social Models

Now we show the results obtained from training and evaluating trend prediction models with the features described in Sect. 4.1. We used Support Vector Machine models for classification, more precisely the SVC implementation from scikit-learn [21], combined with its GridSearchCV class for search of optimal hyperparameters through cross-validation over the training set.

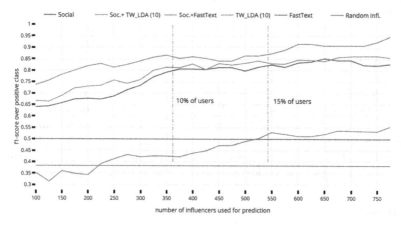

Fig. 5. F1-score on experiments with and without content analysis.

We decided to focus on the experiments considering 10% and 15% of $\mathcal{G}$ as influencers. These values would still return relevant results to our purpose while letting the trained model with enough information. That is the reason why in Fig. 5 we put the vertical lines showing these values. There, we can see that considering 10% of the user space as influencers we have an F_1-score near to 78% over the test data. Details about scoring can be seen in Table 1. In this figure, we can also observe the comparison with the baseline model. Here, we confirm that not all users bring the same information. There is a group that can exert influence over their social environment and another that shows the follower's behavior despite the content.

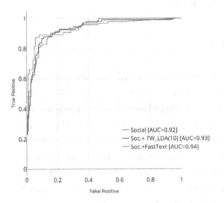

Table 1. Performance evaluations over U_G, using top-10% of U as influencers ($\subset U_I$). TW-LDA(10) refers to model Twitter-LDA with 10 topics.

Model	F_1	Pr.	Rec.
Baseline	42.5%	29.8%	63.8%
TW-LDA(10)	38.3%	30.5%	51.5%
FastText	49.5%	34.3%	92.1%
Social	79.2%	75.8%	82.9%
Soc.+TW-LDA(10)	81.4%	79.1%	83.8%
Soc.+FastText	86.7%	88.7%	84.6%

Fig. 6. ROC curve for social and combined models, using top-10% of U as influencers ($\subset U_I$).

5.3 Social+NLP Models

In this section, we present improved models that add content-based features to the Social Model. Looking for improvements in the scores, we try two alternatives for content analysis: Twitter-LDA [30] and FastText [10]. We apply the first option to discover topics among the tweets and tag each of them by its topic. On the other hand, FastText is used to provide compact dense vector representations of tweets in a way that captures semantic similarities between their content. The feature vectors for combined models are built as follows:

Social+Twitter-LDA: the feature vector of the Social Model is extended by appending the 10-dimensional boolean vectors from Twitter-LDA Model described in Sect. 4.3.

Social+FastText: In this case, the vectors of social features described in the previous section are extended by appending the 100-dimensional vector from FastText Model described in Sect. 4.3.

Even though the purely content-based models performed poorly (even worse than the baseline in some cases), the combined models using content-based and social features obtained the best scores. In Table 1, we compare the baseline with the two new models. The improvement of Twitter-LDA [30] alternative was about 2% over the Social model, obtaining almost the double of performance over the baseline. On the second model, with FastText [10] embeddings we also improved the performance. This time the increase was about 8% over the social model, which makes this model the best fit in our experiments with an 86.7% efficiency ceiling. Also, Fig. 5 shows the performance of combined models using different numbers of influencers from U_I. It is clear that the FastText combined model obtains the best performance. Finally, in Fig. 6 we include ROC curves for social and combined models, which makes it possible to compare our work to the previous content-based work in [17]. In the social and combined cases, we use the full set of influencers U_I for the social features.

6   Conclusions and Future Work

As a general conclusion, we confirm that the information about social connections between Twitter users and their activity can be essential to determine which content becomes popular. We obtained a surprisingly high performance without analyzing the content, which seems to suggest that the source of information has a stronger influence than the actual content when it comes to spreading it across the network. The purely content-based model was far below from the social-based pure model scoring, which reinforces the idea that sometimes our contact lists can provide more information about us than our timeline. Anyway, the combined model with content analysis increased the performance significantly (especially when using FastText word embeddings), which indicates that content still has a level of importance when it is considered within a certain social context. FastText seems particularly well suited for dealing with content from Twitter, specially because of its ability to obtain representations for unseen or misspelled words.

This research opens many doors to evolve the model. The most relevant to us are described next.

A possible improvement is training the model exclusively with tweets published earlier than the tweets used in the test stage. Keeping in mind the temporal variable, using techniques such as Early Prediction [13], we could make a model capable of predicting popularity with the information available on the first minutes of the tweet creation. Later, we can improve this by using Deep-Learning [2]. For influencers detection, alternatives such as [1,16] could be applied to improve the selection of relevant users.

We also propose to conduct research about the aggregation formula for sentence embeddings. We have used a simple average of the vectors of the component words, but there are other more sophisticated functions, such as the weighted average by the inverse document frequency (IDF) [24]. Furthermore, we shall test other embedding models such as Doc2Vec [14] and compare results. Additionally, instead of using the default 100-dimensional pre-trained Spanish model from FastText, we can consider other possibilities such as using a model trained on the Spanish Billion Word Corpus from [5]. To that end, we can train a custom model on our dataset of tweets, or attempt to combine both datasets somehow.

Finally, an interesting line of open research is trying to replicate the experiments for other social networks such as Facebook and Instagram, and see to what extent our conclusions are applicable to those. In particular, the pure social model can be extended to any network of users sharing content, which makes it possible to evaluate it even in image-based networks such as Instagram. However, we are limited by the availability of data to build datasets.

References

1. Azcorra, A., et al.: Unsupervised scalable statistical method for identifying influential users in online social networks. Sci. Rep. **8**, 6955 (2018)
2. Bengio, Y.: Learning deep architectures for AI. Found. Trends Mach. Learn. **2**(1), 1–127 (2009). Also published as a book. Now Publishers (2009)
3. Bryan, K., Leise, T.: The $25,000,000,000 eigenvector: the linear algebra behind google. SIAM Review **48**, 569–581 (2006)
4. Buckley, F., Harary, F.: Distance in Graphs. Addison-Wesley, Boston (1990)
5. Cardelino, C.: Spanish billion word corpus and embeddings. http://crscardellino.me/SBWCE/
6. Celayes, P.G., Domínguez, M.A.: Prediction of user retweets based on social neighborhood information and topic modelling. In: Castro, F., Miranda-Jiménez, S., González-Mendoza, M. (eds.) MICAI 2017. LNCS (LNAI), vol. 10633, pp. 146–157. Springer, Cham (2018). https://doi.org/10.1007/978-3-030-02840-4_12
7. Cossu, J.V., Dugué, N., Labatut, V.: Detecting real-world influence through Twitter. In: 2015 Second European Network Intelligence Conference, pp. 83–90 (2015)
8. Csardi, G., Nepusz, T.: The igraph software package for complex network research. Int. J. Complex Syst. **1695**, 1–9 (2006). http://igraph.org/python/
9. Freeman, L.C.: A set of measures of centrality based on betweenness. Sociometry **40**(1) (1977)
10. Grave, E., Mikolov, T., Joulin, A., Bojanowski, P.: Bag of tricks for efficient text classification. In: Proceedings of the 15th Conference of the European Chapter of the Association for Computational Linguistics, EACL 2017, pp. 427–431, Spain (2017). https://fasttext.cc/
11. Hochreiter, R., Waldhauser, C.: A genetic algorithm to optimize a tweet for retweetability. Mendel, pp. 13–18 (2013)
12. Honnibal, M., Johnson, M.: An improved non-monotonic transition system for dependency parsing. In: Proceedings of the 2015 Conference on Empirical Methods in Natural Language Processing, pp. 1373–1378. ACL, Portugal (2015). https://spacy.io/
13. Smith, J.E., Tahir, M., Sannen, D., van Brussel, H.: Making early prediction of the accuracy of machine learning applications. In: Lughofer, E., Sayed-Mouchaweh, M. (eds.) Learning in Non-stationary Environments: Methods and Applications, pp. 121–151. Springer, New York (2012). https://doi.org/10.1007/978-1-4419-8020-5_6
14. Lau, J.H., Baldwin, T.: An empirical evaluation of doc2vec with practical insights into document embedding generation. In: Proceedings of the 1st Workshop on Representation Learning for NLP, pp. 78–86. Association for Computational Linguistics (2016)
15. Mikolov, T., Chen, K., Corrado, G., Dean, J.: Efficient estimation of word representations in vector space. CoRR abs 1301.3781 (2013)

16. Morone, F., Min, B., Bo, L., Mari, R., Makse, H.A.: Collective influence algorithm to find influencers via optimal percolation in massively large social media. Sci. Rep. **6**, 30062 (2016)

17. Nasir, N., Gottron, T., Kunegis, J., Alhadi, A.C.: Bad news travel fast: a content-based analysis of interestingness on Twitter. In: Proceedings of the 3rd International Conference on Web Science, WebSci 2011 (2011)

18. Naveed, N., Gottron, T., Kunegis, J., Alhadi, A.C.: Bad news travel fast: a content-based analysis of interestingness on Twitter. In: Proceedings of the 3rd International Web Science Conference, WebSci 2011, pp. 8:1–8:7. ACM, New York (2011)

19. Neves, A., Vieira, R., Mourão, F., Rocha, L.: Quantifying complementarity among strategies for influencers' detection on Twitter1. Procedia Comput. Sci. **51**, 2435–2444 (2015). International Conference on Computational Science, ICCS 2015

20. Page, L., Brin, S., Motwani, R., Winograd, T.: The pagerank citation ranking: bringing order to the web. Stanford University, Technical report (1999)

21. Pedregosa, F., et al.: Scikit-learn: machine learning in Python. J. Mach. Learn. Res. **12**, 2825–2830 (2011). http://scikit-learn.org/

22. Pennacchiotti, M., Popescu, A.M.: A machine learning approach to Twitter user classification. In: Proceedings of the Fifth International Conference on Weblogs and Social Media, Barcelona, Catalonia, vol. 11, Spain (2011)

23. Sabidussi, G.: The centrality index of a graph. Psychometrika **31**(4), 581–603 (1966)

24. Arora, S., Liang, Y., Ma, T.: A simple but tough-to-beat baseline for sentence embeddings. In: Proceeding of International Conference on Learning Representations, ICLR 2017, Toulon, France, 24–26 April (2017)

25. Simmie, D.S., Vigliotti, M.G., Hankin, C.: Ranking Twitter influence by combining network centrality and influence observables in an evolutionary model. J. Complex Netw. **2**(4), 495–517 (2014)

26. Uddin, M.M., Imran, M., Sajjad, H.: Understanding types of users on Twitter. CoRR abs/1406.1335 (2014)

27. Vosoughi, S., Roy, D., Aral, S.: The spread of true and false news online. Science **359**(6380), 1146–1151 (2018)

28. Vougioukas, M., Androutsopoulos, I., Paliouras, G.: Identifying retweetable tweets with a personalized global classifier. In: Proceedings of the 10th Hellenic Conference on Artificial Intelligence, SETN 2018, Patras, Greece, 09–12 July 2018, pp. 8:1–8:8 (2018). https://doi.org/10.1145/3200947.3201019

29. Zhang, J., Brackbill, D., Yang, S., Centola, D.: Efficacy and causal mechanism of an online social media intervention to increase physical activity: results of a randomized controlled trial. PM Rep. **2**, 651–657 (2015)

30. Zhao, W.X., et al.: Comparing Twitter and traditional media using topic models. In: Clough, P., et al. (eds.) ECIR 2011. LNCS, vol. 6611, pp. 338–349. Springer, Heidelberg (2011). https://doi.org/10.1007/978-3-642-20161-5_34

Link Prediction in Co-authorship Networks Using Scopus Data

Erik Medina-Acuña[1]([✉]), Pedro Shiguihara-Juárez[1],
and Nils Murrugarra-Llerena[2]

[1] Department of Computer Science, Universidad Peruana de Ciencias Aplicadas,
Lima, Peru
{u201320646,pedro.shiguihara}@upc.pe
[2] Department of Computer Science, University of Pittsburgh, Pittsburgh, PA, USA
nineil@cs.pitt.edu

Abstract. Link Prediction is a common task for social networks and recommendation systems. In this paper, we study the problem of link prediction on Scopus co-authorship networks. We used many well-known relational features, and evaluate them with five different classifiers. Finally,
we perform a feature analysis to determine the most crucial features in
this setup.

Keywords: Data mining · Machine learning · Decision trees ·
Co-authorship network · Link prediction · Supervised learning

1 Introduction

Link Prediction is the technique that forecast links given a partial set of edges
of a graph, and predict new connections or disconnections of edges [12].

In order to achieve a correct link prediction, different studies [10,14,15] have
created their own data structure or model according to their focus or dataset
target. These works focused on social networks, recommendation systems and
how nodes interact with themselves. In these settings, most of the networks
are heterogeneous because there are different ways to link nodes in the same
graph [3,11,15]. In our case, we focus on homogeneous networks. We take data
of research articles and create co-authorship networks. Co-Authorship networks
are compound of researchers as vertices and collaboration articles as edges [11].
This edge relationship can be strengthened by adding references to co-venue
(same affiliation), co-citing and others [9].

In this article, we focus on link prediction for co-authorship using Scopus
data. We extract features from the network and apply supervised learning. Thus,
giving a network of a different period of time, we predict whether the links that

---

The dataset was generated with Elsevier API http://api.elsevier.com and Scopus
http://www.scopus.com.

© Springer Nature Switzerland AG 2019
J. A. Lossio-Ventura et al. (Eds.): SIMBig 2018, CCIS 898, pp. 91–97, 2019.
https://doi.org/10.1007/978-3-030-11680-4_10

were lacking previously are now connected. We used Elsevier[1] for data extraction and many well-known classifiers. We find that decision trees outperform the other classifiers. Also, we find the most relevant features are Jaccard and preferential-attachment.

We review the related work on Homogeneous, Heterogeneous and Social networks in Sect. 2. We describe our link prediction approach to co-authorship networks in Sect. 3. Finally, we discuss our experiments and results in Sect. 4 and conclude in Sect. 5.

2   Related Work

Previous work on link prediction is focused on both homogeneous, heterogeneous networks and social networks.

2.1   Homogeneous and Heterogeneous Networks

Heterogeneous Networks (HN) appear in the domain of internet services (e.g. blogs), citation networks, syntactic networks, and others [8]. These networks usually create a generic structure to handle the different type of links between nodes. The authors in [9] defined their HN of co-authorship networks by merging Co-citing, Co-authorship and co-venue relation between authors. But due to their extreme class imbalance, they couldn't get a good model. We have the same issue, but we take only 10% of the number of negative edges to balance classes.

Zhang [14] uses co-citing, co-authorship, same keywords, and so on to create its HN. Zhou et al. [15] defined the link prediction for co-authorship networks different from social networks like Facebook, where a node can unfriend other nodes. On the contrary, co-author networks grow steadily and there are no much publications per researcher. Thus, its sparsity and dynamics make the task even more challenging. Homogeneous Networks are the ones that use one type of connection. In the link prediction problem, most of the evaluation is based on nodes similarities [13,14]. In our case, our co-authorship network is a Homogeneous Network, which is based on collaboration between authors and its similarity.

2.2   Social Networks

Shall [10,13] proposed the Triadic Closeness (TC) measure, which evaluates a directed graph with three nodes and explains the relations created by them. Yu et al. [13] propose the algorithm Path and Node Combined (PNC) measure, which works on the similarity of nodes in a graph. Esslimani et al. [4] use link prediction methods for social networks to evaluate their novel Densified Behavioral Network Based Collaborative Filtering (D-BNFC). Gong et al., apply Markov Random Fields and Loopy Belief Propagation to accomplish the detection of Sybils [5][2].

---

[1] https://www.elsevier.com.

[2] A Sybil is a user, usually fake, that takes advantage of a system to make negative actions such as steal personal information.

We use the same approach of TC to simulate the way authors recommend their co-authors to each other. This approach takes a sub-graph of three nodes with only two edges. Then, they reduce the search space to evaluate only the missing edges of each TC in the graph.

3 Methodology

The proposed system uses the actual Elsevier Scopus API[3] to retrieve articles data such as references, authors, keywords, author's affiliation, and others. In this section, we describe our data collection procedure, our feature extraction and how our features are feed to a classifier.

3.1 Data Collection

First, a Scopus query is transformed into a required format for Scopus Search API[4,5]. Each request returns a list of 25 articles and their associated Scopus ID (SID). Then, per each SID, the Abstract Retrieval API[6] provides authors, affiliations, publication information, abstract, keywords, etc.

We used a sample query for Bayesian networks. Then, we generate a graph with their vertices as authors and their edges as co-authorship collaborations.

3.2 Feature Extraction

Each edge is represented by two authors and its co-authorship is evaluated using four heuristic features [7]. Each of the features was normalized using min/max normalization, and are summarized on Table 1.

Let $\Gamma(x)$ denote the set of neighbors of node x. Common neighbors measures

$$|\Gamma(x) \cap \Gamma(y)| \tag{1}$$

the likeliness of two authors being introduced by a colleague that they have in common. The Adamic/Adar coefficient [1]

$$\sum_{z \in \Gamma(x) \cap \Gamma(y)} \left(\frac{1}{log|\Gamma(z)|} \right) \tag{2}$$

adds weight to rare features. Neighbors that have fewer authors in common will be more likely to collaborate with a popular author. Then, the Jaccard's coefficient

$$\frac{|\Gamma(x) \cap \Gamma(y)|}{|\Gamma(x) \cup \Gamma(y)|} \tag{3}$$

---

[3] https://dev.elsevier.com/sc_apis.html.
[4] https://dev.elsevier.com/documentation/ScopusSearchAPI.wadl.
[5] https://dev.elsevier.com/tips/ScopusSearchTips.htm.
[6] https://dev.elsevier.com/documentation/AbstractRetrievalAPI.wadl.

is a similarity metric that solves the problem when two authors have many co-authors in common not because they are related but because they have many co-authors. It takes the total number of neighbors in common and divides them by the total of neighbors of both authors. Finally, the preferential attachment coefficient

$$|\Gamma(x) \times \Gamma(y)| \tag{4}$$

evaluates how likely an author will collaborate with new ones considering that if an author has many collaborations, he will continue doing it. We also use as a feature the degree of the author's node.

Table 1. Processed dataset with our employed features.

Feature	Type	Description
Link name (edge between Author A and B)	String	A ↔ B : The edge to be evaluated
Author node A degree	Numeric	Normalized value
Author node B degree	Numeric	Normalized value
Common co-authors	Numeric	Normalized value
Adamic/Adar	Numeric	Normalized value
Jaccard's coefficient	Numeric	Normalized value
Preferential attachment	Numeric	Normalized value
Future link	Nominal	1, if the edge exists in G_n and not in G_0. 0, otherwise

3.3 Data Analysis

We get article data from a query passed through Scopus Search API. Elsevier kindly provide credentials to use their API with the following conditions. One of the restrictions is that you are able to make up to 1000 API calls per week and the queries used should only focus one subject area. Based on that premise, we retrieved 2,454 articles that included a set of keywords related to Bayesian Networks in June 2018. We counted 2044 co-authorships and 524 authors only in 2016.

Each publication had an average of 4 authors. To create our initial graph $G = (V, E)$, we require V^2 iterations between all vertices as seen in Eq. 5.

$$(|Authors_{avg}| * |Articles|)^2 = (4 \times 2454)^2 = 96\ 353\ 856 \tag{5}$$

If we want to iterate between each pair of authors to check whether both have a link we will have to do it 96 353 856 times. This is necessary to create the graph and calculate each edge features. In order to reduce the time of execution we removed vertices that had no impact on the whole graph. To prune the graph, we removed all authors that have a degree less than 5, which are authors with few collaborations. In addition, we take 10% of the missing edges randomly to have our classes balanced as in [7]. We take the 70% of edges and called the resulting graph as G_0. All the positive instances are generated using the TC constraint as follows. We take vertices that have a common vertex but are not currently connected from G_0. Then, we check if authors are connected to the full graph. Let G_n be the new graph used for training. For testing, we will be using the articles from the next year which authors are present in the previous year. We follow the same procedure as described before. Next, we used the features from Table 1.

4   Experiments and Results

4.1   Hardware and Software

We used a Virtual Private Server with Ubuntu 16.04 with 2 GB RAM, Python 3 with Numpy, Pandas, Scikit-learn and Tpot libraries for development.

4.2   Evaluation

We used the data of articles that were published in 2016. We split our data into 70% for training and 30% for testing. We trained our dataset with Logistic Regression (LR), Support Vector Machines (SVM), Decision Trees (DT), Neural Networks (NN) and K Nearest Neighbors (KNN). Our results on the test data are shown in Table 2.

Table 2. Comparison results among five classifiers.

Classifier	Configuration	Accuracy
Logistic regression	loss = log, penalty = l2	50.00
Support vector machines	kernel = rbf	89.29
Decision trees	Default	**91.07**
Neural networks	solver = lbfgs, layers = 5,2	50.00
K-NN	neighbors = 11	89.29

We evaluate the model with two metrics such as accuracy and F1-score. The average accuracy, F1-Score were 91.07% and 91.80% respectively. For more details, the confusion matrix is in Table 3.

Table 3. Confusion matrix for decision tree

		Predicted class		
		Collab.	**Not collab.**	**total**
Actual class	**Collab.**	28	0	28
	Not collab.	5	23	28
	total	33	23	56

4.3 Feature Analysis

We want to understand the success of our method. Thus, we identify the most influential feature. We removed each feature and evaluate the classifier again. The set with less accuracy will highlight the most crucial feature in our approach.

Table 4. Feature relevant analysis. Best features are highlighted.

Feature	Accuracy
Node A edges	91.07
Node B edges	91.07
Adamic/Adar	91.07
Jaccard	**50.0**
Preferential attachment	**78.57**

As a result, we find that Jaccard's is the most relevant followed by preferential attachment in Table 4. We believe that Jaccard is important because it quantifies how non-collaborative authors interact with each other. Also, preferential attachment quantifies possible new collaborations.

5 Conclusion

Link Prediction is a common task used in social networks and recommendation systems. In this article, we predict links on co-authorship networks from Scopus data. We compared different classifiers and identified Jaccard and preferential attachment as the most important features. Decision Tree classifier obtained an accuracy of 91.07%. Also, the lack of Jaccard's feature reduced the accuracy to 50%. Both, Jaccard's and preferential attachment quantify how authors interact and collaborate.

As future work, we plan to test other classifiers such as random forests and deep neural networks. Also, we can expand our features set using author and affiliation Retrieval APIs.

References

1. Adamic, L.A., Adar, E.: Friends and neighbors on the Web. Soc. Netw. **25**, 211–230 (2003)
2. Barabási, A.: Emergence of scaling in random networks. Science **286**, 509–512 (1999)
3. Daud, A., Ahmad, M., Malik, M.S.I., Che, D.: Using machine learning techniques for rising star prediction in co-author network. Scientometrics **102**, 1687–1711 (2014)
4. Esslimani, I., Brun, A., Boyer, A.: Densifying a behavioral recommender system by social networks link prediction methods. Soc. Netw. Anal. Min. **1**, 159–172 (2010)
5. Gong, N.Z., Frank, M., Mittal, P.: SybilBelief: a semi-supervised learning approach for structure-based sybil detection. IEEE Trans. Inf. Forensics Secur. **9**, 976–987 (2014)
6. Jaccard, P.: Etude comparative de la distribution florale dans une portion des Alpes et du Jura (1901)
7. Julian, K., Lu, W.: Application of machine learning to link prediction (2016)
8. Llerena, N.E.M., Berton, L., Lopes, A.D.A.: Graph-based cross-validated committees ensembles. In: 2012 Fourth International Conference on Computational Aspects of Social Networks (CASoN) (2012)
9. Pujari, M., Kanawati, R.: Link prediction in multiplex networks. Netw. Heterog. Media **10**, 17–35 (2015)
10. Schall, D.: Link prediction for directed graphs. Soc. Netw.-Based Recomm. Syst. 7–31 (2015)
11. Singh, H., Tomar, D., Agarwal, S.: Link prediction for authorship association in heterogeneous network using streaming classification. Int. J. Grid Distrib. Comput. **9**, 135–150 (2016)
12. Wang, P., Xu, B., Wu, Y., Zhou, X.: Link prediction in social networks: the state-of-the-art. Sci. China Inf. Sci. **58**, 1–38 (2014)
13. Yu, C., Zhao, X., An, L., Lin, X.: Similarity-based link prediction in social networks: a path and node combined approach. J. Inf. Sci. **43**, 683–695 (2016)
14. Zhang, J.: Uncovering mechanisms of co-authorship evolution by multirelations-based link prediction. Inf. Process. Manag. **53**, 42–51 (2017)
15. Zhou, X., Ding, L., Li, Z., Wan, R.: Collaborator recommendation in heterogeneous bibliographic networks using random walks. Inf. Retr. J. **20**, 317–337 (2017)

Aerial Scene Classification and Information Retrieval via Fast Kernel Based Fuzzy C-Means Clustering

Zhengmao Ye[1(✉)], Hang Yin[1], and Yongmao Ye[2]

[1] College of Engineering, Southern University, Baton Rouge, LA 70813, USA
{zhengmao_ye, hang_yin}@subr.edu
[2] Liaoning Radio and Television Station, Shenyang, China
yeyongmao@hotmail.com

Abstract. Fast kernel based fuzzy C-Means clustering is proposed in this article to accomplish both accurate and robust segmentation, via integration with watershed transform and fast level set schemes. Aerial scenes are inherently linked to (noise and artifact) sensitivities, intensity inhomogeneity, blurry boundary, and information complexity. It is thus necessary to combine the edge or contour based level set method with region based fuzzy C-Means clustering. To achieve fast segmentation, watershed transform is used to secure the initial contour of the fast level set method, so that initial cluster centers of fuzzy C-Means clustering are selected on those closed contour to avoid misclassification and to enhance separability. It reduces time for lengthy computation iteration. Using multiple densely distributed aerial images, robust and fast clustering is observed after comparing between classical and fast kernel based fuzzy C-Means clustering. To further analyze the role of hybrid fast kernel based scheme on scene classification and information retrieval, frequency domain histogram analyses for several clustering cases are conducted on aerial digital images.

Keywords: Fuzzy C-Means clustering · Gaussian kernel · Fast level set · Watershed transform · Histogram analysis

1 Introduction

Segmentation acts as a powerful image processing methodology to reach correct decisions across broad areas including remote sensing and medical diagnosis, using region based or edge (boundary) based schemes [1–3]. The level set method is a geometry oriented active contour approach. A fuzzy level set algorithm has been presented on image segmentation. Spatial clustering has been applied after initial segmentation and robust segmentation can be performed using regularized evolution. Good performance from typical medical images with diverse modalities proves its effectiveness. Initialization and controlling parameter estimations are conducted via fuzzy C-Means clustering. Integration of fuzzy C-Means clustering and fuzzy level set is made on segmentation of the aerial images. Selection of performance index incorporates both local intensity information and spatial information. Based on experimental outcomes on sparse and dense distributed aerial images as well as the typical landscape

© Springer Nature Switzerland AG 2019
J. A. Lossio-Ventura et al. (Eds.): SIMBig 2018, CCIS 898, pp. 98–111, 2019.
https://doi.org/10.1007/978-3-030-11680-4_11

aerial image and skyline aerial image, convincing performance with robustness is obtained at a fast convergence rate. In aerial imagery, the level set method is highly sensitive to parametric initialization. Initialization and control parameter optimization are both crucial to its performance. Control parameters need to be estimated along with dynamic level set approaches with the variational boundary and active contour with respect to space and time. To achieve optimal configuration of controlling parameters, fuzzy clustering is applied for initialization whose spatial information is critical to approximate medical image boundaries of interest and controlling parameters. The fuzzy level set presented is based on the Hamilton-Jacobi function, which is able to suppress boundary leakage and alleviate manual intervention [4, 5]. Watershed transform can be applied to identify initial surface or contour on a basis of gradient information. As an example, it is combined with adaptive contrast stretching for image enhancement. Under conditions of improper illumination, this approach adapts to the intensity distribution and automatically identifies diverse objects using watershed levels to differentiate catchment basins. Operations of erosion and dilation serve as two typical individual processes. To avoid over-segmentation, both foreground markers and background markers are chosen accordingly [6]. Watershed transform is testified for initial partition together with fuzzy C-Means clustering to produce an initial contour. It avoids leaking when the curve propagates along the boundary of level sets. The effectiveness and accuracy of the scheme are verified via numerical simulations on MRI images. Performance enhancement has been observed from the level set evolution [7].

Cluster analysis is concerned with a set of objects to be classified so that the higher similarity occurs within an individual cluster while the lower similarity occurs across multiple clusters. Iterative K-Means clustering represents an unsupervised learning approach which divides the data into a group of hard clusters. It acts as a fundamental method of the region based image segmentation. In contrast to boundary based segmentation, it converges quickly and possibly works independently. Practical implementation has widely been made in remote sensing areas. For instance, remote sensing techniques are crucial for natural environment preservation. To clarify the graphical data accurately in terms of multiple regions and salient objects, similarity criteria should be determined. To distinguish among various objects and regions, K-Means clustering is used to separate digital information into different clusters. For those hard cluster approaches, some facts of practical problems have not been followed where regions and clusters are not unique in general, thus alternative results could also be available. Thus fuzzy C-Means clustering is then carried out to indicate vague belongings of the pixel intensity using well defined fuzzy membership functions in which each pixel of an image processes certain fuzzy degree of belongings to multiple clusters. Fuzzy C-Means clustering defines soft clusters instead where each pixel involves in multiple clusters, which then generates a set of soft clusters for automatic initialization and image partition. Fuzzy C-Means clustering is carried out in Raman spectral analysis for potential decision making on biomedical samples [8]. Quantitative metrics are applied to determine a feasible number of clusters to improve decision accuracy and optimize outcomes from fuzzy C-Means clustering [9]. Fuzzy C-partition also has a wide variety of applications on data interpretation. To speedup convergence rate, a convergence theorem is introduced to generate a biased fuzzy C-Means algorithm with a focal point, which exhibits better scale in data space and less sensitivity to initialization [10]. Fuzzy C-Means algorithms employ information of the color, intensity and texture as well as position to classify the feature space into multiple regions but lack of information on contours. Introduction of fuzzy C-Means clustering helps to locate those vague boundaries such that optimization with respect to both contour information and classical region information is reached [11]. The fuzzy C-Means clustering using a level set model is also deployed in order to recognize the non-ideal IRIS images precisely without over-segmentation or reinitialization. Using spatial clustering and evolutionary feature extraction, it incorporates spatial feature into the curve evolution approach using level sets so as to regularize level set propagation locally. It can combine the sign and magnitude features to improve classification performance [12]. The fast two-cycle model provides rapid level set segmentation. However it is still highly subject to initialization. The spatial kernel fuzzy C-Means clustering is introduced to produce an original contour, so that enhancement on accuracy, convergence and robustness are achieved on both synthetic and real image segmentation [13]. A combination of the level set model with fuzzy C-Means clustering and Lattice Boltzmann method is presented for medical image segmentation that is independent of initial contour. It shows good segmentation performance of speed, effectiveness, accuracy, robustness and efficiency on medical and real world images [14]. In a kernel based Fuzzy C-Means Clustering, a fuzzy factor is dependent on both the space distance among all neighboring pixels and the corresponding gray-level difference as a tradeoff. The factor helps to determine the damping performance of neighboring pixels accurately. To further enhance the robustness against noise and outliers, the kernel distance is adopted in the performance index. The scheme computes adaptively the kernel parameter by applying a fast bandwidth selection rule using the distance variance collected from all data points. Both measures of the weighted fuzzy factor and kernel distance are parameter free. It is shown to be robust against noises and artifacts on both synthetic and real image experiments [15].

2 Watershed Transform for Initialization

Aerial images contain inhomogeneous intensity and blurry boundary which makes accurate segmentation tough. Therefore a state of the art hybrid fast kernel based fuzzy clustering scheme is now proposed to cover both boundary and regional information to enhance accuracy, robustness and efficiency in terms of noise, artifact and outliers. To accelerate the convergence rate, watershed transform is applied for fast initialization of level sets so that the generated optimal closed contour of level sets is applied for cluster center initialization. Quality of information retrieval is analyzed in both spatial domain and frequency domain.

Morphological watershed transform is computed on a basis of mathematical morphology to classify an aerial image in terms of discontinuity. In watershed transform, watershed curves partition separate catchment basins where the gradient of the intensity level represents the altitude. The large gradient regions are referred to as watershed curves and small gradient regions are referred to as catchment basins. Generally watershed transform consists of opening, closing, erosion and dilation. Erosion is to substitute data at each pixel by the minima. Dilation is to substitute data at each pixel by the maxima. Erosion and dilation generate dual operations where the former is to shrink and the latter is to expand. Opening implements dilation on an eroded outcome while closing implements erosion on a dilated outcome. The actual mismatch between the erosion and dilation of a source aerial image results in gradient intensity level. Watershed gradient magnitude is expressed as the mismatch of unit size dilation and unit size erosion from a source image as shown in (1).

$$\text{grad}(X) = (X \circ B) - (X \bullet B) \tag{1}$$

where $X(x, y)$ is defined as the digital image subject to transformation, B is a structuring element, $\circ$ and $\bullet$ indicate dilation and erosion. Downstream watershed segmentation is carried out in context, where erosion is applied ahead of dilation. The watershed transform is applied to capture the initial contour. Due to unavoidable noises, irregularities or other complexity of blurry aerial images, accuracy can barely be guaranteed where both over-segmentation and non-smoothness could occur. If an aerial image is divided into vast number of regions, over-segmentation occurs, controlled markers for foreground and background are used as a tradeoff to compensate for over-segmentation instead, whose results are shown in several numerical results.

3   Fast Level Set Approach

The level set method is on a basis of a geometric deformable model that generates dynamic variational boundaries to represent topological changes on either a level surface or a hypersurface. The active contour is defined as a zero level set function that provides an implicit representation of interface. The typical level set model is expressed as a Signed Distance Function (SDF) to a surface. The initial contour is generated by watershed transform and then the boundary between each pair of neighboring regions has been weighted, in order to define the edge indicator function based on the gradient information. The implicit evolution of existing contours is approximated by chasing the zero level set driven by the partial differential equation. Implicit representation of active contours helps to automatically track topological changes on interfaces and shapes. To evolve a surface towards outer boundaries, external and internal forces are introduced. After parameter selections of initial location, propagation rate and degree of smoothness, the level set model offers a powerful contour based scheme to be applied across splitting and merging processes. For 2D image processing, a level set function represents a closed surface which evolves along with time which can be defined as $\Psi(x, y, t)$. It generates a time dependent PDE function shown in (2).

$$\frac{\partial \Psi}{\partial t} + F|\nabla\Psi| = 0 \tag{2}$$

where $\Psi(x, y, t)|_{t=0} = \Psi_0(x, y)$. Rather than using an arbitrary lengthy initialization process without spatial information, watershed transform is in charge of the contour initiation ahead of the fast level set method, in order to further improve the initialization of fuzzy C-Means and accelerate its convergence accordingly. Evolution of the active contours can be represented by a zero level set $\Gamma(t)$, so that $\Psi(x, y, t) = 0$ whenever (x, y) is placed at $\Gamma(t)$. The contour evolution is then updated using the level set differential Eq. (2). $|\nabla\Psi|$ refers to the gradient operation at the normal direction. F represents a total force function including internal and external forces. The internal force arises from topological geometry while the external one results from the gradient operation or artificial momentum.

Edge indicator g is involved to regulate and terminate the level set evolution around an optima as (3):

$$g = \frac{1}{1 + |\nabla(G_\sigma * I)^2|} \tag{3}$$

where gradient operation is applied to the convolution of the Gaussian kernel with the 2D aerial image I. For each separated sub-image (I_1, I_2, I_3,...), the boundary can be computed such that a composite weighted function serves as the updating edge indicator. The level set model is simplified into (4).

$$\frac{\partial\Psi}{\partial t} = g|\nabla\Psi|[div(\frac{\nabla\Psi}{|\nabla\Psi|}) + v] \tag{4}$$

where the divergence of a normalized term $\nabla\Psi/|\nabla\Psi|$ indicates mean curvature and v acts as the balloon force.

To solve the level set differential equation, numerical schemes can be used to track interfaces and contours. The Fast Level Set method is presented to replace the explicit solution with the implicit solution. A finite difference approach is used to obtain an explicit computation solution of the curvature. Then the fast level set scheme is applied where controlling constant factors μ, λ, v are defined as reflecting penalty, topological conservation and balloon force information, respectively. (4) is then substituted by (5) and (6) after simplification.

$$\frac{\partial\Psi}{\partial t} = \mu\alpha(\Psi) + \beta(g, \Psi) \tag{5}$$

$$\frac{\partial\Psi}{\partial t} = \mu[\Delta\Psi - div(\frac{\nabla\Psi}{|\nabla\Psi|})] + \lambda\delta(\Psi)div(g\frac{\nabla\Psi}{|\nabla\Psi|}) + vg\delta(\Psi) \tag{6}$$

where

$$\alpha(\Psi) = \Delta\Psi - div(\frac{\nabla\Psi}{|\nabla\Psi|}) \tag{7}$$

$$\beta(g, \Psi) = \lambda\delta(\Psi)div(g\frac{\nabla\Psi}{|\nabla\Psi|}) + vg\delta(\Psi) \tag{8}$$

The first term $\mu\alpha(\Psi)$ in (5) acts as the penalty term which is to penalize deviation of Ψ away from a Signed Distance Function upon evolution. Another last term $\beta(g, \Psi)$ of

(5) covers the gradient information which depends on the Dirac function $\delta(\Psi)$. In (8), the role of the gradient operation is to drive the zero level set towards the variational boundaries while the role of the balloon force should be either push or pull the dynamic interface to and from targeting objects. The Dirac function can be defined as an adjustable threshold (9) which is equal to zero when x is outside of $[-\varepsilon, \varepsilon]$.

$$\delta_\varepsilon(x) = \frac{1}{2\varepsilon}[1 + \cos(\frac{\pi x}{\varepsilon})] \tag{9}$$

where ε is a regulating factor of the Dirac function. For a step size τ, an iterative evolution of Ψ is formulated as:

$$\Psi^{k+1}(x, y) = \Psi^k(x, y) + \tau[\mu\alpha(\Psi^k) + \beta(g, \Psi^k)] \tag{10}$$

The dynamic interface evolves towards the optimal boundary adaptively for image segmentation.

4 Kernel Based Fuzzy C-Means Clustering

In practice the cluster boundary of aerial images could be seldom distinct but with certain degree of fuzziness. Fuzzy C-Means clustering becomes suitable for data analysis. The goal of fuzzy C-Means clustering is to classify a finite set of data into C fuzzy clusters to optimize the objective function. An arbitrary pixel is related to multiple clusters with different degrees of belonging, which is interpreted as fuzzy membership function. Using a finite set of data, fuzzy C-Means clustering will generate a group of cluster centers together with the partition matrix that shows the possibility of a pixel being allocated into those clusters. Those pixels near boundary between clusters have less degrees of belonging than those pixels close to cluster centers. Gaussian fuzzy membership function is selected. For any pixel out of C clusters, the sum of degrees of belonging to all clusters is one, as listed in (11).

$$\sum_{i=1}^{c} \mu_{ij}(x) = 1; \mu_{ij}(x) \in [0, 1]; i = \{1, 2, \cdots, C\}; j = \{1, 2, \cdots, N\} \tag{11}$$

where μ represents the fuzzy membership function, C is total number of clusters and N is the total number of pixels subject to clustering. To be more accurate and robust, kernel fuzzy C-Means clustering has been proposed where a nonlinear kernel distance is used instead of the Euclidean distance. A kernel scheme transforms from nonlinear to linear segmentation by mapping source image data into high dimensional kernel space implicitly. Distances defined in the kernel space can be simplified as (12), and Gaussian kernel is shown in (13).

$$\|\Phi(x_i) - \Phi(x_j)\|^2 = G(x_i, x_i) + G(x_j, x_j) - 2G(x_i, x_j) \tag{12}$$

$$G(x_i, x_j) = \exp(\frac{-\|x_i - x_j\|^2}{2\sigma^2}) \tag{13}$$

After initialization of all cluster centers, the degree of belonging is formulated as (14) for every particular fuzzy membership function. It is expressed as the ratio between individual cluster kernel distance measure and the sum of all kernel distances from the pixel to C clusters.

$$\mu_{ij} = \frac{\left\| \Phi(x_j) - \Phi(c_i) \right\|^{-\left(\frac{2}{m-1}\right)}}{\sum_{i=1}^{C} \left\| \Phi(x_j) - \Phi(c_i) \right\|^{-\left(\frac{2}{m-1}\right)}} \tag{14}$$

All the centroids are updated by normalized quantities weighted using Gaussian kernel based fuzzy membership functions. A fuzziness index m ($1 \leq m < \infty$) is applied in kernel fuzzy C-Means clustering.

$$c_i = \frac{\sum_{j=1}^{N} \mu_{ij}^m G(c_i, x_j) x_j}{\sum_{j=1}^{N} \mu_{ij}^m G(c_i, x_j)} \tag{15}$$

To achieve optimization, a feasible objective function must be defined based on spatial information. As a region based scheme, fuzzy C-Means clustering selects the texture, position, intensity and color in the feature space to formulate the objective function of fuzzy clustering as (16). The edge or boundary feature instead is considered in watershed transform and the fast level set model rather than current clustering scheme. Meanwhile, incorporating fuzziness in C-Means clustering helps to locate those vague boundaries that crisp clustering may barely catch.

$$J^{\Phi}(X, C) = \sum_{i=1}^{C} \sum_{j=1}^{N} \mu_{ij}^m \left\| \Phi(x_j) - \Phi(C_i) \right\|^2 \tag{16}$$

where m ($1 \leq m < \infty$) shows fuzziness of segmentation that defined as fuzziness index. $X = \{x_j\}$ indicates a data matrix and each x_j represents the j-th element of this fuzzy matrix. $C = \{c_i\}$ acts as a vector of cluster centers whose c_i represents the centroid of its i-th cluster. $\|.\|^2$ is the L_2 norm. The objective function achieves global optimization in case the pixels close to centers produce largest degrees of fuzzy memberships and pixels far away from centers produce smallest degrees of fuzzy memberships. Initialization of fuzzy cluster centers is conducted using watershed transform with markers followed by fast fuzzy set schemes to accelerate its convergence and to be robust against artifacts, noises and outliers. For every iterative cycle, the set of cluster centers and fuzzy memberships are subject to updating. It evolves until an objective function is minimized and optimal image segmentation can be reached.

5   Densely Distributed Aerial Images

An aerial photo exhibits some unique attributes in image resolution, distance scale and coverage extent. It generally captures more complex information topologically and geometrically than true color still images. Subject to atmospheric dispersion, an aerial image is easily affected by noises and artifacts. 4 typical true color densely distributed aerial images are selected (Fig. 1). Composite color is represented by mixing three primary color components RGB (Red, Green and Blue) at the true color space. The RGB image is described by an intensity level array ($M \times N \times 3$) independent of each other. Rather than classical fuzzy C-Means clustering, fast kernel based fuzzy scheme is selected which also covers 2 initial steps. Morphological watershed transform is applied to capture unique boundaries beforehand without lengthy iteration using gradient magnitude, whose results are applied as the initial level set. Watershed transform itself is sensitive to intensity variations and noises, giving rise to non-smooth boundaries and outliers. The fast level set scheme is thus used next to produce accurate partition based on edge and boundary information whose contours or surfaces are reached rapidly via fast finite difference approach. The weakness of hybrid watershed transform and level set scheme occurs due to lack of regional information. However, closed contour topological information stem from closed surface of the level set has provided perfect spatial information to determine initial cluster centers so as to enhance separability. All initial centroids are chosen from a set of closed contours as constraints to narrow down the region of interest in the unsupervised fuzzy C-Means scheme. Computation time for the enhanced fuzzy C-Means scheme will be reduced remarkably without initial random sampling. Due to nonlinearity and complexity of densely distributed aerial images, kernel based fuzzy C-Means clustering scheme is carried out such that regional information is also involved. Rapid initialization is achieved where much less turnaround time is needed for segmentation in terms of hybrid fast kernel based fuzzy scheme than classical one.

Fig. 1. Typical true color aerial scene images - densely distributed (a. Istanbul; b. Rio de Janeiro; c. Cape Town; d. London) (Color figure online)

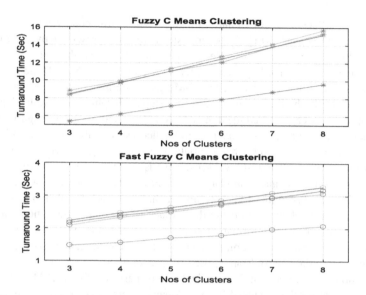

Fig. 2. Turnaround time of fuzzy C-Means clustering in 4 cases (number of clusters: 3–8) (1. Classical fuzzy scheme; 2. Hybrid fast kernel based fuzzy scheme)

In Fig. 2, comparisons of turnaround time between fast kernel based fuzzy scheme and classical fuzzy C-Means clustering are made. For each of 4 aerial photo images, the total cluster number ranges from 3 to 8. The 3 curves on top are associated with 3 cases a, b, and c; while another curve on bottom is associated with the case d. It shows clearly integration of three schemes has significantly reduced the turnaround time, leading to the hybrid fast kernel based fuzzy C-Means clustering scheme. Meanwhile, gradient information, (edge, boundary or contour) information and region information are all taken into account, so that it generates accurate, closed and smooth aerial image segmentation.

6 Frequency Domain Numerical Analysis

The edge, curvature, contour and region information are well captured using hybrid fast segmentation scheme via technology integration being proposed. For complete information retrieval, it is necessary to conduct frequency domain analysis in addition to spatial domain analysis. Numerical simulations on the histogram analysis are conducted in this session to show potential changes occurred in multiple level clustering processes. All selected densely-distributed aerial true color images have three primary color components of red, blue and green. Each individual color component is subject to fast image segmentation via technology integration at multiple levels. In Figs. 3, 4, 5 and 6, three clusters are specified for each of four densely-distributed aerial images. The outcomes of composite color, red, blue and green are listed from the 1st to 4th rows. Clustered images are shown in the 1st column, the corresponding frequency domain histograms are shown in the 2nd column and markers being adopted in watershed transform are shown in the 3rd column.

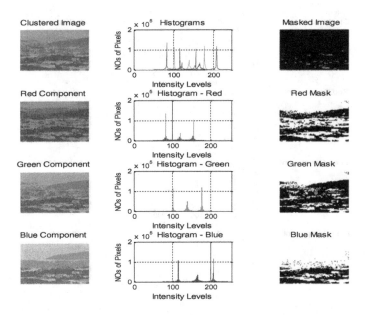

Fig. 3. Histogram of hybrid fast kernel fuzzy C-Means clustering (Case A) (Color figure online)

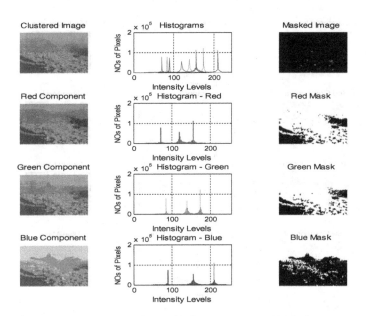

Fig. 4. Histogram of hybrid fast kernel fuzzy C-Means clustering (Case B) (Color figure online)

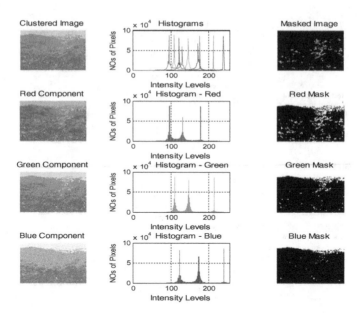

Fig. 5. Histogram of hybrid fast kernel fuzzy C-Means clustering (Case C) (Color figure online)

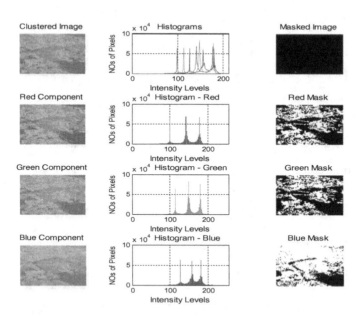

Fig. 6. Histogram of hybrid fast kernel fuzzy C-Means clustering (Case D) (Color figure online)

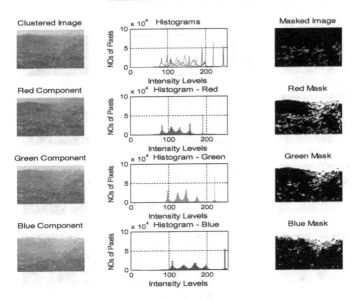

Fig. 7. Histogram of hybrid fast kernel fuzzy C-Means clustering (5-Cluster) (Color figure online)

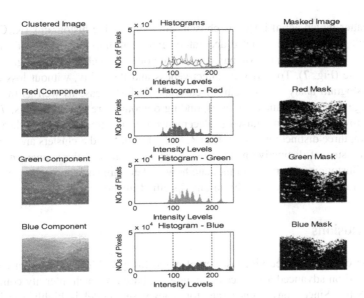

Fig. 8. Histogram of hybrid fast kernel fuzzy C-Means clustering (8-Cluster) (Color figure online)

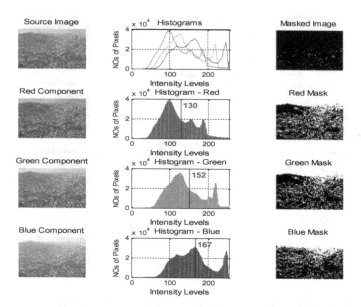

Fig. 9. Histogram of hybrid fast kernel fuzzy C-Means clustering (Source Image) (Color figure online)

No matter which aerial image is chosen, three distinctive histograms (R, G, B) in terms of all 3 clusters are depicted accurately. For each individual case with multiple clusters, the outcomes can be compared with each other as well as the histogram of the source image (Fig. 7). To verify its effectiveness and flexibility, without loss of generality, histogram analyses on 5-cluster case and 7-cluster case as well as the source aerial image are also conducted. Corresponding outcomes are shown in Figs. 7, 8 and 9, whose row-wise and column-wise placements are the same as Figs. 3, 4, 5 and 6. Once again, three distinctive histograms (R, G, B) for all 5 and 8 clusters are accurately shown. The sum of distinctive histograms gives rise to that of the source image. The composite histogram diagram will indicate how to determine an appropriate number of clusters being classified effectively on any specified image.

7 Conclusions

Integration of fast level set with fast kernel based fuzzy C-Means clustering has been implemented on advanced aerial scene recognition based on each intensity component independently. Since convergence rate for a level set model is highly sensitive to initials, watershed transform is applied to provide the fast and rough initialization, near potential best solution where actual level set boundary of objects could be overlapping with results from watershed transform. The fast level set model then helps to deploy active edges or contours and dynamic boundaries, in order to narrow down initial centroid locations for kernel fuzzy C-Means clustering rapidly, making pixel classification and parameter configuration time efficient to handle. The proposed approach has been implemented on multiple aerial true color images with densely distributed attributes. From numerical simulations, no matter how many clusters to be classified in each case, the hybrid approach exhibits much faster turnaround time than classical fuzzy C-Means clustering. Qualitative histogram analyses on the new fast kernel based fuzzy schemes are also conducted to clearly show statistical information retrieved and classified in the frequency domain to completely discover spatial domain and frequency domain information across the clustering processes.

References

1. Duda, R., Hart, P., Stork, D.: Pattern Classification, 2nd edn. Wiley, Hoboken (2001)
2. Gonzalez, R., Woods, R.: Digital Image Processing, 3rd edn. Prentice Hall, Upper Saddle River (2007)
3. Schilling, R., Harris, S.: Fundamental of Digital Signal Processing using MATLAB. Cengage Learning, Stamford (2005)
4. Ye, Z., Mohamadian, H., Yin, H., Ye, Y.: Integration of fuzzy C-Means clustering and fast level set for aerial RGB image segmentation. In: Proceedings of 2015 International Brazilian Meeting on Cognitive Science, Sao Paulo, Brazil, 7–11 December 2015 (2015)
5. Li, B., Chui, C., Chang, S.: Integrating spatial fuzzy clustering with level set methods for automated medical image segmentation. Comput. Biol. Med. **41**, 1–10 (2011)
6. Ye, Z., Mohamadian, H., Ye, Y.: Gray level image processing using contrast enhancement and watershed segmentation with quantitative evaluation. In: Proceedings of 2008 IEEE International Conference on Content-Based Multimedia Indexing, London, UK, 18–20 June 2008, pp. 470–475 (2008)
7. Saikumar, T., Yugander, P., Murthy, P., Smitha, B.: Image segmentation algorithm using watershed transform and fuzzy C-Means clustering on level set method. Int. J. Comput. Theory Eng. **5**, 209–213 (2013)
8. Ye, Z., Ye, Y., Mohamadian, H.: Fuzzy filtering and fuzzy K-Means clustering on biomedical sample characterization. In: Proceedings of the 2005 IEEE International Conference on Control Applications, Toronto, Canada, pp. 90–95 (2005)
9. Ye, Z., Mohamadian, H.: The role of quantitative metrics in enhancing spatial information retrieval via fuzzy C-Means clustering. In: Proceedings of 2011 International Symposium on Remote Sensing of Environment, Australia, 10–15 April 2011 (2011)
10. Fazendeiro, P., Oliveira, J.: Observer-biased fuzzy clustering. IEEE Trans. Fuzzy Syst. **23** (1), 85–97 (2015)
11. Ye, Z., Mohamadian, H.: Enhancing decision support for pattern classification via fuzzy entropy based fuzzy C-Means clustering. In: Proceedings of the 2013 52nd IEEE Conference on Decision and Control, Florence, Italy, 10–13 December 2013, pp. 7432–7436 (2013)
12. Connor, B., Roy, K., Shelton, J., Dozier, G.: Iris recognition using fuzzy level set and GEFE. Int. J. Mach. Learn. Comput. **4**(3), 225–231 (2014)
13. Alipour, S., Shanbehzadeh, J.: Fast automatic medical image segmentation based on spatial kernel fuzzy C-Means on level set method. Mach. Vis. Appl. **25**, 1469–1488 (2014)
14. Arabe, S., Gao, X., Wang, B.: A fast and robust level set method for image segmentation using fuzzy clustering and lattice boltzmann method. IEEE Trans. Cybern. **43**, 910–920 (2013)
15. Gong, M., Liang, Y., Shi, J., Ma, W., Ma, J.: Fuzzy C-Means clustering with local information and kernel metric for image segmentation. IEEE Trans. Image Process. **22**(2), 573–584 (2013)

A Case Study of Library Data Management: A New Method to Analyze Borrowing Behavior

Luis Cano, Erick Hein, Mauricio Rada-Orellana, and Claudio Ortega[✉]

Universidad del Pacífico, Lima, Peru
c.ortegaariza@alum.up.edu.pe

Abstract. The library data management system of academic institutions generates and stores data with information on lends, returns and queries made by users daily. This research aims to provide a useful method to transform this data to obtain relevant knowledge for the management of the library. The process used consisted of a data preprocessing, data transformation and cleaning, and finally the use of privacy and association algorithms. We clarify the proposed method for a university library in Peru where significant results were obtained in the form of association rules. This shows that it is a efficient alternative to have a more detailed knowledge to make better decisions.

Keywords: Pre-processing · Library management · Association rules · Privacy · Knowledge Data Discovery

1 Introduction

Large amounts of data are generated each day across all the interactions between the agents in the world [6]. This phenomena is not only observed at the enterprise level, but also in the education and learning sector [4]. Although educational data is not usually analyzed, this contains relevant information and needs a processing. Among the educational data sets, we are interested in the ones generated in the libraries and archives databases of universities and other academic institutions. To process and analyze the data generated in the transactions, it is necessary to apply methodologies like Knowledge Data Discovery (KDD). This method enable us to use data mining and machine learning algorithms to extract priceless insights [5].

The use of this tool by academic institutions fosters the application of IT not only in the educational process management like student registration or resource allocation, but also in the gathering of high quality data to provide knowledge like as literature preference, rotation and utilization. This knowledge can be used to increase the library usage which can contribute positively to the academic performance as a previous study found at eight UK universities [10] and at Honk Kong Baptist University [12].

© Springer Nature Switzerland AG 2019
J. A. Lossio-Ventura et al. (Eds.): SIMBig 2018, CCIS 898, pp. 112–120, 2019.
https://doi.org/10.1007/978-3-030-11680-4_12

This research aims to provide a useful methodology based on KDD for university libraries to obtain knowledge from their databases and take better decisions for their users. In this way, librarians can decide the allocation of literature, quantity to offer per topic and human resources needed to satisfy the demand. In the next section, we will declare the main techniques we will use, then we will propose the new method with the previous defined techniques; and finally, we will apply the methodology to a case study of a Peruvian university library.

The data we used to perform the research is a semi structured set, captured from one academic semester. The data is composed by four sources: literature lending records, literature returning records, user queries and literature availability. The literature is composed by textbooks, journal articles, newspapers, magazines, reports and audiovisuals.

2   Theoretical Framework

The data mining process consist of collect, clean, process, analyze and obtain knowledge from large amounts of data [1]. The data goes through many subprocesses, called pipeline, to consolidate the information until it reaches a relevant pattern. In this case study, the pipeline aims to improve the knowledge about lending, rotation and stock management of the books.

2.1   Clustering

The clustering process consists in grouping the data in defined sets whose attributes have similar values [2]. We will use k-means algorithm since it is the most widely used from the reviewed literature [2].

This algorithm is a grouping method that aims to partition the data into a defined number k of groups. Observations are included within a group when the distance between their data point and the centroid of the cluster is the least compared to the distance to the centroids of other groups using Euclidean distances. If $\bar{Y} = (Y_1, Y_2, \ldots, Y_n)$ and $\bar{Y} = (Y_1, Y_2, \ldots, Y_n)$ are two observations with n-feature, then the Euclidean distance is given by Eq. 1 [2]:

$$Dist(\bar{X}, \bar{Y}) = \sqrt{\sum_{i=1}^{n} (X_i - Y_i)^2} \tag{1}$$

The procedure of the k-means is as follows: (1) a number of clusters k must be defined, (2) the algorithm generates k centroids randomly and the distances from all data points to each centroid is calculated, (3) this process repeats while it is optimizing the centroids locations by reducing the sum of the squared distances. The break point can be defined as a threshold for the cumulative distance, a number of runs or the experience of the researcher.

2.2   Cluster Validation

The algorithm validation allows us to compare and monitor the performance of potential models. In this work, we used Davies-Bouldin Index and the average within-cluster distance since are the most common indexes in the literature [8].

Davies-Bouldin Index: This index aims to calculate and compare the average within-cluster distance and the average inter-cluster distance. If the algorithms generates clusters with low within-cluster and high inter-cluster distance, the algorithm has a good performance and has an index close to zero. This is calculated with the Eq. 2 [3].

$$DBI(C) = \frac{1}{n_c} \sum_{c_i \varepsilon C}^{n_c} \max_{j \neq i} \frac{S_i + S_j}{M_{i,j}} \qquad (2)$$

Where S is a measure of clusters compactness, M is a measure of clusters separation, C is the set of clusters, c_i is i-th cluster of C and n_c is the number of clusters in C.

Within-Cluster Average Distance: This index calculates the average distance between the centroid of the cluster and all the elements of it [9].

2.3   Association Rules

The classic problem of pattern mining aims to determine the most frequent associations. According to the literature, the most common algorithms are Apriori and FP-Growth. For this research, we will use the FP-Growth since it is a more advanced and computationally efficient technique [7,11].

FP-Growth: This method has a different procedure in the discovery of the frequent *itemsets* from other algorithms. This codify the data in a structure called FP-Tree, which is a representation of the input data in a zipped way. It is built by processing the transactions once at a time and mapping in one of the branches of the tree. After it, FP-Growth generates the most frequent *itemsets* and stores the paths. Then, using the minimum support value, the algorithm chooses the *itemsets* that complies with the value and stores then as "frequents". Finally, it establishes the association rules on the basis of the frequent *itemsets*.

The association rules are statements with the structure "If ... then" that help to discover relationships between not seemingly related data. The rules are created by search patterns and using criteria called "Support" and "Confidence" to identify the most important relationships [11]. "Support" is used to remove rules that do not appear regularly. And "Confidence" measures the number of times when the statements *If ... then* are true.

3   Proposed Method

This work proposes KDD-based method. Figure 1 presents the proposed pipeline and the description of each stage.

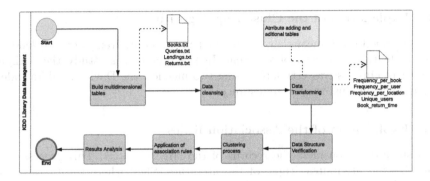

Fig. 1. Proposed model. Own elaboration.

3.1   Build Multidimensional Tables

This stage consists in transforming the semi structured data into structured data. Depending of the technology capacity of the library the input data can be already structured, otherwise, it is needed to perform a manipulation of the tables. It should be noted that each attribute adds more complexity to the analysis.

3.2   Data Cleansing and Enriching

This stage deals with errors found in the data sources, thus several problem-specific decisions must be made. Besides mismatching in data formats, the dataset of literature capacity, which list all literature available to lend, contains repeated titles, or missing ones. There are solution alternatives like deleting the inaccurate data or impute missing values using the tendency or a fix value. For this case study, we opt for delete the records since they were a minimum proportion. To gather more detailed data, this stage generates tables that records the lending frequency and duration according to the dates, literature and users. Nevertheless, one of the challenges in the databases of the libraries of universities and other institutions is that there may be data entry mistakes caused by human errors. For example, some books that appear as returned do not previously appear as borrowed which is a fatal inconsistency.

To tackle this challenge, this stage requires to prepare a timeline for each literature ID and track the registration of lending and returning to determine the duration of lending. In this case study, we found that some books were borrowed many times without prior registration of returning. This is clearly a planning error from the library administration that hamper the knowledge discovery work. Thus, we only calculated the frequencies whose lends had a returning counterpart to keep the data consistency. Finally, we add enriched attributes on the frequencies like categorization of returning times, lending frequencies and binarization to have an ad-hoc structure for the data mining algorithms.

3.3 Explanation of the Clustering Algorithm

We performed the k-means algorithm, that uses lending frequency and lending days as features for each observation. In the present case study the optimal number of clusters was estimated using two metrics: the Davies-Bouldin Index and the within-cluster average distance.

3.4 Explanation of the Association Rules

This stage only considers the records of the largest cluster. This process uses the following attributes: Topic of the literature, Shelf where the literature is available, Literature main keyword, Lending date. The process executes an FP-Growth algorithm generates the *frequent itemsets* with a support level of 30%. Among these *frequent itemsets*, we select the association rules from those *frequent itemsets* with a confidence level of at least of 60%. The chosen support and confidence levels are the most used in the reviewed literature.

4 Case Study: A University Library in Peru

For the present study the university library give us access to the records of their administration system. The main used attributes are indicated in the Table 1, while attributes calculated from data enriching are indicated in Table 2.

Table 1. Descriptive table of attributes

Attribute	Description
Item ID	Literature identifier
Title	Literature title
Location	Place where the literature was available
User ID	User identifier
Lending count	Number of borrowings of the user
Query count	Number of queries executed by the user
Returning count	Number of returning of the user
Lending date	Date when the lending occurred
Returning date	Date when the returning occurred
Query date	Date when the query was executed

- **Main keywords:** We performed a keyword analysis to identify hot topics. Since the university courses are business-oriented, the most relevant keywords are: "economy", "accounting", "Peru", "theory", "analysis", among others. This implies that topics like nature science or arts and media are scarcely searched.

Table 2. Calculated attributes table

Attribute	Description
Daytime	Categorize into dawn, morning, afternoon and night
Available hours	Amount of hours when the literature is available
Weekday	Values between one to seven for each day
Returning time window	Length of lending
Literature topic	Keywords of the literature topic

- **Literature utilization:** We found that 80% of the queries and lends are assigned to only 4% of the literature. This may indicate a misunderstanding from the management to offer suitable literature since there is a gap between what users search and what the library have available.
- **Lend and query schedules:** Most of the traffic query is during the mornings and afternoons, 39% and 46% respectively. During the night, the traffic greatly decreases to 15%, which is explained by the fact that most of the classes are during the morning and afternoon when the students are attending courses and professors are working.
- **Student behavior during the semester:** There are changes of traffic of queries and lends from week to week caused by the behavior of students. We found a high intensity during the first week of the semester (approximately 1300 lends), where students may be more motivated and have more time to immerse into the literature and explore the books referenced by the professors. Then, there are peaks during the mid-term (approximately 1100 lends), which

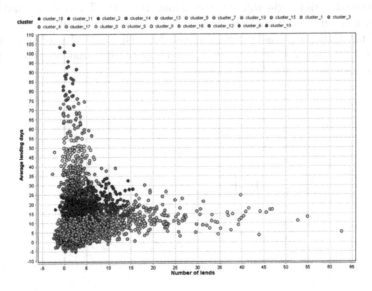

Fig. 2. Clusters generated by the algorithm

is clearly caused by mid-term exams when students are looking up for books. Finally, there are valleys just after the mid-term (approximately 900 lends), when student life organizations are fostering parties and integration events. After that, the traffic increases until reach pre mid-term exams levels. From there, it decreases progressively until final exams week.

– **Clustering algorithm analysis:** Figure 2 shows the twenty cluster model, whose largest agglomerations tend to have a short time window of returning and a low frequency of borrowing. These clusters are conformed by people who borrow a low quantity of literature in the semester and return it in a short while. The other clusters, take clear patterns towards the clusters of students who fail to return in a timely manner (upper left quadrant) and the clusters who are intensive users of the library.

The largest cluster, which represents 15% of the users (cluster 12) is composed by students who return the books in a short time and borrow just a few books. On the other hand, clusters who have a very long time window to return the literature, represent less than 2% of the users each one, as shown in Fig. 3. These clusters return the books after sixty days, when the current library policy is just seven days.

– **Association algorithm analysis:** We identify three rules (from a set of fifty five generated by the algorithm) as the ones of the most importance of this case study, not only because of their confidence level, but also for the insights that they provide.

From the table we can posit three rules:

1. That if the user has borrowed a book from the topic of *Accounting*, then there is a possibility of 84% that he also made the registration at the *Library room 2*. This insight will help the management to offer more accounting literature in the room 2 to increase the rotation and utilization.
2. That if the user has borrowed a book from the topic of *Economics* during the *Afternoon*, then there is a possibility of 80% that he also made the registration at the *Library room 1*.

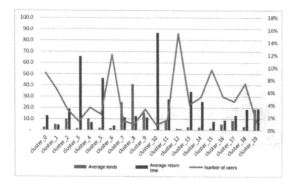

Fig. 3. Analysis of the behavior of the clusters

3. That if the user has borrowed a *Textbook*, during *Afternoon* and at the *Library room 1*, then there is a possibility of 65% that he also borrowed a book of the topic of *Economics*.

5 Conclusions

The use of technology should have a essential role in the educational sector, especially in the libraries of the academic institutions. By doing so, the quality and quantity of knowledge will increase and therefore, the service offered to the users will improve too.

This research offers a modern methodology based on KDD, oriented to libraries and archives databases of academic institutions. The objective of this methodology is to provide relevant insights to the library management to anticipate the literature requirements, optimize the rotation and reduce the unused sources. This applies to the literature and the room spaces, since our methodology incorporates the location of the queries.

The use of clustering algorithms offers a characterization of the users to identify certain profiles. These profiles indicates a specific behavior like defaulting and use intensity, however, a richer database would have offered demographic attributes like age, major, pursued degree or job. The association rules offer the relationship of the user behaviors. The behavior of the main profile provides information about what topics, places or daytime are more associated.

Further research should explore the analysis of panel data to find specific patterns across different libraries. In the case of having more attributes concerning to the user profile, a risk of default model can be elaborated to identify those students who will probably fail to return the borrowed book in the specified time.

References

1. Aggarwal, C.C.: Association Pattern Mining (2015)
2. Bratchell, N.: Chapter 6 cluster analysis. In: Brereton, R.G. (ed.) Multivariate Pattern Recognition in Chemometrics, Illustrated by Case Studies (1992)
3. Davies, D.L., Bouldin, D.W.: A cluster separation measure. IEEE Trans. Pattern Anal. Mach. Intell. **2**, 224–227 (1979)
4. Dunstan, S.M.: El impacto de la informática en la educación (2001)
5. Fayyad, U., Piatetsky-shapiro, G., Smyth, P., Widener, T.: The KDD process for extracting useful knowledge from volumes of data. Commun. ACM **39**, 27–34 (1996)
6. Han, J.: Data Mining: Concepts and Techniques (2005)
7. Han, J., Pei, J., Yin, Y., Mao, R.: Mining frequent patterns without candidate generation: a frequent-pattern tree approach. Data Min. Knowl. Discov. **8**, 53–87 (2004)
8. Kovács, F., Legány, C., Babos, A.: Cluster validity measurement techniques. In: 6th International Symposium of Hungarian Researchers on Computational Intelligence. Citeseer (2005)

9. del Prado Cortez, M.N., Salas, H.A.: Quality metrics for optimizing parameters tuning in clustering algorithms for extraction of points of interest in human mobility. In: SIMBig (2014)
10. Stone, G., Ramsden, B.: Library impact data project: looking for the link between library usage and student attainment. Coll. Res. Libr. **6**, 546–559 (2013)
11. Tan, P.N., Steinbach, M., Kumar, V.: Introduction to Data Mining, 1st edn (2005)
12. Wong, S.H.R., Webb, T.: Uncovering meaningful correlation between student academic performance and library material usage. Coll. Res. Libr. **72**(4), 361 (2011)

Sparkmach: A Distributed Data Processing System Based on Automated Machine Learning for Big Data

Gusseppe Bravo-Rocca⬤, Piero Torres-Robatty⬤, and Jose Fiestas-Iquira⁽⊠⁾⬤

Universidad Nacional de Ingeniería, Rimac Lima 15333, Peru
gbravor@uni.pe,
{piero.torres,jose.fiestas}@uni.edu.pe

Abstract. This work proposes a semi-automated analysis and modeling package for Machine Learning related problems. The library goal is to reduce the steps involved in a traditional data science roadmap. To do so, Sparkmach takes advantage of Machine Learning techniques to build base models for both classification and regression problems. These models include exploratory data analysis, data preprocessing, feature engineering and modeling.

The project has its basis in Pymach, a similar library that faces those steps for small and medium-sized datasets (about ten millions of rows and a few columns). Sparkmach central labor is to scale Pymach to overcome big datasets by using Apache Spark distributed computing, a distributed engine for large-scale data processing, that tackle several data science related problems in a cluster environment. Despite the software nature, Sparkmach can be of use for local environments, getting the most benefits from the distributed processing tools.

Keywords: Semi-automated machine learning · Data Science · Data mining · Statistics · Data engineering · Big data

1 Introduction

Currently, statistics and machine learning are meant to extract new information from patterns found in data. Individually, in Data Science, there are advances in how to automate the end-to-end process of applying machine learning. Automatization offers the advantages of producing more straightforward solutions and an improved way of picking the best learning algorithms and models.

The first published and implemented work was Auto-WEKA in 2013. This software, written in Java, proposed automated algorithms for feature selection and hyperparameter tuning. Later, in 2014, Hyperopt-sklearn was developed and followed a similar focus as Auto-WEKA but using the ML library scikit-learn (2011). This program could only handle small-sized and mid-sized data. The next year MLbase and AutoML were published. The former was the first

© Springer Nature Switzerland AG 2019
J. A. Lossio-Ventura et al. (Eds.): SIMBig 2018, CCIS 898, pp. 121–128, 2019.
https://doi.org/10.1007/978-3-030-11680-4_13

publication in Algorithm and hyperparameters selection involving distributed computing by using Apache Spark-MLlib. The later is based in Auto-WEKA but combines an automated parametric learning platform with Bayesian optimization. Finally, Bravo-Rocca in 2017 developed Pymach, a tool that aims to accelerate the development of machine learning models by looking for the best model available for any data [1]. To improve Pymach scope and performance, we implemented Sparkmach, a distributed successor from Bravo-Rocca's past work. For the transformation, we refactored Pymach library, written in python language, into a Pyspark-based program [9]. Later, we compared both Pymach and Sparkmach programs by using the same case study dataset, based on New York City Bus System and a physic simulation.

The paper's goal is to show the Sparkmach methodology and how it performs in contrast with Pymach by using the same case scenario.

2   Sparkmach: Definition and Structure

This section first summarizes Sparkmach concept. Then it details all the stages working under the proposed system.

2.1   Definition

Sparkmach is an open source library, under MIT licensing, based on Pymach [1] written in Python [3] and deployed in a distributed platform using Apache Spark [5]. Hence, the main difference between both systems is that the former can be executed in a clustered environment for big data solutions by using distributed data processing and the latter locally, facing just small problems regarding size.

The software developed under a chain structure, based on Apache Spark ML pipelines, whose function is to gather different types of estimators and transformers. In consequence, this implementation allows us better modularity, flexibility, and scalability when added new functionality.

Sparkmach can be defined as follows: let $G = (\hat{V}, \hat{E})$ be a directed graph, consisting of the set $\hat{V}$ of nodes and the set $\hat{E}$ of edges. Where $\hat{V} = \{D, A, P, S, E\}$ represents each state in a workflow and $\hat{E} = \{(D, A), (D, P), (D, S), (A, P), (P, E)\}$ each path within that workflow. So that Sparkmach can be define as a pipeline $p^*(G)$, where

$$p(x) = [m_{(x)}^{(0)}, m_{(x)}^{(1)}, \ldots, m_{(x)}^{(k-1)}]$$

and p^* is the best pipeline and $m^{(i)}$ is the ith input's mapper.

The system works in six stages, based on Pymach's seven stages, excluding the Improve stage (grid and random search) that will be supported in a future release. The workflow showed in Fig. 1 describes these stages starting with defining the dataset (metadata) and finishing by presenting the results.

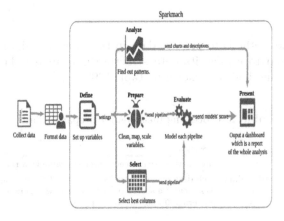

Fig. 1. Sparkmach workflow.

The stages are defined as the following:

- **Define (D)**: Defines the problem to solve.
- **Analyze (A)**: Starts a data exploratory analysis.
- **Prepare (P)**: Modifies the dataset to evaluate.
- **Select (S)**: Chooses the most suitable predictor variables.
- **Evaluate (E)** : Executes the supported models.
- **Present (P)**: Delivers results in a GUI.

2.2 Define

The first stage focuses on defining the input dataset. Table 1 details the required definitions that should be extracted from the starting dataset.

Table 1. Main metadata information extracted from the dataset

Input definition	Description
File path	Absolute file path in DFS, accepts CSV, TXT or PARQUET
Header	Defines header names for each column
Response	Name of the column to be predicted
Problem type	Indicates whether the problem type is regression or classification
Number of features	Number of predictor variables

2.3 Analyze

In the same way as Pymach proceeds with this stage, Sparkmach centrally generates the analysis graphics in the cluster's master node. The resulting distributed dataframe is transformed into a pandas dataframe [7]. Afterward, the latter is plotted by plotly [6] and presented using a web service interface.

2.4 Prepare

Even though the methods to prepare data in this stage are similar to Pymach, these are required to change its data structure into an Apache Spark vector assembler. Some methods are shown in Table 2 [9].

Table 2. Processing methods for preparing data

Processing method	Description
Max & min scaling	This method scales each numeric value in a range from 0 to 1
Normalization, standard scaling	Each data row is re-scaled to a length 1 by using norm 12 and also it can be scaled in order to follow a gaussian distribution of average 0 and deviation 1

2.5 Select

The fourth stage in line is the Select stage, which selects the features that best represent the data to improve the final model. Some functions involve reducing the overfitting, removing highly correlated variables and reducing the space of features [2].

2.6 Evaluate

The actual stage evaluates an algorithm list (Table 3) for the input dataset just as Pymach works. The main difference between both systems is that Sparkmach algorithms are coded to work in a parallel and distributed setting.

Table 3. Algorithms used in the evaluate stage

Problem type	Algorithms
Classification	Logistic regression, decision tree, random forest, gradient boosting, multi layer perceptron, naive bayes, linear support vector machine
Regression	Linear regression, generalized linear regression, decision tree, random forest, gradient boosting

2.7 Present

This stage is about interpreting and communicating the model's results which are an essential step in the data science roadmap. In this stage, the model's results are shown understandably and concisely, which will be the insights that lead to making decisions.

3 Case Study: New York City's Bus Information and a High-Energy Physic Experiment

To assess Sparkmach potential in a real-life situation, we employed information from New York City's Metropolitan Transportation Authority (MTA) and a high-energy physic experiment to search for the signatures of exotic particles (HEPMASS).

As for the first dataset, the source we chose among MTA's data catalog is called "MTA Bus Time Historical Data". Basically, this data consists of GPS output from each bus of the city during service [8]. The sample presented at Table 5 represents preprocessed data from the original source, it takes a 3-month sample from historical Bus data, spanning from August 2014 until the end of October 2014. On the other hand, the second dataset presented at Table 4 is the result of several Monte Carlo simulations of the collisions that produce these particles and the resulting decay products; the goal is to separate particle-producing collisions from a background source [10].

Table 4. HEPMASS simulation dataset sample

f_1	f_2	f_3	...	f_27	Response
1.816216	−0.458454	1.010234	...	0.977542	1
−0.196046	−0.737780	−0.164820	...	−1.448544	0
1.235821	−0.970305	−1.456574	...	1.414154	1
−0.541583	1.284816	−0.645087	...	0.090727	0

Table 5. NY MTA bus dataset sample

BusID	NextStop	Route	Orientation	HourBucket	Response
174	232	273	0	17.5	32
174	232	273	0	18.5	0
174	232	273	0	19.5	64
174	232	273	0	2.5	32

Because Sparkmach has to work in a distributed environment, we built an 8-node cluster with one master node and seven slave nodes. Each one has 4 GB of RAM, a fourth-generation Intel Core I7 processor and CentOS 6.8 as the operating system for each node. The Directed Acyclic Graph showed in Fig. 2 represents the data process from a general perspective. The process starts by ingesting the data and storing it inside an Apache Hadoop HDFS. Then, the data is preprocessed and received by Sparkmach. After processing the data,

Sparkmach will score the most suitable model and then stores it in the HDFS. The system data flow proceeds so that the model built by Sparkmach can be used with real-time inputs.

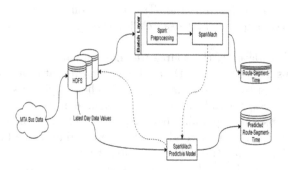

Fig. 2. Data flow architecture showing the MTA dataset as example.

4   Results and Discussion

We made two tests to compare performances from both Sparkmach and Pymach. First, a scalability test is measured on MTA dataset using Sparkmach, as shown in Fig. 3. Second, scalability and predictability tests are measured on HEPMASS dataset using Pymach and Sparkmach, as shown in Fig. 4.

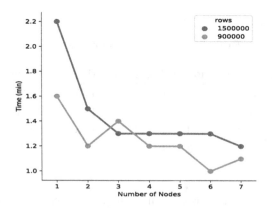

Fig. 3. Scalability test for MTA dataset using sparkmach.

As for the first test, we executed Sparkmach on different size of data and number of nodes. We evaluated a 900,000-row dataset and another one with

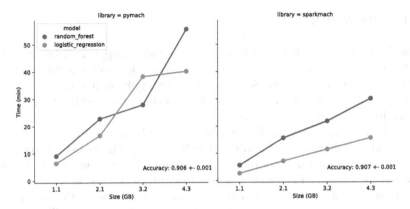

Fig. 4. Scalability and predictibility test for HEPMASS dataset using pymach and sparkmach.

1,458,098 rows. Each evaluation consisted in processing the dataset in a cluster, ranging from 1 node to 7 nodes progressively (Fig. 3).

Furthermore, the results presented a tendency of scalability potential notoriously when the amount of data is more prominent. Also, it demonstrates, for this case, that the scalability achieved by accumulating more data is directly proportional to the number of nodes that belong to a cluster [5]. On the other hand, in Fig. 4 we observed a decrease in time, keeping the same accuracy when dealing with bigger datasets.

5   Conclusions

The automated Data Science process not only involves a sequential methodology in order to solve a specific problem (exploration, preparation, modeling, and optimization) but also it involves the know-how applied to the data for the study. Hence, both Pymach and Sparkmach are meant to be essential tools for building more robust models while complementing the data analysis process.

In the exploratory Analysis stage, we found that valuable information can be extracted from the data as a whole using all of its distribution. In consequence, it dramatically aids preliminary information analysis.

The preprocessing and feature selection are very extensive fields that involve deep knowledge of the The preprocessing and feature selection are very extensive fields that involve in-depth knowledge of the know-how concerning the input data. However, some steps are very general throughout the Data Science cycle. Thus, our proposal integrates these steps in an unsupervised manner to a certain extent that involve the work of "NaN" values, scaling and variable selection.

Finally, we found that the Evaluation stage can model the general problems of classification and regression, providing an initial model base. While it is true that the heuristic logic to find the best model has an exploratory nature, an evaluation of a set of models, our results show that the scores have great importance for an unsupervised process.

6   Future Work

In the Prepare stage, there are Categorical variable-mapping techniques that could improve our model in classification and regression problem situations.

Next, we can improve the Selection stage by applying feature engineering, which not only selects the best features but selects the best extraction and the best modifications from themselves.

Finally, we believe that the Evaluation stage can be upgraded if we implement meta-learning techniques and libraries. For instance, packages like AutoML and AutoSklearn are growing notoriously in the community because of that. In summary, meta-learning as known as "learning to learn," where the program explores a great number of techniques and transfers said knowledge to the solution for future problems, is a new way of representing the data structure. The latter is inspired from the "No Free Lunch" theorem which indicates: "If a solution is good enough to solve a problem, it is not necessarily good for others"[4].

Acknowledgments. The project would have been impossible without the support of *Ciencia Activa* and *Fondo para la Innovación, la Ciencia y la Tecnología* - Innovation, Science and Technology Fund (FINCyT).

References

1. Bravo-Rocca, G.: Pyspark package for getting an overview of a dataset (2016). https://pymach.readthedocs.io/en/latest/readme.html
2. Brownlee, J.: Machine learning mastery with Python (2016)
3. Christensson, P.: Python definition. https://techterms.com. Accessed 7 May 2018
4. Duch, W.: Meta-learning. Nicolaus Copernicus University, Poland
5. Karau, H., Konwinski, A., Wendell, P., Zaharia, M.: Learning Spark, Lightning-Fast Data Analysis. O'Reilly, Sebastopol (2015)
6. Plotly Technologies Inc.: Collaborative data science (2015)
7. McKinney, W.: Data structures for statistical computing in Python. In: Proceedings of the 9th Python in Science Conference, pp. 51–56 (2010)
8. Metropolitan Transportation Authority. MTA — Subway, Bus, L.I.R.R.M.N.: Metropolitan transportation authority. MTA — subway, bus, long island rail road, metro-north (2014). http://web.mta.info/developers/MTA-Bus-Time-historical-data.html
9. Pyspark: Extracting, transforming and selecting features. https://spark.apache.org/docs/latest/ml-features.html. Accessed 7 May 2018
10. Repository, M.L.: Hepmass dataset. UCI, p. 3 (2014). https://archive.ics.uci.edu/ml/datasets/HEPMASS. Accessed 7 May 2018

Deep Dive into Authorship Verification of Email Messages with Convolutional Neural Network

Marina Litvak[(✉)]

Shamoon College of Engineering, Beer Sheva, Israel
`marinal@ac.sce.ac.il`

Abstract. *Authorship verification* is the task of determining whether a specific individual did or did not write a text, which very naturally can be reduced to the binary-classification problem. This paper deals with the authorship verification of short email messages. Hereafter, we use "message" to identify the content of the information that is transmitted by email. The proposed method implements the binary classification with a sequence-to-sequence (seq2seq) model and trains a convolutional neural network (CNN) on positive (written by the "target" user) and negative (written by "someone else") examples. The proposed method differs from previously published works, which represent text by numerous stylometric features, by requiring neither advanced text preprocessing nor explicit feature extraction. All messages are submitted to the CNN "as is," after padding to the maximal length and replacing all words by their ID numbers. CNN learns the most appropriate features with backpropagation and then performs classification. The experiments performed on the Enron dataset using the TensorFlow framework show that the CNN classifier verifies message authorship very accurately.

Keywords: Authorship verification · Binary classification · Convolutional neural network

1 Introduction

Communication through electronic mail is a very basic everyday activity for almost every person these days. This communication can be personal or official and can have different purposes: work, study, commerce, or just chatting with friends. However, because we cannot trust every message that arrives to our account, we use spam filters on a daily basis. Email fraud is one of the most common types of illegal activity enabled by the Internet. Millions of fraudulent messages are sent every day. Statistics[1] say that in Q1 2017, the percentage of spam in email traffic amounted to 55.9%.

Email communication may be misused by various means. An intruder may disguise oneself as a legitimate user by forging messages after breaking into a

---

[1] https://securelist.com/spam-and-phishing-in-q1-2017/78221/.

© Springer Nature Switzerland AG 2019
J. A. Lossio-Ventura et al. (Eds.): SIMBig 2018, CCIS 898, pp. 129–136, 2019.
https://doi.org/10.1007/978-3-030-11680-4_14

mail server and fabricating SMTP messages [18], performing man-in-the-middle attacks [7], hacking an email account, or physically accessing the user's computer. The intruder's purpose can be spying, phishing, or other malicious goals. Therefore, performing authorship verification for suspect email messages may have a crucial role in cybersecurity and forensic analysis. In this paper we introduce an approach to the problem of authorship verification of short messages, which can be accurately applied on a "raw" text of emails and, in contrast to the state-of-the-art works, does not require either enhanced text preprocessing or feature extraction.

2   Related Work

The authorship verification problem has been studied for about decade. Most works used stylometry and relied on shallow classification models. Stylometry aims at reflecting personal writing styles, defined by numerous stylometric features [9]. In general, stylometric features can be categorized into four categories: lexical, syntactic, structural, and content-specific. The total amount of single features in stylometry-based work can reach hundreds, and, therefore, feature selection or dimensionality reduction must be performed prior to classification. Among the most frequently used classifiers in stylometry-based authorship verification models are: k-nearest neighbor (kNN), Naïve Bayes, decision tree, Markov chains, support vector machine (SVM), logistic regression (LR), and neural network. As can be seen from the literature, all authorship verification studies differ in terms of the stylometric features and the type of classifiers employed. An extended survey of stylometric features and authorship detection techniques is given in [8].

The first attempts of authorship verification focused on general text documents and were not realistic for application to online texts, which are usually much shorter, as well as being poorly structured and written. For example, the SVM-based model in [13] obtained 95.70% accuracy for documents containing at least 500 words. Many researchers subsequently investigated the effectiveness of stylometry techniques for authorship authentication on shorter text, including email messages. Their results were not as promising as the results for longer texts. Various classification and regression models with 292 stylometric features yielded an Equal Error Rate (EER) ranging from 17.1% to 22.4% on the Enron email dataset in [9]. Using 150 stylistic features in [5] resulted in an accuracy of 89% for 40 users from the Enron dataset. Authors of [14] combined stylometric representation with 233 features and various classification techniques, obtaining an accuracy of 79.6% on Facebook posts. SVM and SVM-LR classifiers were applied in [3] for authorship verification of short online messages, including email messages. About one thousand stylometric features, enriched by the N-gram model, have been extracted and then selected prior to classification. Experimental evaluation on the Enron email and Twitter datasets produced EER results varying from 9.98% to 21.45%. The SVM model with most frequent words as features [16] achieved 80% accuracy on 50 users from the Enron dataset. A stylometry based authorship verification model based on the Gaussian-Bernoulli deep belief network [4] produced EER results ranging from 8.21% to 16.73% on Enron and Twitter datasets, respectively.

We propose a different approach to the problem of authorship verification of short messages. This approach is based on the deep sequence-to-sequence CNN model, which *does not require either enhanced text preprocessing or feature extraction*. Originally invented for computer vision, CNN models have been lately proven to be effective for various natural language processing (NLP) tasks [6]. For example, a simple one-layer CNN was successfully applied for the sentence classification tasks in [10]. We adapt a similar approach and train the CNN classifier on a two-class training data, composed of positive (written by the "target" user) and negative (writen by "someone else") examples. No pre-trained word vectors are required. This approach, while saving much time and effort that could be invested in feature extraction and selection, produces a very high accuracy.

3   Authorship Verification with CNN

The traditional authorship verification approach, based on stylometry and classification, is usually composed of: (1) extracting a rich set of hand-designed features, (2) selecting the most significant ones, and then (3) feeding them to a standard classification algorithm (for example, SVM). The choice of features is a completely empirical process, mainly based on our linguistic intuition; and to a large extent, this determines the key to success.

Following the idea of application of CNN to NLP tasks [2, 6, 10, 11, 19], we propose a radically different approach: we apply a multilayer neural network (NN), trained in an end-to-end fashion, on a "raw" text, after a very basic preprocessing. The NN architecture takes the input text and *learns several layers of feature extraction* that process the input. The features computed by the deep layers of the network are *automatically trained by backpropagation to be relevant to the task* (of authorship verification in our case).

Typical CNN is composed of several convolutional modules that perform feature extraction. Each module is a sequence of a convolutional and pooling layers. The convolutional layer performs mathematical calculations (filter) to produce features in the feature map. The pooling layer reduces the dimensionality of the feature map. A commonly used pooling algorithm is max pooling, which extracts sub-regions of the feature map and keeps their maximum value, while discarding all other values. The last convolutional module is followed by one or more dense layers. Dense layers perform classification on the features extracted by the convolutional layers and reduced by the pooling layers. In a dense layer, every node is connected to every node in the preceding layer. The final CNN dense layer contains a single node for each target class in the model, with a softmax activation function that generates a probability value for each node. We can interpret the softmax values for a given input as its likelihood to belong to each target class. Figure 1 shows the model architecture adapted to the binary classification of text messages. We explain it in more detail below.

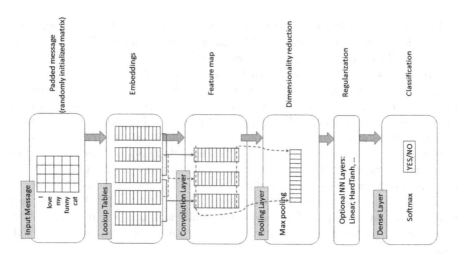

Fig. 1. CNN architecture for email classification task.

Let m_i be the k-dimensional word vector corresponding to the ith word in the message. An email message of length n (zero-padding strategy as in [10] is applied) is represented as $m_{1:n} = m_1 \oplus m_2 \oplus \ldots \oplus m_n$, where $\oplus$ is the concatenation operator.

Let $m_{i:i+j}$ refer to the concatenation of word vectors $m_i, m_{i+1}, \ldots, m_{i+j}$. A convolution operation applies a filter $\varphi \in R^{hk}$ to a window of h words to produce a new feature. For example, a feature c_i is calculated from a window of word vectors $m_{i:i+h-1}$ as $c_i = f(\varphi \cdot m_{i:i+h-1} + b)$, where b is a bias term and f is a non-linear function such as the hyperbolic tangent. Filter φ is applied to each possible window of h word vectors for words in the input text $\{m_{1:h}, m_{2:h+1}, \ldots, m_{n-h+1:n}\}$ to produce a feature map $c = [c_1, c_2, \ldots, c_{n-h+1}]$.

A max-overtime pooling operation [6] is then applied over the feature map and the maximum value $\hat{c} = \max\{c\}$ is taken as the feature corresponding to this particular filter. This stage is aimed at capturing the most important feature—one with the highest value—for each feature map. One feature is extracted from one filter. The model uses multiple filters—with various window sizes—to obtain multiple features. These features are passed to a fully connected dense layer activating softmax function that produces the probability distribution of the input text over target classes.

We use single channel architecture, with one that is fine-tuned via backpropagation. This means that we do not need to provide pre-trained static word vectors (embeddings). All words in our input messages are randomly initialized and then modified during training.

4   Experiments

We evaluated our approach on the Enron email dataset [12]. The Enron dataset was used for different kinds of authorship analysis, including authorship attribution, authorship verification, authorship profiling (characterization), and authorship similarity detection. After discarding users with less than 1000 email messages, we trained our model to 52 remaining users. For each user, 1000 verified email messages were sampled. The parameters of our data are shown in Table 1.

Table 1. Enron data after filtering

# emails	52000
max email length	95 words
# users	52
# messages per user	1000

CNN model[2] was trained for each user. In order to get balanced data, we took the same amount (1000) of positive (all emails that were written by the "target" user) and negative (written by other users and randomly selected) examples. 90% of this data was used for training and 10% for testing. The TensorFlow framework [1] was used in our experiments. In preprocessing, every email was padded to the maximal length (95 words) and encoded by replacing its words by their ID numbers (integers from 1 to V, where V is a vocabulary size).

Figure 2 depicts the accuracy distribution for 52 users with a clustered bar chart. The average overall accuracy is 97%, which is significantly better than what most of the previously published works reported on the Enron dataset.[3] Unfortunately, we could not compare our performance with other works where only EER–that cannot be transformed to accuracy without additional information–was reported or other dataset–even if it is a subset of Enron dataset–was used. CNN performance depends on the amount of epochs (steps) performed during the training.[4] Figures. 3 and 4 show accuracy as a function of epoch number and loss as a function of epoch number, respectively.[5] Blue curves in these figures represent training accuracy and loss, while red curves represent test accuracy and loss.

---

[2] We kept the default settings of the CNN model in the TensorFlow framework, which are as follows: number of embedding dimensions is 128; filter sizes are 3, 4, and 5; number of filters is 128, dropout probability is 0.5, L2 regularization lambda is 0, batch size is 64.

[3] The best accuracy of 89% for 40 users from the Enron dataset was reported in [5].

[4] We ran our model with 500 epochs.

[5] Obtained from training on one of the users.

134   M. Litvak

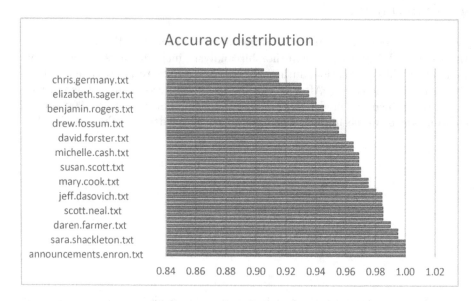

Fig. 2. Accuracy distribution.

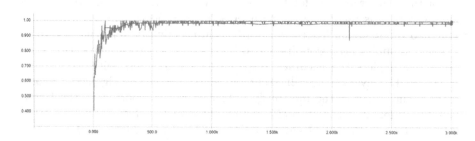

Fig. 3. Accuracy as a function of epochs number. (Color figure online)

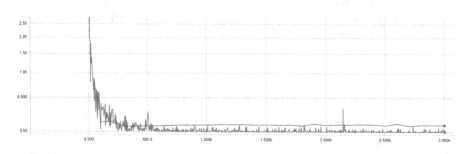

Fig. 4. Loss as a function of epochs number. (Color figure online)

5 Conclusion and Future Work

This paper describes an application of a deep sequence-to-sequence CNN model to the authorship verification task for email messages. In contrast to current state-of-the-art works, our model does not require explicit feature extraction. CNN gets "raw" text as an input, learns features, and performs the classification task on them. The results show that this model verifies email message authorship very accurately. The results can be fine-tuned by performing more epochs, that is generally improves the accuracy of the NN models. In conclusion, the main value of the proposed method is its accurate performance while being applied on a "raw" text. As such, it allows to save time required for the features design, implementation, and extraction and to avoid adding noise to the representation model.

In the future, we intend to experiment with different variations of the CNN architecture, such as the multichannel architecture with several 'channels' of word vectors. These channels will encompass static throughout pre-training with a neural language model [15,17] and non-static that is fine-tuned via backprop-agation, as in [10]. Static vectors can be trained by word2vec [15] or Glove [17]. In such architecture, each filter must be applied to all channels, and the results must be summed to calculate a feature c_i in the feature map. Our experiments can be extended with different baseline methods, additional evaluation metrics (i.e. EER), and real-world (unbalanced) domains. Also, additional task-related features (email message structure and meta-data) can be incorporated into the neural network. In addition, we would like to apply our approach to a different task of authorship analysis—authorship attribution—that can be modeled as a classification task with multiple classes.

Acknowledgments. The author is grateful to Vlad Vavilin and Mark Mishaev for the implementation and running the experiments using the TensorFlow framework.

References

1. Abadi, M., et al.: Tensorflow: a system for large-scale machine learning. In: Proceedings of the 12th USENIX Conference on Operating Systems Design and Implementation OSDI 2016, pp. 265–283. USENIX Association, Berkeley (2016). http://dl.acm.org/citation.cfm?id=3026877.3026899
2. Britz, D.: Understanding convolutional neural networks for NLP (2015)
3. Brocardo, M.L., Traore, I., Woungang, I.: Authorship verification of e-mail and tweet messages applied for continuous authentication. J. Comput. Syst. Sci. **81**(8), 1429–1440 (2015)
4. Brocardo, M.L., Traore, I., Woungang, I., Obaidat, M.S.: Authorship verification using deep belief network systems. Int. J. Commun. Syst. **30**(12), e3259 (2017)
5. Chen, X., Hao, P., Chandramouli, R., Subbalakshmi, K.P.: Authorship similarity detection from email messages. In: Perner, P. (ed.) MLDM 2011. LNCS (LNAI), vol. 6871, pp. 375–386. Springer, Heidelberg (2011). https://doi.org/10.1007/978-3-642-23199-5_28

6. Collobert, R., Weston, J., Bottou, L., Karlen, M., Kavukcuoglu, K., Kuksa, P.: Natural language processing (almost) from scratch. J. Mach. Learn. Res. **12**(Aug), 2493–2537 (2011)
7. Desmedt, Y.: Man-in-the-middle attack. In: van Tilborg, H.C.A. (ed.) Encyclopedia of Cryptography and Security. Springer, Boston (2005). https://doi.org/10.1007/0-387-23483-7
8. El Bouanani, S.E.M., Kassou, I.: Authorship analysis studies: a survey. Int. J. Comput. Appl. (0975 – 8887) **86**(12), 22–29 (2014)
9. Iqbal, F., Khan, L.A., Fung, B., Debbabi, M.: E-mail authorship verification for forensic investigation. In: Proceedings of the 2010 ACM Symposium on Applied Computing, pp. 1591–1598. ACM (2010)
10. Kim, Y.: Convolutional neural networks for sentence classification. In: Proceedings of the 2014 Conference on Empirical Methods in Natural Language Processing (EMNLP), pp. 1746–1751 (2014)
11. Kipf, T.N., Welling, M.: Semi-supervised classification with graph convolutional networks. In: 2017 Proceedings of ICLR (2017)
12. Klimt, B., Yang, Y.: The enron corpus: a new dataset for email classification research. In: Boulicaut, J.-F., Esposito, F., Giannotti, F., Pedreschi, D. (eds.) ECML 2004. LNCS (LNAI), vol. 3201, pp. 217–226. Springer, Heidelberg (2004). https://doi.org/10.1007/978-3-540-30115-8_22
13. Koppel, M., Schler, J.: Authorship verification as a one-class classification problem. In: Proceedings of the Twenty-First International Conference on Machine learning, p. 62. ACM (2004)
14. Li, J.S., Chen, L.C., Monaco, J.V., Singh, P., Tappert, C.C.: A comparison of classifiers and features for authorship authentication of social networking messages. Concurr. Comput.: Pract. Exp. **29**(14), e3918 (2017)
15. Mikolov, T., Sutskever, I., Chen, K., Corrado, G.S., Dean, J.: Distributed representations of words and phrases and their compositionality. In: Advances in Neural Information Processing Systems, pp. 3111–3119 (2013)
16. Nirkhi, S.M., Dharaskar, R., Thakare, V.: Authorship identification using generalized features and analysis of computational method. Trans. Mach. Learn. Artif. Intell. **3**(2), 41 (2015)
17. Pennington, J., Socher, R., Manning, C.: Glove: global vectors for word representation. In: Proceedings of the 2014 Conference on Empirical Methods in Natural Language Processing (EMNLP), pp. 1532–1543 (2014)
18. Polychronakis, M., Provos, N.: Ghost turns zombie: exploring the life cycle of web-based malware. LEET **8**, 1–8 (2008)
19. Zhang, Y., Wallace, B.: A sensitivity analysis of (and practitioners' guide to) convolutional neural networks for sentence classification. arXiv preprint arXiv:1510.03820 (2015)

Monitoring of Air Quality with Low-Cost Electrochemical Sensors and the Use of Artificial Neural Networks for the Atmospheric Pollutants Concentration Levels Prediction

Ana Luna$^{(\boxtimes)}$, Alvaro Talavera, Hector Navarro, and Luis Cano

Universidad del Pacífico, Av. Salaverry, 2020 Lima, Peru
{ae.lunaa,ag.talaveral,h.navarrob,lcanovasquez}@up.edu.pe

Abstract. This paper shows the preliminary results of the monitoring and estimation of air pollutants at a strategic point within the district of San Isidro, Lima - Peru. Low-cost, portable, wireless and geo-locatable electrochemical sensors were used to capture reliable contamination levels in real-time which could be used not only to quantify atmospheric pollution exposure but also for prevention and control, and even for legislative purposes. For the prediction of CO_2 and SO_2 levels, computational intelligence algorithms were applied and validated with experimental data. We proved that the use of Artificial Neural Networks (ANNs) has a high potential as a tool to use it as a forecast methodology in the area of air pollution.

Keywords: Electrochemical sensors · Air pollution ·
Artificial Neural Networks

1 Introduction

In the last decades, several research papers have been published showing the correlation between the decrease in the quality of life and the increase of respiratory and cardiovascular diseases with air pollution [5,11,18]. According to the 2014 report of the World Health Organization (WHO), Lima is one of the cities with the most polluted air in Latin America. For this reason, the monitoring of air quality should be one of the priorities of our current society. The last report of the Agency of Evaluation and Environmental Enforcement elaborated in 2015, shows that only 29% of the existing municipalities in Metropolitan Lima supervises and controls the air pollution in their jurisdiction.

Currently, there are ten fixed air quality monitoring stations in Metropolitan Lima and Callao that permanently evaluate the atmospheric pollution. The Meteorology National Service and Hydrology of Peru (SENAMHI) is the entity that monthly reports the results of these automatic stations.

© Springer Nature Switzerland AG 2019
J. A. Lossio-Ventura et al. (Eds.): SIMBig 2018, CCIS 898, pp. 137–150, 2019.
https://doi.org/10.1007/978-3-030-11680-4_15

The acquisition of pollution levels is usually carried out through static monitoring stations that are high precision devices which require permanent maintenance and calibration and whose cost far exceeds tens of thousands of dollars each [15]. Consequently, this brings with it a significant limitation in the number of possible monitoring networks. However, the growing advance of technology has allowed the development of a new generation of sensors that are more economical and small.

Low-cost air pollution sensors are attracting attention because they offer air pollution monitoring possible in many more locations. On the other hand, those kinds of sensors can be sensitive to weather conditions (temperature and humidity variations, and wind speed) or can have difficulties distinguishing pollutants, and their respective calibrations are not always easy to carry out. Besides, measurements with low-cost sensors are often of lower and more questionable data quality than the results from official monitoring stations. Nevertheless, in certain well-defined situations, the measurement uncertainty of these devices may approach the level of 'official' measurement methods, that is why low-cost gas sensors have been used in several air quality campaigns at rural and urban sites [10,16,17].

This paper presents the design, equipment, and start-up of a portable and low-cost air pollution monitoring station in the district of San Isidro, Lima, Peru. It also details how the acquired values were processed, and the graphs of the levels of the pollutants are shown according to the registered time slot. Next, some observations are outlined, and artificial intelligence is used, specifically neural networks not only for the prediction but also for the validation of the critical pollutants measured. Finally, we analyzed the obtained results, and in the last section, we presented the conclusions.

2   Artificial Neural Networks

Neural Networks are a field within Artificial Intelligence that is inspired by the connections between neurons of the human brain, trying to create artificial models that use conventional algorithmic techniques to find the solution of a particular problem. Neural networks are, in short, processing units that exchange data or information and are used to recognize patterns, such as financial trends. They can also learn and improve their functioning [9].

The artificial neuron model consists primarily of a set of inputs of a certain number of components, in the form of a vector, which generates a single output. There is also a set of synaptic weights that represent the interaction between pre-synaptic and post-synaptic neurons. The propagation rule provides the post-synaptic potential, and the activation function offers the activation state of the neuron as a function of the previous state and the post-synaptic value. Finally, the output function is obtained from the activation state.

An artificial neural network (ANN) is composed of nodes (processing units) and connections. Each node has a transfer function. The inputs are external, and the outputs are obtained from the outgoing links of the ANN.

The network used in this paper is a multilayer perceptron (MLP) that has a specific structure that allows visualizing how the synaptic connections of the network are. In the MLP, the neurons are usually grouped into layers, one is the input, which receives the data from the environment; the other one is the output that provides the response of the network to the input stimulation, and there are also some hidden layers that are part of the internal processing of the network and have no direct contact with the external environment (Fig. 1).

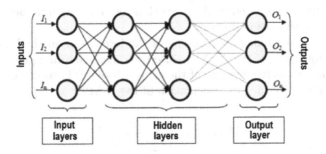

Fig. 1. Example of a multilayer perceptron.

The learning process of an ANN consists of the update of the weights through learning algorithms, which is a numerical procedure for adjusting weights to minimize the committed error.

Any machine learning model has to be trained, and this process starts by defining the loss function. The idea is to minimize this objective function to learn the parameters that better adjust the prediction model. Equation (1) defines the loss function.

$$loss = G(\theta) = \frac{1}{N} \sum_{i=0}^{N} loss(f(x^i; \theta), y^i) \tag{1}$$

Where $G(.)$ is the function to be minimized, i is the number of examples, $f(.)$ is the predicted output and y is the real value and θ is the weights to be adjusted [9].

Then, to minimize this value and to find the correct parameters θ needed for the model, a technique called stochastic gradient descent is used [9,11]. Its particular algorithm (see [9,12]) is:

The training algorithm used in this work is a variation of the stochastic gradient descent called Levenberg-Marquardt, which is applied mainly to neural networks with several layers because the speed of convergence is breakneck, and it quickly minimizes the error function.

After the training process, a validation step must be done. For this purpose, another set of data is required, called the validation set. Each example of the evaluation set contains the values of the input variables, with their corresponding solution taken; but the solution is not granted to the neural network.

Algorithm 1. Stochastic Gradient Descent (SGD)

1: Initialize θ randomly
2: For N epochs
3: For each training example (x, y)
4: Compute Loss Gradient: $\frac{\partial G(\theta)}{\partial \theta}$
5: Update θ with the update rule: $\theta = \theta - n \frac{\partial G(\theta)}{\partial \theta}$

Then, the solution estimated by the network is compared with the actual solution to complete the validation process.

To measure the prediction accuracy of the forecasting method and to evaluate the performance of the ANN, we used the mean absolute percentage error (MAPE) which measures the size of the error (absolute) in terms of percentage. Equation (2) was used for the calculation of the MAPE:

$$MAPE = \frac{1}{N} \sum_{i=1}^{N} \frac{|x_i - \hat{x}_i|}{|x_i|} \tag{2}$$

where N is the total number of values; x_i and $\hat{x}_i$ are the actual and predicted values for the network, respectively.

We also used the root mean squared error (RMSE) to quantify the performance of the forecasting method. RMSE is an estimator that measures the average of squared errors, that is, the difference between the estimator and what is being estimated, evaluating the quality of a set of predictions as to its variation and the degree of bias. The calculation formula of the RMSE is shown in Eq. 3:

$$RMSE = \sqrt{\frac{1}{N} \sum_{i=1}^{N} (x_i - \hat{x}_i)^2} \tag{3}$$

the variables of this equation are described just below Eq. 2.

3 Materials and Methods

The measurements were carried out in the facilities of the Municipality of San Isidro, Lima, Peru as shown in Fig. 2. The geographic coordinates of San Isidro district where we located the sensors are: -12.1167 latitude, -77.05 longitude and 109 m.a.s.l. San Isidro has an extension of 9.78 km^2 and a population of approximately 58.056 inhabitants.

3.1 Low-Cost Electrochemical Sensors

We employed the *Alphasense* outdoor sensors (Essex - United Kingdom) for the air quality measurement [1–3]. They are wireless, powered by a battery and a solar panel (Fig. 3a). They have Wi-Fi connection (GSM cellular interface) as

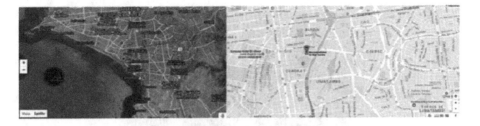

Fig. 2. The map shows the point where the atmospheric's sensors were installed.

well as being geo-locatable (built-in GPS). The data are sent to the cloud and can be read from any PC with an internet connection (M2M). The electrochemical sensors have four electrodes designed to measure gas levels in nmol/mol units. The first of the electrodes is called the "Work Electrode", the second is called the "Reference Electrode", the third one is the "Electrode Counter", and the last one is the "Auxiliary Electrode", which is used to correct the zero before the changes of current, and as is not in contact with the gas; therefore, it provides useful information about the effect of ambient temperature. Each of the sensors provides two signals, WE and AE. The constants WE_0 and AE_0 are the respective background signals, and the manufacturer supplied their values. Having those data, the levels of the pollutants could be obtained using Eqs. 4 and 5. The air pollutants recorded were: CO_2, VOC (alcohols, aldehydes, aliphatic hydrocarbons, amines, aromatic hydrocarbons, CH_4, LPG, ketones and organic acids), CO, SO_2, O_3, and NO_2. Besides, the simultaneously meteorological variables acquired were the ambient temperature, the relative humidity and the barometric pressure (Fig. 3). The cost of M2M air quality sensors is approximately \$1,350 and includes a GSM cellular interface, GPS sensor, meteorological sensor, micro USB cables, the eight atmospheric pollutants sensors previously mentioned, housings for the sensors, one year of application cloud SaaS and sensors's APIs, in addition to licenses, software updates, and product support.

3.2 Field Measurements

The data stored for the different air pollutants were recorded from 8:00 a.m. to 12:30 p.m. during the days of Wednesday 5 - Friday 7 April 2017 and Monday 10 - Wednesday 12 April 2017. We recorded one data point per minute, giving a maximum of 270 daily registered values. The information acquired for indirect measurement pollutants, CO, SO_2, O_3 and NO_2, must be processed. To do this, we generated several codes in the Python programming language; the first one performs the conversion of the recorded data into voltage units. So, the pollutant levels provided by the $Valarm$ dashboard were converted in parts per million (ppm) values, using Eq. 4,

$$Value(ppm) = \frac{(WE - WE_o)(AE - AE_o)}{Sensitivity} \tag{4}$$

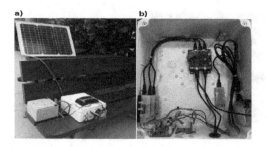

Fig. 3. (a) Air quality sensors (CO_2, VOC, SO_2, O_3, NO_2 and CO) and meteorological (ambient temperature, relative humidity, and barometric pressure) kit. The solar panel, a solar regulator, and the battery give the autonomy to the module. (b) Pollutant and meteorological sensors' housing, length: 29 cm, width: 25 cm, height 13 cm (including the cover).

where WE are the data recorded by the "Work Electrode", AE are the values of the "Auxiliary Electrode", and WE_o and AE_o both correct the background signals of the respective electrodes. The sensitivity value for each pollutant is obtained from the following Eq. 5:

$$Sensitivity = WE_{sensitivity}(\frac{nA}{ppm}) * Gain(\frac{mV}{nA}) \tag{5}$$

its values' factors were data provided by the manufacturer of the sensors.

The next stage involves the processing of the data, which are now expressed in units of ppm. The new program eliminates both negative values and zeros and also those values that deviate three times from the value of the standard deviation corresponding to the average of each one of the respective pollutants. The discarded values are replaced by those obtained by linear interpolation between the previous and subsequent data of the registered value. In the case of the first and last values that have been eliminated, they are completed with the data of the average of the acquisition.

Finally, all the treated data are processed with a third program that graphs the levels of contamination as a function of time, and inserts in each figure the value of the meteorological variables (pressure, humidity, and temperature) and the value of the *Air Quality Standard* (AQS) for each pollutant.

The values of the concentration levels of the measured pollutants are shown only for one of the acquisition days, as an example in Fig. 4. The rest of the days measured are qualitatively analogous, and the respective results of their mean values are summarized in Fig. 5.

Table 1 reports the meteorological variables of each of the days in which the measurements were made.

In Fig. 5, we show the average levels of air pollutants registered in the time slot from 8:00 a.m. to 12:30 p.m. during the days specified in the graphs.

During the measurements made in the district of San Isidro, we observed that the average value of CO_2 exceeds by almost 32% the maximum values within its air quality standards (Figs. 4 and 5). There is a direct relationship between global

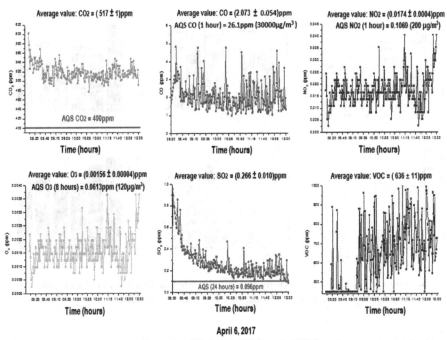

April 6, 2017
Temperature: 28°C, Humidity: 65%, Pressure: 1000.7 hPa

Fig. 4. Time series of pollutant levels measured in the time slot from 8:00 a.m. to 12:30 p.m. on Thursday, April 6th, 2017.

Table 1. Measured values of temperature, relative humidity and pressure in the days and hours in which the concentration levels of atmospheric pollutants were recorded

Weather variables	April 5th 2017	April 6th 2017	April 7th 2017	April 10th 2017	April 11st 2017	April 12th 2017
Temperature	30 °C	28 °C	32 °C	32 °C	29 °C	32 °C
Relative humidity	62%	65%	60%	60%	64%	59%
Pressure	999 hPa	1000.7 hPa	999.8 hPa	1000 hPa	1000 hPa	999 hPa

warming or climate change and the increase in greenhouse gas emissions caused by human societies, both industrialized and developing ones. Recent studies have shown that the concentration of CO_2 in the atmosphere is increasing steadily due to the use of fossil fuels as an energy source [7]. This phenomenon causes essential changes in living beings and ecosystems, as well as economic, social and environmental consequences of great magnitude (the average temperature of the earth's surface has increased, erosion and salinization phenomena in coastal areas increased, infectious diseases, etc.).

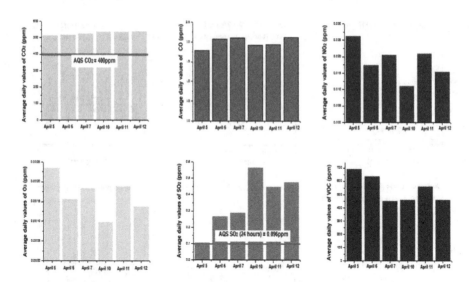

Fig. 5. Average values of levels of daily air pollutants (CO_2, CO, NO_2, O_3, SO_2 and VOC). Registration time slot: 8:00 a.m. to 12:30 p.m. Year 2017.

Both, CO_2 and SO_2 are primary gaseous pollutants and are emitted directly into the atmosphere. The SO_2 also registers values above the quality standards (Figs. 4 and 5), the average of the measured values exceeds by more than 270% the amount reported by the AQS.

The sources that produce these chemical compounds are varied, but the main artificial sources are the burning of fossil fuel and that coming from factories, industries, power plants, and automobiles.

When the sulfur dioxide is in the atmosphere, it reacts with moisture and forms aerosols of sulfuric and sulfurous acid that are then part of the so-called acid rain and its components are highly harmful to the vegetation.

High levels of SO_2 also attack building materials that are usually made up of carbonate minerals. Besides, sulfur dioxide forms sulfates, and exposure to these as to acids derived from SO_2, is of an extreme health risk because they enter directly into the human circulatory system through the respiratory tract, being very irritating to lungs. Also, the excess of SO_2 brings with it severe consequences at the health level, such as, for example, the increase in the number of hospitalizations due to chronic obstructive pulmonary diseases (COPD) as recorded in the reference [7].

The article [4] also shows an increase in the incidence of allergic rhinitis, bronchitis, and asthma in children as a consequence of SO_2 contamination. Moreover, in the work [8] different effects of long-term exposures to SO_2 are analyzed, among others, especially in young adults who are severely affected by asthma. Undoubtedly, what we breathe affects our health, and can increase morbidity and mortality in older people and children; therefore it should be a priority to reduce air pollution to improve health.

3.3   Artificial Neural Networks Applied to Atmospheric Pollutants

In the work of [13], neural networks are used for the estimation of CO between the years 1996–2006, in the city of Delhi. The production of CO and its dispersion in the atmosphere follow a non-linear and complex dynamic, so the authors showed that the use of the neural network algorithm is an efficient method for the study of this type of problem. On the other hand, in the article [6], autoregressive neural network model is used for forecasting CO_2 emission in the cereal sector in Apulia region (Italy), for the improvement and the decision-making policy. In the publication [14], Support Vector Machines regression (SVMr) algorithm and neural networks were used for the prediction of hourly ozone values in Madrid urban area. Solar radiation, temperature, and different windows of measured ozone hours were collected from the stations of 27 districts and were all taken into account for the forecast. In this sense, the use of computational intelligence techniques, such as neural networks, are algorithms that actively contribute to the modeling of a non-linear relationship, for example, using non-linear autoregressive models that could face, among others, a problem to predict an atmospheric variable. Because of that, in this current work, we used artificial neural networks to forecast the concentration levels of the two atmospheric pollutants that exceed the established ECA values, CO_2 and SO_2 (see Figs. 4 and 5). The network used in this work is an MLP neural network, which provides a non-linear model. In particular, in our study it relates the variables of atmospheric pollutants for the estimation of contamination levels of CO_2 and SO_2. For the prediction of CO_2, we first build the neural network. The optimal topology found for the network was 14 neurons in the first hidden layer and 3 neurons in the second hidden layer (Fig. 6). This result was obtained using a search algorithm where all possible combinations between neurons in a structure of 2 hidden layers and a maximum of 20 neurons in each layer were tested. The network training data used was the concentrations of atmospheric pollutants measured during the 5 days prior to the prediction day, these data were divided into 3 sets, 70% was used for training, 20% was considered for validation and the remaining 10% was taken into account for the test. The training algorithm was Levenberg-Marquardt. The convergence occurred in iteration number 14, in which there was no significant variation in the training error of the network between the previous step and the one after this last iteration.

Figure 7 shows the graph of the measured values and the prediction of CO_2 concentration levels for Wednesday, April 12nd, 2017. The latter were obtained from the previous training of the model with the incorporation of the measured values of the rest of the pollutants except CO_2. The data of temperature, relative humidity and atmospheric pressure were also considered. The RMSE obtained was 20 ppm and the MAPE reached almost 3%, and the Pearson correlation coefficient (PCC) between the prediction and the measured values was 0.65. Comparing this result to a more straightforward and more traditional approach, like linear regression, the PCC is worse and equal to 0.42.

146   A. Luna et al.

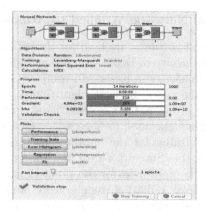

Fig. 6. Architecture of the neural network used for the prediction of CO_2. It can be seen that the convergence occurred in the 14th iteration, and there was no significant variation between the performance obtained in a previous iteration and the iteration immediately after.

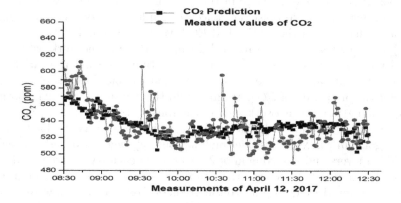

Fig. 7. Carbon dioxide forecast.

The electrochemical sensors provide real-time monitoring and generate an output signal which is directly proportional to the concentration of the target gas. Although there is a compromise between the sample rate and the number of samples of our data acquisition system, the electronics components sensors can be affected by some surface turbulence effect or by a lack of sensors time stabilization. For that reasons, it is possible to observe some fluctuations in the respective pollutant time series that deviate from the expected value. In this way, we could explain the few sudden peaks that differ from the predicted values in Figs. 7 and 9.

To find the optimal architecture of the neural network used for the prediction of SO_2 concentration levels, the search algorithm was again used, testing all possible combinations between neurons in a structure of 2 hidden layers and a maximum of 20 neurons in each layer. The result was a network composed of 8 neurons in the first hidden layer and 5 neurons in the second hidden layer (Fig. 8).

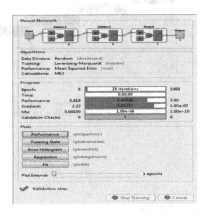

Fig. 8. Architecture of the neural network used for the prediction of SO_2.

The training data of the network were the pollutant concentrations of 5 days before: Wednesday 5 - Friday 7 April 2017 and Monday 10 - Tuesday 11 April 2017, and the respective meteorological values. The values measured with the low-cost sensors were divided into 3 sets (following the same procedure for the CO_2), 70% was used for training, 20% was used for validation and the remaining 10% for the test. The chosen training algorithm, as it was explained in Sect. 2, was the Levenberg-Marquardt. The convergence was achieved in the iteration number 12.

Table 2 presents the results of six models of neural networks, all of them used to estimate the ppm levels of the pollutant SO_2 on Wednesday, April 12th, 2017. The values used for the inputs in each of the networks are in the "Variables" column and are marked with a cross (X). Besides, in all the cases, the meteorological variables: temperature, relative humidity and atmospheric pressure were considered in the training. Finally, Table 2 shows the results of the MAPE and the RMSE obtained for the predictions of the concentration levels of SO_2 on Wednesday, April 12th, 2017 between 8:30 a.m. and 12:30 p.m.

As it can be seen in Table 2, the prediction of SO_2 that showed the lowest values of MAPE and RMSE was the one in which O_3 was not contemplated (Fig. 9). That is because only the primary pollutants (CO, CO_2, NO_2, VOC), that come from the automotive fleet burning of fossil fuel give a better prediction for the pollutant SO_2 which is of the same type. On the other hand, the O_3 is a secondary pollutant (it is not emitted directly from a source), and it does not contribute to the prediction of SO_2.

Table 2. Results of MAPE and RMSE for each prediction of SO_2.

Variables	Prediction 1 of SO2	Prediction 2 of SO2	Prediction 3 od SO2	Prediction 4 of SO2	Prediction 5 of SO2	Prediction 6 of SO2
SO2						
CO	X		X	X	X	X
CO2	X	X		X	X	X
NO2	X	X	X		X	X
O3	X	X	X	X		X
VOC	X	X	X	X	X	
MAPE	45%	49%	51%	51%	29%	54%
RMSE (ppm)	0.30	0.36	0.43	0.36	0.27	0.42

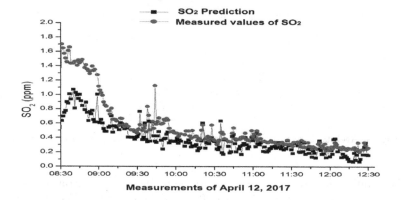

Fig. 9. Sulfur dioxide forecast.

The best linear correlation coefficient between the prediction, using ANN, and the measured values of SO_2 was 0.89. But, using linear regression we obtained a PCC of 0.55. So, for both predictions, the CO_2 and SO_2 emissions, classical linear methods are not considered to be adequate for model complex phenomena because they usually make too rigid approximations. On the other hand, ANN gives a more flexible architecture to implement a short-term emission forecasting tool in Lima. Therefore, as a first approximation, the use of artificial intelligence having low cost atmospheric sensors is a potential tool for predicting reliable concentration levels of pollutants.

4   Conclusions

Field experiments are essential to understand certain aspects of weather and climate of a region. In this paper, the data acquired from various air pollutants in the district of San Isidro were shown and analyzed.

The low-cost sensors used in this work have great potential to be used as a complement to the scarce static monitoring stations; increasing, in this way, the spatial resolution of the measurements. Although measurements of electrochemical sensors do not necessarily have the accuracy or precision of traditional on-site instruments, they are versatile and economical. So, they could be used in the acquisition of air pollution values and obtain preliminary information on emissions from fixed and mobile sources. Even more, critical points of air pollution could be identified in real time, that is, areas with the deterioration of air quality. In consequence, measures could be proposed to reverse this environmental situation or mitigate the critical episodes of air pollution on the population settlements.

From these preliminary results obtained in this work, it is planned to carry out field measurements during a whole year, differentiating seasons and then applying different machine learning techniques for the prediction of the levels pollutant studied.

Acknowledgment. The authors would like to thank the Universidad del Pacífico and in particular the Department of Engineering for the purchase of the air and sound pollution sensors; and of the SIM cards used for data transmission. The authors would also like to express their gratitude to Mrs. Jimena Sánchez Velarde, Director of datosabiertosperu.com and Pamela Olenka Peña, Manager of Sustainability of the Municipality of San Isidro, who authorized the use of the facilities within the Municipality, allowing the realization of this research.

References

1. Alphasense: CO-B4 Carbon Monoxide Sensor 4-Electrode (2015). http://www.alphasense.com/WEB1213/wp-content/uploads/2015/04/COB41.pdf. Accessed 4 June 2018
2. Alphasense: OX-B431 Oxidising Gas Sensor Ozone + Nitrogen Dioxide 4-Electrode (2017). http://www.alphasense.com/WEB1213/wp-content/uploads/2017/07/OX-B431.pdf. Accessed 4 June 2018
3. Alphasense: SO2-B4 Sulfur Dioxide Sensor 4-Electrode (2017). http://www.alphasense.com/WEB1213/wp-content/uploads/2017/05/SO2-B4.pdf. Accessed 4 June 2018
4. Chiang, T.Y., Yuan, T.H., Shie, R.H., Chen, C.F., Chan, C.C.: Increased incidence of allergic rhinitis, bronchitis and asthma, in children living near a petrochemical complex with SO2 pollution. Environ. Int. **96**, 1–7 (2016)
5. Day, O., et al.: Air Pollution and Health (2013)
6. Gallo, C., Contò, F., Fiore, M.: A neural network model for forecasting CO2 emission. Agris On-line Pap. Econ. Inf. **6**(2), 31 (2014)
7. Ghozikali, M.G., Mosaferi, M., Safari, G.H., Jaafari, J.: Effect of exposure to O3, NO2, and SO2 on chronic obstructive pulmonary disease hospitalizations in Tabriz, Iran. Environ. Sci. Pollut. Res. **22**(4), 2817–2823 (2015)
8. Greenberg, N., et al.: Different effects of long-term exposures to SO2 and NO2 air pollutants on asthma severity in young adults. J. Toxicol. Environ. Health, Part A **79**(8), 342–351 (2016)
9. Haykin, S.: Neural networks, a comprehensive foundation. Technical report, Macmilan (1994)
10. Jiang, Q., et al.: Citizen sensing for improved urban environmental monitoring. J. Sens. **2016**(2), 1–9 (2016)
11. Landrigan, P.J.: Air pollution and health. Lancet Public Health **2**(1), e4–e5 (2017)

12. Mitchell, T.M.: Machine Learning, 1st edn. McGraw-Hill Inc., New York (1997)
13. Nigam, S., Nigam, R., Kapoor, S.: Forecasting carbon monoxide concentration using artificial neural network modeling. Int. J. Comput. Appl. (0975 – 8887), International Conference on Current Trends in Advanced Computing "ICCTAC-2013", pp. 36–40 (2013)
14. Ortiz-García, E., Salcedo-Sanz, S., Pérez-Bellido, Á., Portilla-Figueras, J., Prieto, L.: Prediction of hourly O3 concentrations using support vector regression algorithms. Atmos. Environ. **44**(35), 4481–4488 (2010)
15. Piedrahita, R., et al.: The next generation of low-cost personal air quality sensors for quantitative exposure monitoring. Atmos. Meas. Tech. **7**(10), 3325 (2014)
16. Spinelle, L., Gerboles, M., Villani, M.G., Aleixandre, M., Bonavitacola, F.: Field calibration of a cluster of low-cost available sensors for air quality monitoring. Part A: Ozone and nitrogen dioxide. Sens. Actuators B: Chem. **215**, 249–257 (2015)
17. Sun, L., et al.: Development and application of a next generation air sensor network for the Hong Kong marathon 2015 air quality monitoring. Sensors **16**(2), 211 (2016)
18. Wong, G.W.K.: Air pollution and health. Lancet Respir. Med. **2**(1), 8–9 (2014)

Data Mining Algorithms for Risk Detection in Bank Loans

Alvaro Talavera$^{(\boxtimes)}$, Luis Cano, David Paredes, and Mario Chong

Universidad del Pacífico, Av. Salaverry 2020, Lima, Peru
{ag.talaveral,lcanovasquez,db.paredesm,m.chong}@up.edu.pe

Abstract. This article proposes a new approach on detection of fraudulent credit operations applying computational intelligence techniques. We use a dataset of historical data of customers from a financial entity and we split it to train a classification and clustering algorithm. We train a radial basis function network to classify clients that commit or not credit fraud. Then, we build a Fuzzy c-means clustering to group data points to create customer profiles. This algorithm has the capacity of grouping the data inside clusters and assigning a degree of membership to the points outside the clusters. Subsequently, the trained classification algorithm is applied to the clusters to provide additional information about customer profiles. We demonstrate good performance for fraudulent credit operations and identification of customer profiles.

Keywords: Risk detection · Fuzzy C-means ·
Radial basis function networks · Finance profiles

1 Introduction

The rapid growth of technologies has allowed financial entities, such as banks, to generate new strategies on making profit, as well as the creation of credit cards. Although technologies can help to improve users' lives, it could also generate a negative impact on the economy. In [5], the authors explain how the misuse of credit cards can impact negatively on the inflation, disfavoring the economy. It is very important to handle such matters efficiently. A question arises on how credit risk must be detected rapidly, in order to take preventive measures.

When financial entities make decisions about granting credit to customers, they perform a probability weighting to find out if customers can accomplish the credit payment. This assessment can be based on the evaluator's judgment or data mining techniques. Data mining techniques used in this area has increased because of the large amount of applicants and attributes to be considered [2].

The most common computational intelligence techniques in detection of fraudulent credit operations are neural networks (NN), support vector machines (SVM), bayesian networks (BN), decision trees (DT), among others [14]. In [7], a fraud detection model is developed with neural networks, applying a model to create an association of rules of "if then" type, so the performance of the

© Springer Nature Switzerland AG 2019
J. A. Lossio-Ventura et al. (Eds.): SIMBig 2018, CCIS 898, pp. 151–159, 2019.
https://doi.org/10.1007/978-3-030-11680-4_16

neural network can improve and at the same time obtain relevant information of fraudulent cases. Similarly, [8] presents an approach comparing the performance of three fraud detection model based on NN, BN and DT in a 10-fold cross-validation setting. Moreover, [10] compares two fraud detection models based on neural and bayesian networks. Both models are trained under the same settings and tested with features and patterns that have not been seen before in order to analyze the performance later.

In order to avoid credit fraud, fraud prevention mechanisms must be established. Nonetheless, in [11], the authors present 3 difficulties to create these mechanisms. First, fraudulent credit operations may perfectly fit in a normal credit operation. Second, privacy and safety concerns limit the exchange of ideas for fraud detection, and third, the datasets and models to be evaluated are rarely available or non-anonymized. Methods to overcome these difficulties have been developed in the literature, using supervised and unsupervised learning. In the case of the first one, there are methods using DT [4,11], NN [6] and SVM [11]. In the case of unsupervised learning, clustering methods are based on k-means [12] and self-organized maps (SOM) [15].

In [1], an approach based on unsupervised techniques is proposed, which offers an alternative to studies using supervised learning. It shows that credit fraud could be detected through the detection of abnormalities, it means identifying outliers and classifying them as credit fraudulent candidates through peer group analysis techniques (PGA). PGA aims to generate clusters with common patterns and produce statistics inside of every cluster in order to identify outliers and classify the set of potential fraudulent credit operations. This methodology is applied in this study to find local abnormalities in the dataset and identify these possible illegal operations.

We propose a methodology which historical data of customers are grouped using a fuzzy clustering algorithm [13]. The aim of this phase is to find customer profiles, followed by a classification algorithm to discover which group is more likely to present operations as credit default. The rest this paper is organized as follows: Sect. 2 explains the theoretical foundations of the classification and clustering algorithms. Section 3 provides details on the methodology and its phases. Section 4 presents the results of the application in the database, which contains fraudulent credit operations. Section 5 presents the final conclusions.

2   Background

This section explains the algorithms applied to this study, such as radial-basis function networks (RBFN), which is used for classification. These networks have a good performance in addressing approximation issues. The other algorithm used is Fuzzy c-means (FCM), which is a clustering method.

2.1   Radial-Basis Functions Networks

In the context of artificial neural networks, radial-basis function networks are approximating universal functions. They are also computationally simple because they have fewer layers than traditional neural networks.

The structure of a radial-basis function network consists of three layers. The first layer is the input of the network and it has m units, where m is the dimension size of the training set. The second layer is composed of hidden units where non-linear transformations are applied from the input values space to a feature space and the number of units in this layer is equal to the size of the training set. The last layer is a linear layer of values as output, which has less units than in the hidden layer. They key difference with traditional multilayer perceptron networks (MLP) are radial-base functions as non-linear transformations in the second layer. This functions are represented as follows:

$$\varphi_j(x) = \varphi(x - x_i) \tag{1}$$

On this equation, "φ" represents a radial function. An example of this type of function is the Gaussian function, which is presented below:

$$\varphi_j(x) = exp(-\frac{1}{2\sigma^2}||x - x_i||^2) \tag{2}$$

The parameters for this function are "σ" and "x_j", where "σ" is the dispersion measure of the j-th Gaussian function with "x_j" as center. The dispersion and center of the Gaussian functions of the hidden layers are determined in the first stage of the training. It is important to mention that a critical factor for this technique is selecting an appropriate amount of Gaussian functions in the hidden layer. Low values of Gaussian functions does not allow to fit the data properly, since the flexibility of the model would be limited. On the other hand, high values of Gaussian functions would cause a wrong generalization of the data and poorly adjusted noise due to the high flexibility of the model.

It should be noted that the functioning of radial-base functions is justified through the Cover's theorem [3]. This theorem ensures that a complex classification problem that is inside a non-linear separable space of a high dimensional, is more likely to be linearly separable than in a low dimensional space. Thus, the way how radial-base function networks solve a classification problem is transforming the data into a non-linear, high dimensional and separable space and then separating the data by their respective classes.

2.2   Fuzzy C-Means

The objective of clustering is to split a dataset Z into c clusters. In general, a partition A_i from Z follows the next properties: the union of partitions must contain all the data, clusters must be disjoint with each other, the partitions cannot be either empty or the entire data set.

A partition can be represented by a partition matrix $U = [u_{ik}]_{c \times N}$, where the i-th row of the matrix contains the membership functions of A_i from Z. In the case of fuzzy clustering and fuzzy partitions, $[u_{ik}]$ can take values between 0 and 1. In this way, the properties of a fuzzy partition are the following:

$$[u_{ik}] \in [0,1], 1 \le i \le c, 1 \le k \le N, \tag{3}$$

$$\sum_{i=1}^{c} u_{ik} = 1, 1 \le k \le N, \tag{4}$$

$$0 < \sum_{k=1}^{N} u_{ik} = 1, 1 \le i \le c. \tag{5}$$

Fuzzy C-Means is a fuzzy clustering algorithm [13] based on the minimization of the following objective function:

$$J(Z; U, V) = \sum_{i=1}^{c} \sum_{k=1}^{N} (u_{ik})^m ||z_k - v_i||_A^2, \tag{6}$$

Where U is the matrix of fuzzy partitions, V is a vector of centroids of clusters, D_{ikA}^2 is a distance measure and m is a fuzzy exponent that goes from 1 to ∞. The parameters of this algorithm are the following: *Number of Clusters*: The parameter c is the most important, since the algorithm will check for clusters c even if these do not exist in the data. *Parameter of Fuzziness*: The exponent m influences on the fuzziness of the partitions. If m goes to infinity, the partition becomes completely fuzzy. *Completion Criteria*: The algorithm finishes when the norm of the difference of U in two successive interactions is less than a threshold ϵ. *Norm-Inducing Matrix*: Determines the shape of the clusters depending on matrix A that is chosen for the distance measurement. *Initial Partition Matrix*: A partition matrix with initial random degree of membership.

3   Methodology

The aim of this study tries to build a model to classify fraudulent credit operations, as well as identifying relevant information about customer profiles. The database DB is divided in two partitions DB_1 and DB_2, where DB_1 is used to train a classification algorithm and DB_2 is for building the clustering algorithm.

In the initial stage, a classification algorithm is trained using the database DB_1. In the second stage, Fuzzy c-means detects candidate clusters $CC = \{cc_1, cc_2, ..., cc_n\}$ using the database DB_2. Then, the profile of each candidate cluster is analyzed. Finally on stage 3, using the trained classification algorithm, the cluster with the highest amount of individuals who committed credit fraud cc^* is identified from the group of candidate clusters. All stages are shown in the Fig. 1.

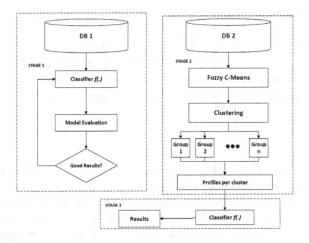

Fig. 1. Block diagram

This methodology helps to find potential customers for whom a bank should not have to risk granting credits. The benefits of the FCM clustering method are detecting customers with a certain degree of membership could belong to different clusters. It allows high flexibility in the identification of customer profiles.

4   Results

We evaluate our work on "Statlog Database" [9], a dataset of fraudulent credit operations which consists of 1000 instances with 20 attributes (7 numeric and 13 categorical) and a binary label which represents good or bad credit risk for the bank.

Following the first stage of the methodology (Fig. 1), a RBFN network model is trained on (80% of the total). The best value for "σ" is 84 when the models are trained under the same configuration. Adicionally, we trained a MLP model on the same data under the same configurations. From DB_1, we have a training set and validation set of 600 and 200, respectively. Table 1 shows the confusion matrix of the results of the models.

The equations of the metrics of *accuracy* and *recall* are the following:

$$Accuracy = \frac{tp + tn}{tp + tn + fp + fn} \tag{7}$$

$$Recall = \frac{tp}{tp + fn} \tag{8}$$

Where *tp* means true positives, *tp* true negatives, *fp* false positives and *fn* false negatives. The results from the RBFN model has *accuracy* of 72% and *recall* of 77.7%, while MultiLayer Perceptron MLP model has 69% *accuracy* and 75.3% recall. These metrics were obtained from the confusion matrix, as Table 1 shows.

Table 1. Confusion matrix of RBFN and MLP models

Real\Predicted	RBFN		MLP	
	Good risk	Bad risk	Good risk	Bad risk
Good risk	115	23	113	25
Bad risk	33	29	37	25

In the stage 2, a Fuzzy C-Means clustering model was trained with DB_2 (20% of the total) using three clusters. This number of clusters was selected by the best performance of intra and extra cluster measurement was achieved with 3 clusters. The scatter plot of the variables "credit amount" and "duration of the credit in months" is depicted in Fig. 2. In more details, the cluster 1 (red points) has 21 data points, the cluster 2 (green points) has 110 data points and the cluster 3 (blue points) has 62 data points. Descriptive statistics were performed on each variable of all the clusters to discover profiles. The three variables that generate the largest differences between clusters are credit purpose, duration of credit and credit amount and can be observed in Tables 2, 3 and 4. Table 2 includes the percentage of people who answered to each value of each variable and the cluster to which they belong. Cluster 1 was named "risky" because according to the statistics, the most representative characteristics for this cluster are: the credit purpose is mostly for household appliances over 36 months, the average credit amount is 9972 currency units and the percentage of the denominated customers as bad risk is higher than 40%. Regarding to cluster 2, it was named "good credit payers" because on these characteristics: the credit purpose is mostly to buy a new car over 16.5 months, the average credit amount is 1609 currency units and the percentage of the denominated customers as bad risk is 27%. Finally, cluster 3 was named "opportunity" because of these characteristics: the credit purpose is mostly to buy new furniture over 25 months, the average credit amount is 4229 currency units and the percentage of the denominated customers as bad risk is 30%, thus it could potentially be a good credit opportunity for the bank.

In relation to the credit purpose variable of Table 2, the responses relating to 0, 1, 2, 3, 4, 5, 6, 7, 8, 9 and 10 are: new car, used car, furniture, radio and television, household appliances, reparations, education, holidays, recycling, businesses and others, respectively. For the variable duration of credit (months) from Table 3, the average, maximum and minimum values for each cluster are presented. For the credit amount variable from Table 4 the average, maximum and minimum values for each cluster are also presented.

In the next stage 3, clustered data was fed in the best classification algorithm from stage 1. Tables 5 and 6 shows that identified customers that with bad risk in each cluster using the RBFN model. Table 5 presents the proportion of predicted customers as "bad risk", where cluster 1 has more customers classified as bad risks than the others. The results of the classification of the RBNF model in the clusters are shown in Table 6 with a 66.5% *accuracy* and 76.1% *recall*.

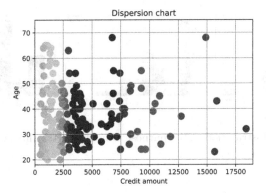

Fig. 2. Scatter plot (Color figure online)

Table 2. Credit purpose

Cluster\Purpose	0	1	2	3	4	5	6	7	8	9	10
Cluster 1	7%	0%	10%	25%	35%	14%	0%	0%	7%	0%	0%
Cluster 2	41%	1%	15%	25%	0.9%	9%	1%	2%	0%	0.9%	0%
Cluster 3	14%	4%	27%	24%	17%	9%	1%	0%	0%	0%	0%

Table 3. Duration of credit (months)

Cluster	Mean	Max	Min
Cluster 1	36.04	60	6
Cluster 2	16.55	45	4
Cluster 3	25.03	60	9

Table 4. Credit amount

Cluster	Mean	Max	Min
Cluster 1	9972.75	18424	7166
Cluster 2	1609	2848	454
Cluster 3	4229.82	6872	2859

Figure 3 shows the distribution of the purpose of credit variable for cluster 1, which includes the majority of customers who asked for credit to purchase household products, radio or television. In the same figure, the distribution of the purpose of credit variable for cluster 2 includes the majority of customers who asked for credit to purchase a new car, a radio or television. Additionally, the distribution of the purpose of credit variable for cluster 3 includes the majority of customers who requested credit to purchase new furniture, a radio or television.

Table 5. Confusion matrix to clusters

Real/Predicted	Good risk	Bad risk
Good risk	105	34
Bad risk	33	28

Table 6. Good and bad risk - cluster

Cluster	Good risk	Bad risk
Cluster 1	50%	50%
Cluster 2	70%	30%
Cluster 3	75.8%	24.2%

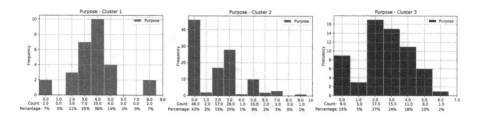

Fig. 3. Distribution of the purpose of credit on the three clusters

5 Conclusions

The aim of this paper is to propose a methodology to identify fraudulent credit operations by using the RBFN classification algorithm and the Fuzzy C-Means clustering. Three different types of profiles were found, which were defined as risky, good credit payers and opportunity. The first profile has a higher amount of customers classified as bad risks than the others. An advantage of this methodology is to increase the knowledge about the descriptions of profiles of clusters. The algorithm has showed a good performance in the metric of recall, a convenient indicator in the credit scoring issue. The reason for this is that is worse to grade a client as a good risk when in fact it is considered to be a bad risk, than to qualify a client as a bad risk when in fact it is considered to be a good risk.

References

1. Bolton, R.J., Hand, D.J.: Unsupervised profiling methods for fraud detection. In: Proceedings of Credit Scoring and Credit Control VII (2001)
2. Bravo, E., Talavera, A., Serra, M.R.: Prediction and explanation in credit scoring problems. In: Data Analytics Applications in Latin America and Emerging Economies. Auerbach Publications, New York (2017)
3. Cover, T.M.: Geometrical and statistical properties of systems of linear inequalities with applications in pattern recognition. IEEE Trans. Electron. Comput. **14**, 326–334 (1965)
4. Delamaire, L., Abdou, H., Pointon, J.: Credit card fraud and detection techniques: a review. Banks Bank Syst. **4**(2), 57–68 (2009)
5. Geanakoplos, J., Dubey, P.: Credit cards and inflation. Games Econ. Behav. **70**, 325–353 (2010)
6. Ghosh, S., Reilly, D.L.: Credit card fraud detection with a neural-network. In: Proceedings of the Twenty-Seventh Hawaii International Conference on System Sciences (1994)
7. Guo, T., Li, G.Y.: Neural data mining for credit card fraud detection. In: 2008 International Conference on Machine Learning and Cybernetics, vol. 7 (2008)
8. Kirkos, E., Spathis, C., Manolopoulos, Y.: Data mining techniques for the detection of fraudulent financial statements. Expert Syst. Appl. **32**(4), 995–1003 (2007)
9. Lichman, M.: UCI machine learning repository (2013). http://archive.ics.uci.edu/ml

10. Maes, S., Tuyls, K., Vanschoenwinkel, B.: Credit card fraud detection using Bayesian and neural networks (2015)
11. Sahin, Y., Duman, E.: Detecting credit card fraud by decision trees and support vector machines. In: Vector Machines, International Multiconference of Engineers and Computer Scientists (2011)
12. Vaishali, V.: Fraud detection in credit card by clustering approach. Int. J. Comput. Appl. **98**(3), 29–32 (2014)
13. Yang, M.S.: A survey of fuzzy clustering. Math. Comput. Model. **18**(11), 1–16 (1993)
14. Zareapoor, M., Seeja, K.R., Alam, M.A.: Analysis on credit card fraud detection techniques: based on certain design criteria. Int. J. Comput. Appl. **52**(3), 35–42 (2012)
15. Zaslavsky, V., Strizhak, A.: Credit card fraud detection using self-organizing maps. Int. J. Inf. Secur. **18**, 48 (2006)

DETECTOR: Automatic Detection System for Terrorist Attack Trajectories

Isaias Hoyos, Bruno Esposito, and Miguel Nunez-del-Prado[⊠]

Universidad del Pacífico, Av. Salaverry 2020, Lima, Peru
{i.hoyoslopez,bn.espositoa,m.nunezdelpradoc}@up.edu.pe

Abstract. To guarantee national security against terrorist attacks or organized crime, states must implement homeland security solutions based on ubiquitous systems to know in advance the number of suspects involved in an attack. This work proposes a method, which combines popular trajectory similarity metrics to estimate the number of attackers participating in a malicious act through the analysis of the trajectories described by the attacker's cell phone connection to antennas (i.e. Call Detail Records). Therefore, measuring trajectory similarity in CDRs generates different challenges compared to those similar metrics applied over GPS and video datasets.

Keywords: Terrorist · Trajectory · Similarity

1 Introduction

In the last two decades, we have been witnesses of the increment of terrorist acts against states. These malicious acts cost many lives, destroy public infrastructure and create an atmosphere of fear. This kind of attacks are committed by groups. Anticipating in advance the number of attackers of these groups could be useful for public forces and could reduce the damage to society. Therefore, it is important to be technologically prepared for terrorist attacks.

We aim to identify the number of criminals quicker through the use of Call Detail Records data and lead a faster intervention by the police authorities. This will lead to a future and progressive reduction of citizen insecurity. From the citizens' point of view, this project could improve their perception of security if they are aware of a faster criminal identification mechanism. In consequence, the confidence of the citizen in public institutions could raise.

This effort aims to solve this problem by taking advantage of the high amount of data generated through the use of cell phones. In the present study, we aim to determine the number of aggressors before the terrorist act.

The rest of the paper is structured as follows: Sect. 2 describes related works on trajectory similarity metrics. Section 3 defines theoretical concepts on trajectory distance. Then, Sect. 4 introduces the experimentation process and reports the results, while Sect. 5 discusses the findings and limitations of the present study. Finally, Sect. 6 concludes the work and presents future research avenues.

© Springer Nature Switzerland AG 2019
J. A. Lossio-Ventura et al. (Eds.): SIMBig 2018, CCIS 898, pp. 160–173, 2019.
https://doi.org/10.1007/978-3-030-11680-4_17

2   Related Works

There are two different approaches to measure trajectory similarity. The first approach is to work with the whole trajectory [4,5,7,8,16–19] and another approach is to work with a part of the trajectory (*i.e.*, a sub-trajectory) [13–15,17,18,20,22]. In Table 1, we summarize (i) the kind of trajectory, (ii) the distances metrics and (iii) the dataset characteristics of the related works found in the literature.

Table 1. Distance summary where trajectory (TRA), video (VID)

Authors	trajectory	sub-trajectory	Jaccard	DTW	EUCL	Hausdorff	LCSS	relative	SDTW	CDTW	Edit	sigmoidsim	UMS	PCA-EUCL	# Trajectories	Technology	Category	Update rate (s)
[17]	✓	✓	✓	✓	✓	✓	✓								200000	GPS	TRA	15
[5]	✓							✓							17621	GPS	TRA	-
[16]		✓	✓				✓								182	GPS	Geolife	16.2
[16]		✓	✓				✓								49	Video	VID	-
[22]		✓	✓	✓		✓			✓						30,000	GPS	TRA	60
[19]	✓		✓							✓					324	GPS	TRA	-
[21]	✓												✓		12	Video	VID	-
[18]		✓	✓												104	GPS	People	-
[20]		✓	✓			✓					✓				50	GPS	TRA	-
[8]	✓													✓	536	GPS	TRA	61.8
[8]	✓													✓	182	GPS	Geolife	16.2
OA		✓		✓	✓	✓								✓	30	GSM	CDR	900

In the work of Rayatidamavandi *et al.* [17], they propose two different strategies such as Locality-Sensitive and Distance-Based Hash function to group trajectories. The former uses two different hash functions. One relies on p-Stable distribution [24] and takes as input the trajectory and the Hash function outputs a bucket index probability. Consequently, the closer two trajectories to a given index the more similar they are. Another locality-sensitive hash function uses Min-wise Independent Permutation [2]. This distance needs the trajectories to be in discrete form by mapping the location points into a pre-defined grid to apply Jaccard similarity. The latter strategy relies on dynamic time warping (DTW), Euclidean, Hausdorff and larges common subsequence distances (LCSS) [23].

In the work of Fan and Yao [5], they propose a function that takes as input two discrete trajectories F_x and F_y of the same length generated by a road-based location sign. Thus, the function compares whether two points of each trajectory are within a ϵ threshold to output one or zero otherwise. Therefore, each pair of points is measured and normalized to have a distance between 0 and 1.

Wang *et al.* [22] analyze and compare six widely used measures of trajectory similarity: (i) Euclidean distance, (ii) dynamic time warping, (iii) piece-wise dynamic time warping, (iv) edit distance with real penalty, (v) edit distance on real sequence, and (vi) longest common sub-sequence. The stress tests implemented were three: re-sampling trajectory, point shift and the addition of noisy points. They find that there is no overall measure superior. Each one has specific benefits depending on the characteristics of the dataset.

Toohey and Duckman [20] implement and compare four common trajectory similarity measures: (i) longest common subsequence, (ii) Frechet distance, (iii) dynamic time wrapping, and (iv) edit distance. They provide a CRAN implementation for each one. Later, Mao et al. [15] evidence the lack of similarity measures which are robust to noisy data. They introduce three distances that cope with this problem: (i) the point-segment, (ii) the prediction and (iii) the segment-segment distances. These consider the traditional dynamic time warping algorithm (DTW) and suggest a new segment-based dynamic time warping algorithm (SDTW). This approach exhibits better accuracy and lower sensitivity to noisy than (i) the longest common sub-sequence algorithm, (ii) edit distance on real sequence algorithm, and (iii) DTW. Similarly, Vlachos, Gunopulos, and Kollios [21] suggest a distance function for data with a high amount of outliers. The function proposed is based on the longest common sub-sequence.

Trajectory similarity measures usually do not consider the uncertainty introduced by the sampling process, as shown by Furtado et al. [8]. The authors propose a new distance function, which is robust to movement uncertainty generated by the heterogeneity in the sampling. Further experiments show that this approach is superior to a more straightforward approach, such as linear interpolation. Sharif and Alesheik [19] and, later, Sharif, Alesheikh, and Tashayo [18] offer a different approach, where, instead of using the spatial footprint, the similarity of the trajectories is measured by a notion of internal and external contexts. The authors show that the conjunction of both techniques delivers better results in an applied experiment.

In the present effort, we take into account the most popular distance metrics, such as Euclidean distance, principal component analysis with Euclidean distance, dynamic time warping (DTW), and larges common subsequence (LCSS) distance to combine them for taking advantage of the different benefits of each metric to capture similarity. It is worth noting that the works in the literature apply similarity metrics to trajectories issued from Global Positioning System (GPS) or Video recording, which have an outstanding location update rate. In the context of our work, we analyze popular similarity metrics to deal with trajectories from Call Detail Records, which have a poor update rate compare to GPS and video (*c.f.*, Table 1).

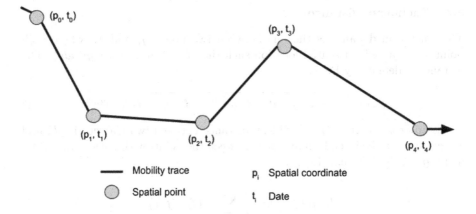

Fig. 1. Trajectory example.

3   Background

In this section, we define the concepts of (i) trajectory, (ii) distance measures namely, Euclidean, principal component analysis, distance time wrapping and longest common sequence, (iii) trajectory alignment, and (iv) the ranking quality measure *i.e.*, normalized discounted cumulative gain (nDCG).

3.1   Mobility Trace

A mobility trace is a tuple f that contains the following elements, according to [9]:

- Identifier: can be the real identifier of the equipment, a pseudonym or an unknown value. A pseudonym is generally used to protect the privacy of the user and maintain linkability between activities done by the same user.
- Spatial coordinates: location of the mobility trace. It can be a latitude and longitude tuple, a geographic area or the label of a specific place.
- Date: time frame in which the movement was made. It can be hours, days or a time interval.
- Additional information: optional data such as directions, velocity, the existence of users around a certain radius.

We can represent a mobility trace as a tuple $f_i = \{identifier, latitude, longitude, timestamp, epsilon\}$.

3.2   Trajectory

A trajectory is a set $F = \{f_1, f_2, ..., f_t, ..., f_T\}$ of T mobility traces, which describe the movement of an object in every unit of time t. Generally, each mobility trace is represented by spatial coordinates and a date [6], as shown in Fig. 1.

3.3   Euclidean Distance

The Euclidean distance is the line between two points f_i and f_j, where each point f has a defined spatial location such that $f = \{latitude, longitude\}$. The distance is defined by Eq. 1.

$$d_E(f_i, f_j) = \sqrt{(f_i.lat - f_j.lat)^2 + (f_i.lon - f_j.lon)^2} \tag{1}$$

In order to measure the Euclidean distance between two trajectories F_i and F_j, we average the Euclidean distance between the points within each trajectory, as proposed by [10] (c.f., Eq. 2).

$$d_{EM}(F_i, F_j) = \frac{1}{L} \sum_{k=1}^{L} d_E(f_{i,k}, f_{j,k}) \tag{2}$$

Where $d_E(f_{i,k}, f_{j,k})$ is the euclidean distance between the spatial points $f_{i,k}$ and $f_{j,k}$, which are extracted from trajectories F_i and F_j at position k. L is the length of trajectories F_i and F_j. Figure 2 provides a visual summary of this distance metric.

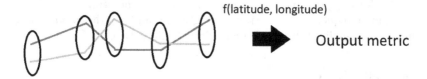

f(latitude, longitude)

Output metric

Fig. 2. Euclidean distance example

3.4   Principal Component Analysis (PCA)

This technique helps us to find patterns in high dimensional data by transforming a set of data possibly related into a set of data linearly independent. These are known as principal components.

For the trajectory case, we can reduce the dimensionality of the original spatial trajectories f_i by keeping the principal component with the highest eigenvalue. The coordinates of the trajectories are placed in a one-dimension vector and, then, projected into a lower dimensional space f_i' [1].

After applying PCA, the distance measure between two trajectories is obtained through the Euclidean distance between the principal components a_1 at each position among the trajectory, as seen in Eq. 3. Trajectories must have equal length.

$$d_{PCA}(F_i, F_j) = \frac{1}{N_\lambda} \sum_{l=1}^{N_\lambda} d_E(f_{i,l}', f_{j,l}') \tag{3}$$

Where $d_E(f'_{i,l}, f'_{j,l})$ is the euclidean distance between points $f'_{i,l}, f'_{j,l}$, which belong to a lower dimensional space than $f_{i,l}, f_{j,l}$. N_λ is the number of spatial data points in each transformed trajectory. Figure 3 provides a visual summary of this distance metric.

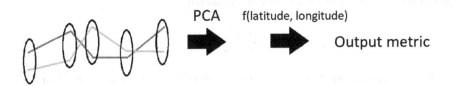

Fig. 3. PCA distance example

3.5 Trajectory Alignment (TA)

The aforementioned distances require trajectories of the same length. Hence, we propose a trajectory alignment technique, which takes as input two trajectories F_i, F_j and a time interval.

Figure 4 depicts an example of trajectory alignment. First, TA finds the begin and end time for both F_i and F_j. In our example, the begin and end time are *4:15* and *5:45*, respectively for a 15 min time window. Then, we keep only time slots that have events from both F_i and F_j namely, time slots 2, 3, 7 and 8. Therefore, those are the aligned trajectories. Once trajectories are aligned, we are able to use Euclidean or PCA distances.

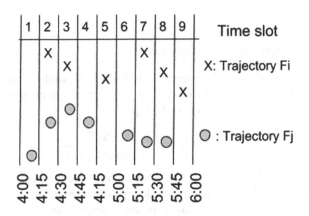

Fig. 4. Trajectory alignment example

3.6 Distance Time Warping (DTW)

This technique measures the similarity between two temporal sequences, which can have velocity changes. Additionally, it compares sequences of different length because it identifies a temporal jump that minimizes the total distance between paired points [12]. Equation 4 details the metric.

$$d_{DTW}(F_i, F_j) = \frac{1}{T_i} \sum_{t=1}^{T_i} d_E(\phi_{i,t}, \phi_{j,t}) m_t / M_\phi \qquad (4)$$

Where ϕ_i and ϕ_j are two temporal jump functions which minimize the distance between the paired points, m_t is a weight coefficient, and M_ϕ is a normalization factor. T_i is the number of data points in temporal jump function ϕ_i.

Since this measure is not symmetric, we use Eq. 5 to average the measure from f_i to f_j and from f_j to f_i to have a symmetric distance. Figure 5 provides a visual summary of this distance metric.

$$D_{DTW}(F_i, F_j) = \frac{d_{DTW}(F_i, F_j) + d_{DTW}(F_j, F_i)}{2} \qquad (5)$$

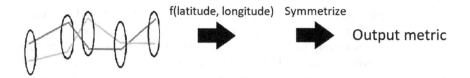

Fig. 5. Distance time warping example

3.7 Longest Common Sub-sequence

This technique aligns data with different length. Not all points have to be paired. Instead of matching mobility traces one by one, if there is no suitable pair for the mobility trace, it is ignored [3]. (c.f., Eq. 6)

$$D_{LCSS}(F_i, F_j) = 1 - \frac{LCSS(F_i, F_j)}{min(T_i, T_j)} \qquad (6)$$

Where $LCSS(F_i, F_j)$ shows the number of paired data points between the two trajectories. $min(T_i, T_j)$ outputs the length of the smallest trajectory. This technique can be implemented through dynamic programming.

3.8 Normalized Discounted Cumulative Gain

The Normalized Discounted Cumulative Gain (nDCG) is a measure of a ranking quality. It measures the information gain of one trajectory based on (i) its position in a list or ranking and (ii) the relevance associated with the trajectory (*c.f.*, Eq. 7). The information gain is higher in the first positions of the ranking, while the last positions are heavily discounted.

$$nDCG_p = \frac{DCG_p}{IDCG_P} \tag{7}$$

nDCG is a modification of the Eq. 8, Discounted Cumulative Gain (DCG) [11], which assumes that trajectories with higher relevance are more useful when their position in the ranking makes them appear first. Additionally, trajectories highly relevant that appear on the last positions in the ranking should be heavily penalized.

$$DCG_p = \sum_{i=1}^{p} \frac{2^{rel_i}}{log_2(i+1)} \tag{8}$$

Where rel_i represents the relevance of the i-th trajectory in the ranking.

DCG is not consistent for different length rankings. Therefore, the cumulative gain in position p must be normalized [22]. This is done by finding the ratio between (i) DCG of the evaluated ranking and (ii) the DCG of the ranking ordered by relevance, which is an ideal IDCG (*c.f.*, Eq. 9).

$$IDCG_p = \sum_{i=1}^{|REL|} \frac{2^{rel_i} - 1}{log_2(i+1)} \tag{9}$$

Where $|REL|$ represents the list of trajectories ordered by relevance.

In the next section, the aforementioned concepts will be applied to compute trajectory similarity.

4 Experiments

In the present section, we detail the process depicted in Fig. 6 to extract the list of the n most similar trajectories, and we present the results of the analysis performed with our dataset.

The process proposed in the present work has five steps, ranging from data input, distance computing, relevance calculation, ranking quality measurement and output of the most similar trajectories.

The first step is the data input. In this step, the process receives a suspicious trajectory F_s, which in this case is the *user34* and a dataset containing all possible similar trajectories. The data used for this work is the call detail records of a European telephone provider. For each user, we have the time and location in which the user enters the coverage area of a telephone antenna. These events represent 50 535 cell events of 30 users (*i.e.*, cell phones) in an area covered by 643 000 antennas. The data was gathered during July 2014.

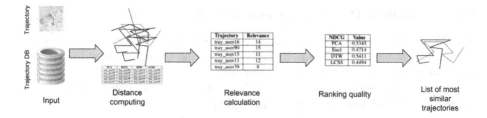

Fig. 6. Detector process

The second step, which is the distance computing takes the suspicious trajectory and a set of trajectories to compare to. Then, using the four metrics mentioned before in Sect. 3, we measure the distance of the suspicious trajectory with each one of the trajectories in the dataset of trajectories. For example, Table 2 summarizes the ranking of the experiment, where each column shows the ranking of trajectories for each distances metric namely, principal component analysis (PCA), Euclidean distance (EUCL), Distance Time Warping (DTW), and Longest Common Sub-Sequence (LCSS). The position of each trajectory could change according to the metric used.

Table 2. Distances matrix

PCA	EUCL	DTW	LCSS
user16	user90	user15	user90
user90	user16	user11	user11
user39	user11	user16	user15
user15	user39	user39	user16
user11	user15	user90	user39

Once the distances are computed, *the third step is the relevance calculation.* Therefore, for each metric, the process establishes a trajectories ranking (R_{PCA}, R_{EUCL}, R_{DTW}, and R_{LCSS}) by a distance ranging from the nearest one to the farthest one. Thus, in the first element, we have the most similar trajectories with the highest relevance. In the second row, we have the second most similar trajectories with the second highest relevance and so on. Hence, a relevance value is based on the trajectory position in the ranking as shown in Table 3.

For instance, in Table 3 we have a column with the relevance weight, the first trajectories in the second row weights five, the second set of trajectories in the third row have a weight of four and so on. Therefore, relevance is the arithmetic sum of the trajectories' weight assigned by each metric. Hence, the less distance between a given trajectory F_i and the suspicious one F_s, the more relevance F_i will have. The relevance of each trajectory considers its position in the different rankings.

Table 3. Distances matrix

PCA	EUCL	DTW	LCSS	Relevance weight
user16	user90	user15	user90	5
user90	**user16**	user11	user11	4
user39	user11	**user16**	user15	3
user15	user39	user39	**user16**	2
user11	user15	user90	user39	1

The relevance values are shown in Table 4. For example, to compute the relevance of *user16*, the process adds the relevance weights of the corresponding place in the ranking. Therefore, for the PCA, EUCL, DTW and LCSS metrics the relevance weights are five, four, three and two, respectively (*c.f.* Table 3)

Table 4. Relevance ranking

Trajectory	Relevance
user90	15
user16	14
user11	12
user15	11
user39	8

After obtaining the relevance values, *the four step computes the ranking quality*. Hence, we establish a trajectory ranking by relevance to apply nDCG in order to determine the *ranking quality* of the obtained ranking. The ideal DCG (iDCG) is represented by the Table 4. Combining the weights of all distance metrics we obtain the ideal ranking sorted by relevance. Thus, taking into account the iDCG of Table 4, we obtain the nDCG value for each distance metric as shown in Table 5.

Finally, *the fifth step, the list of the most similar trajectories* selects the ranking R_{PCA}, R_{EUCL}, R_{DTW} or R_{LCSS} with the highest nDCG and, from it, the process proposes the trajectory or the set of trajectories with the lowest distance to the initial F_s trajectory. In our experiment, the selected distance metric is *DTW* as illustrated in Table 5.

From the analysis, the process outputs the trajectory of the *user15*, who has the highest relevance in this case (*c.f.* column DTW in Table 3). To perform the experiments, we have implemented a graphical user interface (*c.f.*, Fig. 7), which defines (i) the suspicious trajectory, (ii) the list of proposed trajectories, (iii) the number n of trajectories to be compared and (iii) a time windows. The code is publicly available in Github[1].

---

[1] DETECTOR Website: github.com/bitmapup/detector.

Table 5. Ranking quality

NDCG	Value
R_{PCA}	0.5345
R_{Eucl}	0.4714
R_{DTW}	0.5411
R_{LCSS}	0.4494

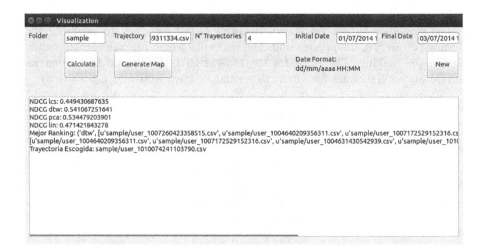

Fig. 7. DETECTOR graphical user interface

The suspicious and the proposed trajectory are graphed in Google Maps through its API service in JavaScript. Figure 8 illustrates the whole events of the suspicious and the most similar. It is worth noting that the set of antenna events are complex and sometimes sparse in time. Nevertheless, the trajectories displayed can be restricted by time windows, which initial and final dates are defined in the graphical interface. Hence, we present the analyzed trajectories as a set of events within the black box.

5   Discussion

In the present section, we mention the contribution of this study to the literature and discuss the validation of the results.

We propose a novel methodology to identify the number of agents participating in certain event. In addition, we present results that display the capabilities of the algorithm. The four metrics used are commonly used in the literature. Nonetheless, we are the first to combine these in order to detect similar spatial trajectories within a temporal window. We believe this can be applied to terrorism detection in real time.

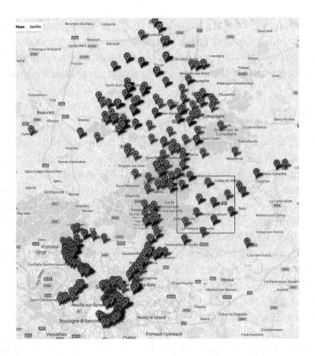

Fig. 8. Example of analyzed trajectories.

It is worth mentioning that we are unable to display accuracy results for the dataset used due to the absence of labeled data. We are unable to truly recognize groups of attackers and evaluate the predictive power of our proposal. Even though the reach of our predictive tests is limited, this downfall is mitigated by the use of multiple distance metrics. In this manner, we minimize the risk of selecting a distance that is not capturing the patterns of the sample.

In the present effort, we have validated our method using heuristics and visualization. The former validation consist in the duplication of the suspicious trajectory A into A'. Thus, the A' trajectory is inserted in the dataset containing all the other candidate trajectories. Therefore, when computing the trajectory similarity for A, A' must be the most similar trajectory. We have repeated this process for five randomly chosen trajectories. In all cases the methodology choosed A' as the most likely candidate to trajectory A. The latter method is also used by Mao *et al.* [15] and Sharif and Alesheikh [19]. Besides, mobility traces issued from a telecommunication operator are not trivial to analyze due to the location update rate. Thus, we can have few events within the time windows under study. Furthermore, the attach policy between cell phones and antennas affect the number of location updates depending if the location is updated when the cell phone makes a handover or receives a call, sms or use data. Besides, the precision is affected by the load balancing algorithm over the antennas, which send cell phone connections to contiguous antennas increasing artificially distances when comparing two trajectories. To solve this problem, in the future we plan to generated a ground truth dataset containing trajectories of a set of cell phones, which will travel together. This dataset will enable us to study the impact of the aforementioned factors in the distance similarity quantification.

6 Conclusions

In the present effort, we propose a process to compute the most similar trajectories using call detail records as input data. The methodology uses multiple metrics to determine which distance generates the best ranking of trajectory similarity by using the Normalized Discounted Cumulative Gain. The used distance metrics are principal component analysis (PCA), Euclidean distance (EUCL), Distance Time Warping (DTW), and Longest Common Sub-Sequence (LCSS).

This project has citizen security applicability because the system will identify the most similar trajectory to a suspicious trajectory. Therefore, it will determine which trajectories belong to a set of terrorist partners or group.

We believe that there are a few additional research possibilities to complement our work. First, the spatial trajectories described could be modeled as a part of a mobility model through Markov Chain Models [9]. Second, additional distance metrics based on hash functions would improve and add robustness to the results shown. Third, a solution for the absence of labeled data would be to artificially generate one by selecting an arbitrary suspicious trajectory and create a group of targets around it by adding random distortions to the initial trajectory.

References

1. Bashir, F.I., Khokhar, A.A., Schonfeld, D.: Object trajectory-based activity classification and recognition using hidden Markov models. IEEE Trans. Image Process. **16**(7), 1912–1919 (2007)
2. Broder, A.Z., Charikar, M., Frieze, A.M., Mitzenmacher, M.: Min-wise independent permutations. J. Comput. Syst. Sci. **60**(3), 630–659 (2000)
3. Buzan, D., Sclaroff, S., Kollios, G.: Extraction and clustering of motion trajectories in video. In: Proceedings of the 17th International Conference on Pattern Recognition, ICPR 2004, vol. 2, pp. 521–524, August 2004
4. Ester, M., Kriegel, H.P., Sander, J., Xu, X.: A density-based algorithm for discovering clusters in large spatial databases with noise. In: KDD, vol. 96, no. 34, pp. 226–231 (1996)
5. Fan, H., Yao, W.: A trajectory prediction method with sparsity data. In: 2017 IEEE International Symposium on Parallel and Distributed Processing with Applications and 2017 IEEE International Conference on Ubiquitous Computing and Communications (ISPA/IUCC), pp. 1261–1265. IEEE (2017)
6. Feng, Z., Zhu, Y.: A survey on trajectory data mining: techniques and applications. IEEE Access **4**, 2056–2067 (2016)
7. Ferreira, N., Klosowski, J.T., Scheidegger, C.E., Silva, C.T.: Vector field k-means: clustering trajectories by fitting multiple vector fields. In: Computer Graphics Forum, vol. 32, pp. 201–210. Wiley Online Library (2013)

8. Furtado, A., Alvares, L., Pelekis, N., Theodoridis, Y., Bogorny, V.: Unveiling movement uncertainty for robust trajectory similarity analysis. Int. J. Geogr. Inf. Sci. **32**, 1–29 (2017)
9. Gambs, S., Killijian, M.O., del Prado Cortez, M.N.: Show me how you move and i will tell you who you are. In: Proceedings of the 3rd ACM SIGSPATIAL International Workshop on Security and Privacy in GIS and LBS, SPRINGL 2010, pp. 34–41. ACM, New York (2010)
10. Hu, W., Xie, D., Fu, Z., Zeng, W., Maybank, S.: Semantic-based surveillance video retrieval. IEEE Trans. Image Process. **16**(4), 1168–1181 (2007)
11. Järvelin, K., Kekäläinen, J.: Cumulated gain-based evaluation of IR techniques. ACM Trans. Inf. Syst. **20**(4), 422–446 (2002)
12. Keogh, E.J., Pazzani, M.J.: Scaling up dynamic time warping for datamining applications. In: Proceedings of the Sixth ACM SIGKDD International Conference on Knowledge Discovery and Data Mining, KDD 2000, pp. 285–289. ACM, New York (2000)
13. Lee, J.G., Han, J., Whang, K.Y.: Trajectory clustering: a partition-and-group framework. In: Proceedings of the 2007 ACM SIGMOD International Conference on Management of Data, pp. 593–604. ACM (2007)
14. Liu, L.X., Song, J.T., Guan, B., Wu, Z.X., He, K.J.: Tra-dbscan: a algorithm of clustering trajectories. In: Applied Mechanics and Materials, vol. 121, pp. 4875–4879, Trans Tech Publications (2012)
15. Mao, Y., Zhong, H., Xiao, X., Li, X.: A segment-based trajectory similarity measure in the urban transportation systems. Sensors **17**(3), 524 (2017)
16. Nanni, M., Pedreschi, D.: Time-focused clustering of trajectories of moving objects. J. Intell. Inf. Syst. **27**(3), 267–289 (2006)
17. Rayatidamavandi, M., Zhuang, Y., Rahnamay-Naeini, M.: A comparison of hash-based methods for trajectory clustering. In: 2017 IEEE 15th International Conference on Dependable, Autonomic and Secure Computing, 15th International Conference on Pervasive Intelligence & Computing, 3rd International Conference on Big Data Intelligence and Computing and Cyber Science and Technology Congress (DASC/PiCom/DataCom/CyberSciTech), pp. 107–112. IEEE (2017)
18. Sharif, M., Alesheikh, A., Tashayo, B.: Similarity measure of trajectories using contextual information and fuzzy approach, January 2018
19. Sharif, M., Alesheikh, A.A.: Context-awareness in similarity measures and pattern discoveries of trajectories: a context-based dynamic time warping method. GIScience Remote Sens. **54**(3), 426–452 (2017)
20. Toohey, K., Duckham, M.: Trajectory similarity measures. SIGSPATIAL Spec. **7**(1), 43–50 (2015)
21. Vlachos, M., Gunopulos, D., Kollios, G.: Robust similarity measures for mobile object trajectories. In: Proceedings of 13th International Workshop on Database and Expert Systems Applications, pp. 721–726, September 2002
22. Wang, Y., Wang, L., Li, Y., He, D., Liu, T.Y., Chen, W.: A theoretical analysis of NDCG type ranking measures. CoRR abs/1304.6480 (2013)
23. Zhang, Z., Huang, K., Tan, T.: Comparison of similarity measures for trajectory clustering in outdoor surveillance scenes. In: 18th International Conference on Pattern Recognition, ICPR 2006, vol. 3, pp. 1135–1138. IEEE (2006)
24. Zolotarev, V.M.: One-Dimensional Stable Distributions, vol. 65. American Mathematical Society, Providence (1986)

Car Monitoring System in Apartments' Garages by Small Autonomous Car Using Deep Learning

Leonardo León-Vera[1,2]([envelope]) and Felipe Moreno-Vera[1,2]([envelope])

[1] Universidad Nacional de Ingenieráa, Lima, Peru
[2] Information Technology and Communications Center, Lima, Peru
{lleonv,felipe.moreno.v}@uni.pe
http://www.uni.edu.pe
http://www.ctic.uni.edu.pe

Abstract. Currently in Peru, people prefer to live in apartment instead of houses but in some cases there are troubles with belongings between tenants who leave their stuffs in parking lots. For that, the use of an intelligent mobile mini-robot is proposed to implement a monitoring system of objects, such as cars in an underground garage inside a building using deep learning models in order to solve problems of theft of belongings. In addition, the small robot presents an indoor location system through the use of beacons that allow us to identify the position of the parking lot corresponding to each tenant of the building during the route of the robot.

Keywords: Object detection · Localization · Self-driving · Low energy · Bluetooth

1 Introduction

Currently in Peru, families prefer to live in apartments (see Fig. 1(a)), since among the advantages obtained are a more welcoming space and less effort in cleaning unnecessary rooms. In the same way, the size of the apartments is reduced more and more, therefore, families opt for the option of keeping their belongings in their parking lots next to their cars or instead of thems, as long as they are in an underground garage, but exist a problem when some objects disappear.

A mini robot is proposed to constantly supervise cars of the people in their respective parking lots during a schedule that does not bother the tenants of the building. Through autonomous driving techniques and object detection, a computational vision system based on deep learning algorithms is proposed in order to achieve the navigation of the robot and identify mainly the cars in question. Also, The location system construct with Beacons that determine the mini robot relative position in real time.

© Springer Nature Switzerland AG 2019
J. A. Lossio-Ventura et al. (Eds.): SIMBig 2018, CCIS 898, pp. 174–181, 2019.
https://doi.org/10.1007/978-3-030-11680-4_18

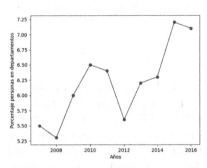 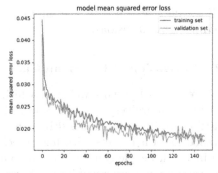

(a) Growth of people living in apartments. (b) Graphics of training and validation
Source: Private households of Peruvians performance of the model of [1]. Minimum
2007-2016. INEI.   quadratic error loss function vs Epochs.

Fig. 1. Graphics of demand and training

The navigation of the mini car in [13] is based on images using deep learning
and done for more complicated tracks than the garage of an apartment and using
different sensor for better accuracy in indoor scene.

2   Structure of the System

Implementation of a mini-robot that runs through an underground garage and
verify that the belongings of the tenants are in their respective parking lots. It
is proposed to start the solution to the problem by detecting objects to identify
the position of the car and its license plate to obtain information from the owner
of the car.

2.1   Data Collected from Tenants of the Building

Data is collected from the tenants of the building such as the apartment number,
owner of the apartment, ID of parking lot, car models, license plate of their car,
objects they stored in the parking lot, etc. (See Fig. 2). These will be compared

ID Depart.	Name	ID parking lot	Car model	License Plate	Car	Bicycle	...
1101	Pedro	23	Toyota	ACM-123	Si	No	...
1502	Juan	34	Gol	ABC-345	No	Si	...

Fig. 2. Data of tenants.

with the data obtained by the mini car and shown to the caretaker so that he can draw his conclusions.

2.2   Navigation of the Small Car

To solve this task, we use webcams which are much cheaper than the Lidars and Radars which are very used for self-driving due to the important data they can contribute to the autonomous driving model, between 3D maps and calculate the distance towards objects with high precision. In contrast, cameras emulate the way in which people can see the environment giving a better classification and interpretation of textures of images in comparison to the previous ones [4]. Their future will be strongly dependent on the development of the software algorithms controlling the self-driving and how can they process the massive amount of data generated.

A Convolutional Neural Network based on the NVIDIA model [1] for regression task is used to predict the angle of rotation of the wheels of the mobile robot by only obtaining an image of the path (See Fig. 1(b)).

Data Collection. Being supervised learning it was necessary to collect a video recording of the route that the mini car must follow to obtain an optimal performance. Being a flat floor, we did not get additional problems which would give problems when testing the system. However, when obtaining an environment where the walls are very similar, we look for a correct positioning of the camera and cut out the image so that it only focus on taking images that correctly identify the curves of straight roads. The input data are the images of three front cameras separated from each other by a few centimeters to collect more data from the road and speed of the small car at each moment, and values of angles between $0°$ and $180°$ as output values for training the model.

Neural Network Architecture. A deep sequence model of layers is used in the following manner, with its respective number of filters: Conv24-Conv36-Dropout-Conv48-Dropout-Conv64-Dropout-Conv64-Dropout-FC-FC-FC-FC with nonlinear activation function ReLU, which bring good results in computational vision tasks in the Convolutional layers. Regularization methods such as Dropout with a probability of 0.5 are added, due to the overfitting that occurs with the NVIDIA model when iterations are increased considerably.

Neural Network Training. The training of the data was done with the board NVIDIA P4000 with a partition of 20% validation set and 80% training set. The MSE loss function and the ADAM optimizer method are used without the need to manually set the speed of the learning rate.

Neural Network Testing. Testing of the convolutional neuronal network is performed on the Udacity simulator, obtaining a clear outstanding performance when driving autonomously as can be seen in the video [5] and later applied in underground garage.

2.3   Cars Detections

Cars detection task is in charge of a Convolutional Neuronal Network using Tiny YOLO model [3] that is the best model with the best performance between accuracy and inference time in detection objects in real time such as cars, bicycles and others objects (see Fig. 3(b)), model also detects void spaces if probability to find any object is lower than the threshold.

Dataset. The dataset used to train the model is "Visual Object Classes Challenge 2012" (VOC 2012). That present 20 classes, among them is cars (in a future work, we train chairs, tables, bicycles, etc.). For future work, the model is going to be trained using ILSVRC [11] 2014 dataset which have better quality on images such object scale, level of image clutterness, 200 classes of objects, among others.

Neural Network Architecture. Tiny YOLO model consists in a convolutional neuronal network with 9 convolutional layers of 16, 32, 64, 128, 256, 512, 1024, 512, 425 filters each one. This model is lightweight in comparison with "YOLO9000: Better, Faster, Stronge" [3] and accors.

(a) Mini-Robot for testing in the garage.

(b) Car detection in subway garage using tiny YOLO model.

Fig. 3. Mini robot and car detection

Neural Network Test. Test of the neural network was performed on an NVIDIA P4000 GPU server with images taken in the garage and on the NVIDIA TX1 board which we used in the mini-robot in the garage with a C920 camera obtaining 15 fps which works without any problem with the tiny YOLO model.

Results of Cars Detection. Detection of objects is very well identifying cars of general classes (See Fig. 3(b)), however for images where it is required to identify brand and model of the car is not get results since there is no dataset of images of front or back of the cars with annotations of their marks from which the neural network can learn. The nearest set of images found is Stanford dataset [12], but majority of images are not taken from front or back view. We trained with this dataset, but the images used for test were unlabeled. So, we discard this dataset for the model.

In contrast, identify digits of plates can get a better result because it is enough to locate the plate and perform a digit recognition [10]. ALPR is solved by [2] as well. Dataset for these tasks are SSIG dataset [14], a commercial dataset or UFPR-ALPR dataset [2] recently made with more fully annotated images and more vehicles in real-world scenarios for academic purposes.

After recognizing the digits of the plates, SUNARP online consultation [9] is accessed and information is obtained from the owner of the car (see Fig. 4).

Fig. 4. Query of the owner of license plate in SUNARP online application [9].

2.4   Mini-robot Mechanism

Mini-robot is built with base of a Monster Truck 1/18 which includes a motor and 2 servos for the movement of front and rear tires. An Arduino Uno is added that allows communication between the engine and the NVIDIA TX1 board, as you can see in Fig. 3(a). The Arduino sends instructions to the engine through its GPIOs where indicates speed of the wheels and angle of rotation of the front wheels. Communication of the board TX1 to the Arduino is serial through a USB which allows to send integer values encoded in characters to indicate the rotation of the front wheels and the instructions to go forward, stop and rewind. Through python program on the TX1 board, instructions are sent to the Arduino and this send to the motor which angle have to turn.

(a) Detection of objects, walls and signals of beacons. See the colors: yellow (beacons), red (object detection field), green (direction of route) and celestial (Bluetooth signal).

(b) Positioning of the beacons in the parking lot. See the colors: yellow (beacons), red (object detection field), green (direction of route) and celestial (Bluetooth signal)

Fig. 5. Environment and Beacon's location map (Color figure online)

2.5 Tenant Parking Mapping by Beacons

Summary. In this section, we present the way that robot knows where is it in real time. This action can be regulated with Beacons that is useful for indoor location in places where the Internet is not enough to make a connection [6], even allowing tracking in real-time applications [7].

Detection of Beacons. In garage studied, there is a separation distance of approximately 8 m between walls and approximately 2.45 or 7.35 m separation between vehicles, as shown in the Fig. 5(a) and (b).

As can be seen, Beacons have been distributed efficiently (see Fig. 6) that allows to determinate where it is located in real time, as well as identifies who owns that area due to the match between image recognition and the current position, based on previous work [8].

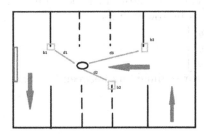

Fig. 6. Locations of the beacons in the garage and variables to be taken for the calculation and identification of the beacons.

Determination of the Position of the Mini Robot. In this section, we present the equations that define system to solve, by forming imaginary circles around each beacon, the respective radii are obtained. Then we proceed to calculate the relative position with respect to the global system of beacons based on distance equations forming a system of equations of the form:

$$E_i : (x - x_i)^2 + (y - y_i)^2 = d_1^2$$
$$\text{for i} = 1, ..., 3$$

Where (x, y) is the current position of the mini-robot and the indexes i correspond to the beacons and $r_i, ..., r_3$ correspond to the respective vectors.

For this system formed proceeds to solve: Take (x_i, y_i) as coordinates of each beacon, we deduce $r_i = r_c + d_i$.

Where r_c is the current position of the mini-robot, for all equations the pair of indices i, j are not equal $(i \neq j)$.

So, The module is taken:

$$\|r_i\|^2 = \|r_c\|^2 + 2(r_c)(r_i) + \|d_i\|^2.$$

Calculating:

$$\|r_i\|^2 - \|r_j\|^2$$

Obtains:

$$r_c(d_i - d_j) = \|d_j\|^2 - \|d_i\|^2 + \|r_i\|^2 - \|r_j\|^2 = Y_i$$

By which we would have:

$$x_c(x_i - x_j) + y_c(y_i - y_j) = Y_i$$

Which forms a new linear system that is solved by numerical methods and represented as:

$$Y = AX.$$

Where:

$x = (x_c, y_c)^t$: es the column vector of the positions of mini-robot.
A: is the matrix forms by row vectors.

$$A = \begin{bmatrix} x_1 - x_2 & y_1 - y_2 \\ x_2 - x_3 & y_2 - y_3 \\ x_3 - x_1 & y_3 - y_1 \end{bmatrix}$$

Y: is the column vector of differents between beacons.

3   Conclusions

We see that during the implementation of the robot car, it has to take into account many things that had not been foreseen to be able to handle the autonomous car.

In the work procedure, an optimal performance is obtained for the object detection task using Artificial Intelligence algorithms.

For the determination of positions of the mini robot based on the terrain delimited by columns and walls, it is necessary to specify the locations and the numerical system that solves the system of equations generated in such a way that a minimum error is obtained, which serves to determine and to identify the place where it is when we detect cars and objects of a certain car park, generating the relation object detected - location.

References

1. Bojarski, M., et al.: End to end learning for self-driving cars, arXiv preprint arXiv:1604.07316 (2016)
2. Laroca, R., et al.: A robust real-time automatic license plate recognition based on the YOLO detector. CoRR, vol. abs/1802.09567 (2018)
3. Redmon, J., Farhadi, A.: YOLO9000: Better, Faster, Stronger, CoRR, arXiv preprint arXiv:1612.08242 (2016)
4. Santo, D.: Autonomous Cars' Pick: Camera, Radar, Lidar? https://www.eetimes.com/author.asp?section_id=36&doc_id=1330069. Accessed 7 June 2018
5. CarND Self-Driving Nanodegree Udacity - Behavioral-Cloning. https://youtu.be/e7D23Kdy72Q. Accessed 7 June 2018
6. Li, J., Guo, M., Li, S.: An indoor localization system by fusing smartphone inertial sensors and bluetooth low energy beacons. In: Frontiers of Sensors Technologies (ICFST), Shenzhen, China (2017)
7. Gorovyi, I., Roenko, A., Pitertsev, A., Chervonyak, I., Vovk, V.: Real-time system for indoor user localization and navigation using bluetooth beacons. In: 2017 IEEE First Ukraine Conferenceon Electrical and Computer Engineering (UKR-CON) (2017)
8. Momose, R., Nitta, T., Yanagisawa, M., Togawa, N.: An accurate indoor positioning algorithm using particle filter based on the proximity of bluetooth beacons. In: IEEE 6th Global Conference on Consumer Electronics (GCCE) (2017)
9. SUNARP: Servicio Gratuito de Consulta Vehicular con datos relacionados a las características de los vehículos registrados a nivel nacional. https://www.sunarp.gob.pe/ConsultaVehicular. Accessed 7 June 2018
10. Hsieh, C.T., Chang, L.-C., Hung, K.M., Huang, H.-C.: A real-time mobile vehicle license plate detection and recognition for vehicle monitoring and management. In: Joint Conferences on Pervasive Computing (JCPC) (2009)
11. Russakovsky, O., et al.: ImageNet large scale visual recognition challenge. Int. J. Comput. Vis. (IJCV) **115**(3), 211–252 (2015)
12. Krause, J., Stark, M., Deng, J., Fei-Fei, L.: 3D object representations for fine-grained categorization. In: 4th IEEE Workshop on 3D Representation and Recognition, at ICCV 2013 (3dRR-13) (2013)
13. Zhou, C., Li, F., Cao, W.: Architecture design and implementation of image based autonomous car: THUNDER-1. In: Multimedia Tools and Applications (2018). https://doi.org/10.1007/s11042-018-5816-9
14. Goncalves, G.R., da Silva, S.P.G., Menotti, D., Schwartz, W.R.: Benchmark for license plate character segmentation. J. Electronic Imaging **25**(5), 053034 (2016)

A Framework for Analytical Approaches to Combine Interpretable Models

Pedro Strecht(✉) , João Mendes-Moreira , and Carlos Soares

INESC TEC/Faculdade de Engenharia, Universidade do Porto,
Rua Dr. Roberto Frias, 4200-465 Porto, Portugal
{pstrecht,jmoreira,csoares}@fe.up.pt

Abstract. Analytic approaches to combine interpretable models, although presented in different contexts, can be generalized to highlight the components that can be specialized. We propose a framework that structures the combination process, formalizes the problems that can be solved in alternative ways and evaluates the combined models based on their predictive ability to replace the base ones, without loss of interpretability. The framework is illustrated with a case study using data from the University of Porto, Portugal, where experiments were carried out. The results show that grouping base models by scientific areas, ordering by the number of variables and intersecting their underlying rules creates conditions for the combined models to outperform them.

Keywords: Knowledge generalization · Interpretable models ·
Prediction of performance · Decision tree merging · C5.0

1 Introduction

Present-day challenges of automatic *Knowledge Discovery from Databases* (KDD) are going beyond the goal of transforming data into knowledge. Despite the importance of having an accurate prediction, most decision makers are also quite interested in perceiving the rationale behind it. Therefore, nowadays, it is becoming essential to find a suitable way to present knowledge to aptly support decision making. The interpretability of models has been a key quality towards that direction by providing a familiar and appealing language to decision makers. This has been materialized with easy to read models, describing action axioms, i.e., a set of rules and alternatives, leading to a specific outcome. Common examples of interpretable models are decision trees and decision rules [8].

The number of organizations using interpretable models has been increasing, however, generating such models to predict or describe a phenomenon in organizations with a decentralized activity presents new challenges. In this paper we address this topic by devising a framework based on the concept of combining a set of interpretable models and exemplify it with a case study using decision trees.

© Springer Nature Switzerland AG 2019
J. A. Lossio-Ventura et al. (Eds.): SIMBig 2018, CCIS 898, pp. 182–197, 2019.
https://doi.org/10.1007/978-3-030-11680-4_19

An example of an organization with decentralized activity is a company that does its sales through subsidiaries or even by authorized individual distributors. Another is of a university offering numerous courses to its students. This organizations have their problem domain broken down into what can be seen as several units that operate concurrently. This parallelism makes it increasingly common to generate not a single model but multiple models, each relating to a unit. In the company example, each subsidiary can have a model to describe/predict its monthly sales level. Likewise, in the university context, each course can have a model to describe/predict the performance of the students enrolled in it. Yet, the fact that these models are associated with only one unit makes it hard to find global knowledge in the perspective of the whole organization. In the aforementioned examples, this would render as gaining insight of the overall monthly sales level behavior of the organization or the overall performance behavior of the students of the university.

The need to bring together models emerged essentially from two contexts. The first was to create models for systems based on distributed environments, i.e., where the data sources were scattered across different locations. The problem was presented as "mining data that is distributed on distant machines, connected by low transparency connections" [2]. The second was a consequence of the growth in the amount of data collected by information systems. It became necessary to create models that could manage large datasets [11]. At the time there was a lack of available resources to handle the task, being described as "a very slow learning process sometimes overwhelming the system memory" [5] or "the emergence of datasets exceeding available memory" [1].

The concept of combining interpretable models has been described in different problem domains with distinct jargon as well. Still, going through research, it is clear that there are common patterns in the intermediate phases of the process, even if named differently. Hence, although combining models itself can be viewed as a generic process, it nonetheless presents challenges that can be addressed in various forms. From this perspective, we designed a framework for analytical approaches to combine interpretable models. Its major procedures are the creation and evaluation of interpretable models (called base models), their organization into groups, their subsequent combination within each group and the evaluation of the resulting combined models.

The remainder of this paper is structured as follows. Section 2 presents related work on combining models. Section 3 describes the framework. Section 4 presents an application of the framework to a case study with a few experiments and discussion of results. Section 5 concludes with final remarks.

2   Related Work

Before looking at the topic of combining interpretable models, it is worthwhile to differentiate it from ensemble learning, which, at first glance may look similar. Ensemble learning [12] consists of using the predictions made by a number of base models to make a single prediction. In contrast, combining models consists of using a set of base models to create a generalized model, which is the only one making a prediction. The goals of each technique are also quite different. While in ensemble learning it is focused on improving accuracy, in combining models it is concerned in obtaining aggregated models without significantly affecting accuracy. Moreover, model interpretability is a goal per se for combining models but not for ensemble learning.

Approaches to combining models fall into two major categories: analytical and mathematical. Essentially, all analytical approaches consist of breaking down a set of models into rules and then assemble them in order to create a combined model. On the other hand, mathematical approaches consist in applying a mathematical function to a group of models which results in the combined model.

Analytical approaches were first introduced by Williams [18] and over the years other researchers have made contributions, coming up with different ways of carrying out the process. The motivations have been either to create a model from distributed data or from large datasets.

In problems with naturally distributed data, every location has its own local dataset with identical format and structure. These are moved over a channel to a centralized location where they are joined into a monolithic dataset, i.e., a non-distributed dataset stored in a single location. A generalized model is then created using all available data. Still, such scenario presents a major problem: moving data may be unsafe, expensive or simply impossible due to its large volume. An alternative of moving data is to move the models instead. This implies the creation of a model in each location and then moving all local models through a channel to a centralized location. Once there, they are combined into a single generalized model [2,13].

In problems with the need to create models from large datasets, it is essential to artificially create distributed data. This is achieved by breaking down a large dataset into as many individual datasets as necessary until it becomes possible to create a model for each [1]. Under such circumstances, all base models are combined into a generalized one.

Mathematical approaches are quite different from each other and were designed to solve specific problems. Kargupta and Park [7], motivated by the need to analyse and monitor time-critical data streams using mobile devices, proposed an approach to combine decision trees using the Fourier Transform. This mathematical operation decomposes a function of time (a signal) into its frequencies yielding the frequency domain representation of the original signal. According to the authors, mining critical data streams requires on-line learning that produces a series of decision trees models, which may have to be compared with each other and aggregated, if necessary. Transmitting these models over a wireless network is presented as a problem. As the decision tree is a function, it can be represented in a frequency domain, resulting in the model spectrum. Combining models becomes a matter of adding their spectra, a trivial task in the frequency domain. If required, the combined model can be transformed back to the decision tree domain by the Inverse Fourier Transform.

Gorbunov and Lyubetsky [4] devised a mathematical approach to combine models by exploring the problem of constructing a tree which is the "nearest" on average to a given set of trees, also referred to as *supertree*. The notion of nearest is a function that counts the number of events that occur by comparing each tree to the average one. Examples of events are divergence, duplication, loss and transfer. The method is tested on the domain of analysis of the evolution trees of different species. In this context, the problem is to map a set of gene trees into a species tree (the average tree). However, the approach can be used for any kind of tree in which there are operations which can be mapped to the four types of events indicated above. It is worthwhile noting that the authors present an approach to convert non-binary trees to binary ones in order to allow the algorithm to be used in those contexts as well. The algorithm has also been applied in the context of molecular biology [10].

Shannon and Banks [15] presented an approach, called *Maximum Likelihood Estimate* (MLE) to combine a set of classification trees into a single final tree by finding a *central tree*. It uses a probability distribution on the set of classification trees and a distance metric based on structural differences between trees. This metric allows weighting to penalize tree differences occurring near the root differentially compared with less serious differences occurring near the terminal nodes. The approach was applied to a set of classification trees obtained from biomedical data. Specifically, 13 classification trees were created that predict the presence or absence of cancer based on immune system parameters. The authors conclude that the resulting model retains the interpretability of a single tree model and has excellent generalizability.

3   Framework Description

This section presents a framework for analytical approaches to combine interpretable models by sequencing the generalized parts and identifying the ones that can be specialized. The goal is to, given a set of datasets, train the corresponding base models and obtain a set of combined models representing the knowledge contained in groups of base models. Algorithm 1 summarizes the workflow and the next subsections describe some of the procedures in detail.

The process begins with the creation of folds for each dataset (D). Due to its relatively low bias and variance, 10-fold cross validation is recommended for estimating accuracy [3]. Each fold (f) contains approximately the same number of examples. Each combined model has to be evaluated using unseen data, i.e., data not used in the creation of base models. As each base model is to provide rules to a combined model, one fold of its associated dataset is put aside destined to incorporate a test dataset to evaluate that same combined model. As a consequence, to ensure it remains new data, this fold (denoted as λ) is never included in the data for creating or evaluating base models. At this point, the question arises of which fold to use for this purpose. Instead of choosing a specific fold, the process of combining models and subsequent evaluation is performed 10 times, each using a different fold. Each fold maps in an iteration (λ) of the evaluation

Algorithm 1. Framework to combine interpretable models

Input: Datasets $= \{D_1, \ldots, D_n\}$
Output: Improvement scores $= \{\varepsilon_1, \ldots, \varepsilon_k\}$
for i such that $1 \leq i \leq n$ **do**
    $\{f_i^1, \ldots, f_i^{10}\} \leftarrow$ CreateFolds(D_i)
end for
for λ such that $1 \leq \lambda \leq 10$ **do**
    for i such that $1 \leq i \leq n$ **do**
        $M_i^\lambda \leftarrow$ TrainModel$(D_i \setminus f_i^\lambda)$
        $\eta_i^\lambda \leftarrow$ EvaluateModel$(D_i \setminus f_i^\lambda)$
    end for
    $\{G_1^\lambda, \ldots, G_k^\lambda\} \leftarrow$ CreateGroups$(\{M_1^\lambda, \ldots, M_n^\lambda\})$
    for j such that $1 \leq j \leq k$ **do**
        $\Omega_j^\lambda \leftarrow$ CombineModels(G_j^λ)
        $\sigma_j^\lambda \leftarrow$ EvaluateCombinedModel$(\Omega_j^\lambda, \{f_1^\lambda, \ldots, f_p^\lambda\}, \{\eta_1^\lambda, \ldots, \eta_p^\lambda\})$
    end for
end for
for j such that $1 \leq j \leq k$ **do**
    $\varepsilon_j \leftarrow \frac{1}{10} \sum_{\lambda=1}^{10} \sigma_j^\lambda$
end for

---

cycle. Base models (M_i^λ) are created and then evaluated using the data in all folds except the λ fold $(D_i \setminus f_i^\lambda)$. Hence, the evaluation of base model uses a 9-fold cross-validation set-up based on k-fold cross-validation [16]. The result is an evaluation score, conceptually denoted as η_i, measuring the predictive quality of the model. Examples of such metrics are accuracy, area under curve (AUC) or the F1-score [6].

The base models are then organized into groups (G_j^λ), each to yield a combined model (Ω_j^λ). Next, the evaluation test folds are assembled as test dataset for the combined model $(\{f_1^\lambda, \ldots, f_p^\lambda\})$. The evaluation procedure takes the combined model, the test dataset and the base models performances $(\{\eta_1^\lambda, \ldots, \eta_p^\lambda\}$ (p denoting the number of models in the group) resulting in an *improvement score* of the combined model (σ_j^λ). This metric estimates the average gain (if positive) or loss (if negative) in predictive quality relative to the base models. Finally, as the evaluation cycle is replicated 10 times, the improvement scores of each combined model are averaged across all iterations yielding the *overall improvement score* (denoted as ε_j).

3.1   Create Groups

In this procedure, the base models are gathered into groups and then ordered within each group (Eq. 1).

$$G_j^\lambda = \{M_1^\lambda, \ldots, M_p^\lambda\} \tag{1}$$

Models can be grouped reflecting a business driven criterion. For example, if a company is interested in knowing the performance of sales of its subsidiaries, it may want to group the models by geographic zone. Alternatively, there are applications where the creation of groups may be completely automated, for example, by criteria related to the complexity of the model (e.g., the number of rules). In such cases, clustering techniques can be used to assist the creation of the groups. There may be applications where there is no need to create groups. However, in order to keep the framework generic, it is considered that there is a single group with all the models.

The order by which the models are to be combined plays a role in the process, as the combining operation may not be associative. After assigning models to groups, they are sorted according to some criterion. Usually, this accounts for the similarity of models (e.g. the variables used).

3.2   Combine Models

In this procedure, the base models in each group are combined resulting in a new model, as described in Algorithm 2. Initially, all base models (M) are converted to decision tables, denoted as T, which allows the process to be language independent. These are then combined sequentially following the pre-established order. Depending on the approach chosen to combine decision tables, there may be circumstances that generate an empty decision table. If so, the procedure skips that attempt and carries on selecting the next decision table to combine with the last one that succeeded (T_ω). After all the decision tables in the group are scanned, the final combined decision table is converted back into the same language as the base models, yielding the combined model (Ω). The next subsections detail these operations.

Algorithm 2. Combine models

Input: Group of base models $\{M_1, \ldots, M_p\}$
Output: Combined model Ω
$T_\omega \leftarrow$ ExtractRules(M_1)
for i such that $2 \leq i \leq p$ **do**
    $T_\theta \leftarrow$ CombineRules(T_ω, ExtractRules(M_i))
    if $T_\theta \neq \varnothing$ **then**
        $T_\omega \leftarrow T_\theta$
    end if
end for
$\Omega \leftarrow$ BuildModel(T_ω)

Extract Rules. Each rule is a conjunction of conditions on independent variables (x_i), which, if true, predict a class in a target variable ($\hat{y}$). Rules map

to regions in a multidimensional space which, in turn, become rows in a decision table, while columns specify the variables. Figure 1 presents an example
of extracting the rules of a decision tree model and presenting them in a decision table. A special case is the one of an "empty" model with only one leaf
node, corresponding to a single decision rule covering the whole decision space.
The decision table includes only one row with the target variable column filled
with the value in the leaf node, being devoided of columns for the independent
variables.

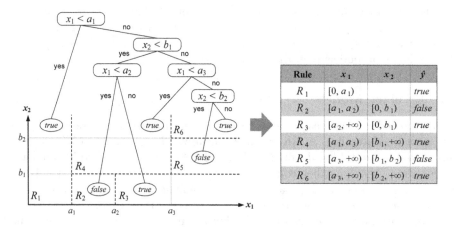

Rule	x_1	x_2	$\hat{y}$
R_1	$[0, a_1)$		true
R_2	$[a_1, a_2)$	$[0, b_1)$	false
R_3	$[a_2, +\infty)$	$[0, b_1)$	true
R_4	$[a_1, a_3)$	$[b_1, +\infty)$	true
R_5	$[a_3, +\infty)$	$[b_1, b_2)$	false
R_6	$[a_3, +\infty)$	$[b_2, +\infty)$	true

Fig. 1. Extracting the rules of a decision tree model into a decision table

Combine Rules. This operation attempts to combine the rules in a pair of
decision tables (T_1 and T_2) into one (T_θ). It encompasses three sequential tasks
as presented in Algorithm 3, each yielding a decision table.

Algorithm 3. Combine rules

Input: Decision tables T_1 and T_2
Output: Combined decision table T_θ
$T_\alpha \leftarrow$ SynthesiseRules(T_1, T_2)
if $T_\alpha \neq \emptyset$ **then**
  $T_\beta \leftarrow$ ResolveConflicts(T_α)
  $T_\theta \leftarrow$ JoinRules(T_β)
else
  $T_\theta \leftarrow \emptyset$
end if

The task *Synthesise rules* implies a specific approach to derive the rules of the
combined table. An intuitive example is the intersection of the inner product of

the rules of both tables [1,2,17]. In Eq. 2, rule $R_{ij}^{T_\alpha}$ results from the intersection of R_i from decision table T_1 with R_j from T_2. The operation is replicated until all rules from both tables are combined. A possible consequence is that none of the rules of both tables overlap, resulting in non-interceptable tables. If this occurs, the combined table is empty and the process stops.

$$R_{ij}^{T_\alpha} = R_i^{T_1} \cap R_j^{T_2} \tag{2}$$

Intersection is an example of a *Combination function*. It is important to note that in the framework definition it is generic, i.e., it may be specialized with other functions. Therefore, the *Synthesise rules* task can be described by Algorithm 4 ($|T_1|$ and $|T_2|$ denoting the number of rules in T_1 and T_2 respectively).

Algorithm 4. Synthesise rules

Input: Decision tables T_1 and T_2
Output: Combined decision table T_α
for i such that $1 \leq i \leq |T_1|$ **do**
    for j such that $1 \leq j \leq |T_2|$ **do**
        $R_{ij}^{T_\alpha} \leftarrow$ CombinationFunction$\left(R_i^{T_1}, R_j^{T_2} \right)$
    end for
end for

A conflict exists if a pair of overlapping rules of T_1 and T_2 do not agree on the target variable value. The task *Resolve conflicts* selects, for each conflict found, which value should be set to the target variable of the new rule. For example, an approach is to assign the target value of the rule/region with larger volume in the multidimensional space [1]. Another is to select the one created with more examples [17]. After this task, the resulting decision table T_β has no conflicts. If there are no conflicts in the first place, then $T_\beta = T_\alpha$.

The task *Join rules* attempts to decrease the number of rules by identifying adjacent rules in multidimensional space sharing the same class in the target variable. These can be joined together, thus reducing the number of rules. If none is found, then $T_\theta = T_\beta$.

To illustrate these tasks let us consider the examples of decision tables T_1 and T_2 in Fig. 2. Decision table T_1 has two independent variables x_1 and x_2 while T_2 only has one independent variable x_1. These tables will be combined using intersection as the combination function. There is also a column N specifying the number of examples used to create each rule. This number is required as part of the strategy for conflict resolution.

The result of the *Synthesise rules* task is the decision table T_α in Fig. 3. Each row represents an attempt to combine each rule of T_1 with each rule of T_2 by applying Eq. 2 over each variable. Each row has a *Status* column for the result of the combination attempt (*combined*, *conflict* or *disjoint*). There are conflicts in rows $R_{11}^{T_\alpha}$ and $R_{22}^{T_\alpha}$ as the underlying combined rules, although overlapping,

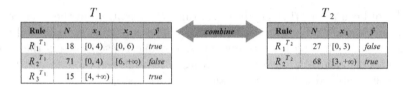

Fig. 2. Two decision tables to be combined

do not agree on the target variable value. Row $R_{31}^{T_\alpha}$ is disjoint as the rules $R_3^{T_1}$ and $R_1^{T_2}$ do not overlap on x_1.

After dropping the disjoint rule, all conflicts must be resolved. Using as criterion to select the target variable value of the rule created using more examples, the *Resolve conflicts* operation results in the decision table T_β (Fig. 3). For example, in $R_{11}^{T_\beta}$, *false* is selected for $\hat{y}$ as $N_1^{T_2}$ (27) is larger than $N_1^{T_1}$ (18).

T_α

Row	x_1	x_2	$\hat{y}$	Status
$R_{11}^{T_a}$	[0, 3)	[0, 6)	true/false	conflict
$R_{12}^{T_a}$	[3, 4)	[0, 6)	true	combined
$R_{21}^{T_a}$	[0, 3)	[6, +∞)	false	combined
$R_{22}^{T_a}$	[3, 4)	[6, +∞)	false/true	conflict
$R_{31}^{T_a}$	∅		n/a	disjoint
$R_{32}^{T_a}$	[4, +∞)		true	combined

T_β

Row	x_1	x_2	$\hat{y}$
$R_{11}^{T_\beta}$	[0, 3)	[0, 6)	false
$R_{12}^{T_\beta}$	[3, 4)	[0, 6)	true
$R_{21}^{T_\beta}$	[0, 3)	[6, +∞)	false
$R_{22}^{T_\beta}$	[3, 4)	[6, +∞)	false
$R_{32}^{T_\beta}$	[4, +∞)		true

T_θ

Rule	x_1	x_2	$\hat{y}$
$R_1^{T_\theta}$	[0, 3)	[0, 6)	false
$R_2^{T_\theta}$	[3, 4)	[0, 6)	true
$R_3^{T_\theta}$	[0, 4)	[6, +∞)	false
$R_4^{T_\theta}$	[4, +∞)		true

Fig. 3. Example of combining two decision tables via intersection of rules

The result of the *Join rules* task is the decision table T_θ in Fig. 3. Firstly, all rows in T_β are renumbered as rules in T_θ. After examination, there are two that can be joined together. In fact, $R_{21}^{T_\beta}$ and $R_{22}^{T_\beta}$ are contiguous over x_1, completely overlap over x_2 and share the target variable value. Rule $R_3^{T_\theta}$ is the union of both.

Build Model. This operation converts a decision table back to the base model representation. For example, if the base models are decision trees, then the combined model should also be a decision tree. However, if they are presented as a set of rules, then that should be the language of the combined model.

The operation presents, however, unexpected challenges. An inevitable consequence of repeatedly changing and removing decision regions along the combination process is a final decision table frequently failing to cover the entire multidimensional space. An approach for decision tree models is depicted in Fig. 4. It consists in artificially generating examples falling into each decision region of the final combined decision table T_w [17]. The examples of all regions are gathered in a dataset D^{T_w} from which a model is trained (Ω).

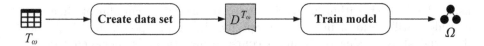

Fig. 4. Building a model from a decision table

3.3 Evaluate Combined Model

In this procedure, a combined model is evaluated following the steps in Algorithm 5. The predictive quality of the combined models is measured by an *improvement score* (σ).

Algorithm 5. Evaluate combined model

Input: Combined model $= \Omega$, Test folds $= \{f_i, \ldots, f_p\}$, Performance of base models $= \{\eta_i, \ldots, \eta_p\}$

Output: Improvement score of combined model $= \sigma$

for i such that $1 \leq i \leq p$ **do**

  $\mu_i \leftarrow \text{EvaluateModel}(\Omega, f_i)$

  $\Delta_i \leftarrow \mu_i - \eta_i$

end for

$\sigma \leftarrow \frac{1}{p} \sum_{i=1}^{p} \Delta_i$

The fold that was put aside in each base model is now used as test data to evaluate the combined model. Evaluation consists in using the combined model to make predictions on the test data and then comparing them with the true values of the target variable. The evaluation metric (μ_i) has to be the same as the one used to evaluate base models. For example, if the F1-score was chosen to evaluate base models, then it should also be used to evaluate the combined ones.

As the aim is to estimate the variation in predictive quality of replacing the base models with a combined one, the difference of performances (Δ_i) is calculated. If positive, the combined model performs better than the base model, otherwise, its performance is worse. The cycle is replicated for all folds coming from each dataset of the base models associated with the combined model. The improvement score of the combined model is the average of the differences of performances relative to all base models in the original group.

4   Framework Application

4.1   Case Study

This section presents the application of the framework to a case study. As part of the analysis of academic data in the University of Porto, decision tree models, using the C5.0 algorithm [9], were created to predict at enrollment time how a student will perform in a course. Specifically, at this point, models predict if a student is passing or failing a course using student's socio-demographic information and previous academic performance.

Being able to explain why some students perform better than others in a course is valuable knowledge for the teachers lecturing and assessing it. For this reason, a separate decision tree model was created for each course. The availability of these models has created a context which paved the way to the possibility of combining them in order to create global models that can be useful for other decision-levels of the university, such as department directors or even faculty deans.

In the case study, datasets relate to courses that occurred in the year of 2012/2013 and were obtained from the institution academic database. Each example in datasets represents a student enrollment in the corresponding course, described by the following variables: age, sex, marital status, nationality, displaced, scholarship, special needs, type of admission, type of student, type of dedication, debt situation, status of student, years of enrollment, delayed courses and approval. As the goal of the models is to predict the success or failure of a student in the course, the target variable is approval. A total of 5779 datasets (from 391 programmes) were extracted. The size of each dataset varies significantly as the number of enrolled students is very different across courses. It ranges from 1 student (in PhD programmes) to 950 (in a multi-programme course). A decision tree model was created for each course with at least 100 enrolled students, which resulted in 730 models. All models were evaluated with the F1-score as performance metric.

4.2   Experiments

In a first set of four experiments, all base models were combined without creating groups. In other words, it was considered a single group containing all models. In each experiment, a different ordering criteria was applied.

In the first criterion all models in a group are sorted randomly, i.e., no specific criteria is used. The second criterion is to order the models within each group by the number of variables used. The reason for this criterion is to find out if combining the simplest models first while leaving the complex ones last improves the quality of the combined model. The number of variables in the models ranges from zero up to seven. Models having the same number of variables are sorted randomly. In the third criterion, models are ordered according to the number of examples in the corresponding training dataset. Unlike the previous criterion, increasing the number of examples to train a model does not necessarily translate into a more complex one. Instead, this criterion focuses on finding out if combining models trained with less examples first while leaving models trained with a greater number of examples last, plays a role in the performance of the combined model. Models trained with the same number of examples are sorted randomly. The fourth criterion consists of using the order provided by a hierarchical clustering method [6]. As in the third criterion of grouping, we used the importance score [14] of the four topmost used variables in models to calculate the euclidean distance between them. Then, hierarchical clustering, besides its main goal of creating clusters (not used for this purpose), also sorts the models according to their distance.

Decision tables were combined using intersection as combination function. Still, every combined model was created more than one time, each relating to a different specialization of the framework. Four parameters were defined to control the operations *Combine rules*, *Resolve conflicts* and *Build model*. In this case study, rules are combined based on the number of examples used to create them. The weight is a metric introduced to normalize the number of examples in a decision rule relative to the dataset associated with the model.

The first parameter is *weight assignment*, controlling if the combined rule weight should be the maximum, the minimum or the average of the pair. The second parameter is *conflict resolution*. When a conflict is found, it controls the combined rule target value as whether the maximum or the minimum weight of the combined pair.

The combined decision trees were created using artificially generated examples extracted from the decision rules. A problem that may arise with this approach is the creation of a very large number of examples, which can make it hard to create a model. Therefore the third parameter is *examples generated* controlling whether there should be an example for every value, or one for a predefined step between the limits. Another issue is how to ensure that the weight of the rules is reflected in the decision tree. The fourth parameter, *repeat examples*, addresses it by controlling whether examples should be repeated to account for weight.

Covering all parameters creates 24 possibilities of specialising the framework. Therefore, in each experiment there are 24 versions of each combined model. Figure 5 depicts the combined models' improvement score distribution, separating grouping criteria from ordering criteria. Following one of the assumptions of the framework, all the combined models were evaluated with the F1-score (as it was used for the base models).

4.3   Results and Discussion

The first set of experiments (ungrouped models) is the baseline. In each set, random order is the baseline for the ordering criteria. All cases with $\varepsilon > 0$ are combined models that outperform the base ones. It is immediately clear that grouping models by scientific areas leads to combined models that overall have a better predictive quality than their ungrouped counterparts. This consistency is interesting from a business perspective as it supports the hypothesis that there are cases in which a combined model can replace base models without compromising predictive power. Although the results are far from outstanding and entirely empirical they reveal that the particular combination of grouping models by scientific areas while ordering then by the number of examples yields the best results. This setting achieves the highest number of cases in which the combined models outperform the base models (about half of the total).

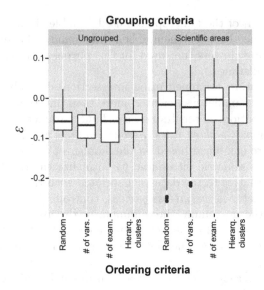

Fig. 5. Combined models' improvement score distribution

The main strength of the approach presented in this framework is the flexibility by which parts of the process can be altered. The criterion to group models, how to order them, the performance metric, the combination function, the strategy to resolve conflicts between rules, and even how to create a model at the end of the process are not hard-coded. There are no constraints on the models themselves either as long as they can be converted to decision tables. It can be applied to both binary or multiclass problems as well as models with any number or type of variables.

The experimental set-up to evaluate a combined model is relatively straightforward. It consists essentially of reserving a portion of the datasets associated with each base model and using them to make predictions with the combined model, employing the same performance metric used to evaluate to base models. The result is an estimate of whether the combined models can replace the base models they are intended to represent.

The usefulness of a combined model is to generalize the knowledge of a set of models, but without substantially decreasing the predictive power with respect to those base models. On one hand, it is important to guarantee interpretability, which is achieved by not changing the language of the combined models. On the other hand, it is important to ensure that the predictive performance of the combined model is at least close to the one observed in the base models. The results obtained from the case study illustrate both goals. Each combined model created is a decision tree, thus holding the promise of being interpretable by those who can read the base models. Additionally, the results show that there are specific cases in which the predictive ability is higher relative to the base models.

It is also important to discuss the weaknesses of the approach. In fact, the very same flexibility that was pointed out as a strong point can also be perceived as an obstacle to the application of the framework. Because there are so many configurable parts, it is difficult to apply it directly to a new problem. Intrinsically, it is required to know somewhat well the base models to decide how to best tune the framework. The first problem is how to group them. Intuitively, it is reasonable to go for a business-oriented criterion. For example, if the models relate to geographically dispersed business units, an aggregate subset might reflect this (e.g., base models relating to cities grouped in countries). The example of courses grouped into scientific areas from the case study also fits in this pattern. These are the kind of decisions that often require an in-depth knowledge of the business. On the contrary, if the goal is to obtain a model that represents a set of models, without any specific business logic underneath, then a criterion based on meta-features of the models or their datasets may be make more sense. Under these circumstances the grouping process may even be automated thus removing the need to decide on the criterion to group models.

In the case study, the combination function is the intersection of rules which implies preserving the rules that overlap the most over several models. Hence, the resulting combined models contain the most common rules found in the respective base models. In this context, the combination process is more likely to succeed in creating high quality combined models, the more similar the models in each group are. However, such concern is directly related to the chosen combination function. This hints the need for an alignment between the criterion to group models and the combination function. Other applications, with different criteria to group models possibly lead to devise specific combination functions to generalize knowledge. Consequently, choosing a combination function is not a straightforward decision.

The strategy to resolve conflicts in the target variable is instrumental in keeping the rules generic over groups of models. The example presented in the case study is to assign the value of the target variable corresponding to the rule created with the largest weight of the rule in the context of the model. This allows to compare models with few examples with others having many examples. Other complex and creative strategies can be found in the literature. As in the combination function, this can be an intricate decision, requiring careful consideration of the implications of each possible alternative.

The issue of creating a combined model in the same language as the base models is paramount to ensure its interpretability. In the case study, it is exemplified with the most common method found in the literature, consisting on generating examples from a decision table and then training a decision tree model.

The generality of the framework creates the conditions for comparing various methods for each part such as Bursteinas and Long [2], Andrzejak et al. [1] and Strecht et al. [17]. However, comparing with mathematical approaches is not as plain as the combination functions are not always available. Nor is the process broken down into steps. Instead, models are combined in a single operation hard to generalize.

5   Conclusions

Combining interpretable models has emerged over the years as an approach to solve different problems in particular contexts. Analytical approaches, which separate the rules of a set of models and then recombine them, although presented in a variety of forms, can be abstracted to a generic method. The main contribution of this paper is to devise a framework that sequences the main procedures and then identifies the operations that can be deployed in different ways.

The core of the process lies in the representation of models as a set of decision rules (or regions), listed in a decision table, which helps the process of combining them. Then, the sub-problems of how to combine decision regions, resolve class conflicts of the target variable in overlapping regions, or create a combined model remain open to different approaches, without loss of generality. Combined models are evaluated by assessing their ability to replace the base models. Although the set-up to evaluate combined models is part of the framework, the evaluation metric itself is generic.

Overall, the framework creates conditions for a systematic study to fine tune the combination of models process by assessing the impact of selecting different alternatives to solve the sub-problems. Its application is exemplified in a case study in education. The results reveal that there are particular circumstances in which the combined models beat the base models' predictive performance.

Using intersection as the combination function implies that the base models should be grouped according to similarity. Obviously the question arises of which are the metrics or features that best measure similarity in order to maximize the results of the process. In the case study, the criterion of grouping models by scientific areas is business-oriented and probably one that makes sense to decision makers. However, there are contexts where criteria such as the number of rules, the depth of the underlying tree or others related to the meta-features of the datasets may be worth investigating.

It is clear that the purpose of this framework is not to answer the question of which is the best tuning for a specific problem of combining models. On the contrary, it aims to acknowledge that there are many open problems for research and that can be solved in several ways. The biggest challenge is to discern what is the best combination of strategies that can maximize the combination process in the context of each particular problem.

Acknowledgments. This work is funded by projects "NORTE-07-0124-FEDER-000059" and "NORTE-07-0124-FEDER-000057", financed by the North Portugal Regional Operational Programme (ON.2 - O Novo Norte), under the National Strategic Reference Framework (NSRF), through the European Regional Development Fund (ERDF), and by national funds, through the Portuguese funding agency, Fundação para a Ciência e a Tecnologia (FCT).

References

1. Andrzejak, A., Langner, F., Zabala, S.: Interpretable models from distributed data via merging of decision trees. In: Proceedings of the 2013 IEEE Symposium on Computational Intelligence and Data Mining. IEEE (2013)
2. Bursteinas, B., Long, J.: Merging distributed classifiers. In: Proceedings of the 5th World Multiconference on Systemics, Cybernetics and Informatics (2001)
3. Efron, B., Tibshirani, R.J.: An Introduction to the Bootstrap. Chapman & Hall, London (1993)
4. Gorbunov, K., Lyubetsky, V.: The tree nearest on average to a given set of trees. Probl. Inf. Transm. **47**(3), 274–288 (2011)
5. Hall, L., Chawla, N., Bowyer, K.: Combining decision trees learned in parallel. In: Working Notes of the KDD-97 Workshop on Distributed Data Mining, pp. 10–15 (1998)
6. Han, J., Kamber, M., Pei, J.: Data Mining: Concepts and Techniques. Morgan Kaufmann, San Francisco (2011)
7. Kargupta, H., Park, B.: A Fourier spectrum-based approach to represent decision trees for mining data streams in mobile environments. IEEE Trans. Knowl. Data Eng. **16**, 216–229 (2004)
8. Kohavi, R., Quinlan, R.: Data mining tasks and methods: classification: decision-tree discovery. In: Handbook of Data Mining and Knowledge Discovery, pp. 267–276. Oxford University Press Inc., New York (1999)
9. Kuhn, M., Weston, S., Coulter, N., Quinlan, J.: C50: C5.0 decision trees and rule-based models. R package version 0.1.0-16 (2014)
10. Lyubetsky, V., Gorbunov, K.: Fast algorithm to reconstruct a species supertree from a set of protein trees. Mol. Biol. **46**(1), 161–167 (2012)
11. Maimon, O., Rokach, L.: Data Mining and Knowledge Discovery Handbook, 2nd edn. Springer, Boston (2010). https://doi.org/10.1007/978-0-387-09823-4
12. Opitz, D., Maclin, R.: Popular ensemble methods: an empirical study. J. Artif. Intell. Res. **11**, 169–198 (1999)
13. Provost, F.J., Hennessy, D.N.: Scaling up: distributed machine learning with cooperation. In: Proceedings of the 13th National Conference on Artificial Intelligence, pp. 74–79 (1996)
14. Quinlan, J.: C4.5: Programs for Machine Learning. Morgan Kaufmann, San Mateo (1993)
15. Shannon, W.D., Banks, D.: Combining classification trees using MLE. Stat. Med. **18**(6), 727–740 (1999)
16. Stone, M.: Cross-validatory choice and assessment of statistical predictions. J. R. Stat. Soc.: Ser. B **36**(2), 111–147 (1974)
17. Strecht, P., Mendes-Moreira, J., Soares, C.: Merging decision trees: a case study in predicting student performance. In: Luo, X., Yu, J.X., Li, Z. (eds.) ADMA 2014. LNCS (LNAI), vol. 8933, pp. 535–548. Springer, Cham (2014). https://doi.org/10.1007/978-3-319-14717-8_42
18. Williams, G.: Inducing and combining multiple decision trees. Ph.D. thesis, Australian National University (1990)

Processing Quechua and Guarani Historical Texts Query Expansion at Character and Word Level for Information Retrieval

Johanna Cordova[1,2(✉)], Capucine Boidin[2], César Itier[3], Marie-Anne Moreaux[1], and Damien Nouvel[1]

[1] INALCO ERTIM, 2 rue de Lille, 75007 Paris, France
`johanna.cordova@caramail.fr`
[2] Paris 3 IHEAL, 28 Rue Saint-Guillaume, 75007 Paris, France
[3] INALCO CERLOM, 2 rue de Lille, 75007 Paris, France

Abstract. The LANGAS project provides an online database containing historical (16th–19th) texts in Quechua, Guarani and Tupi, for sociolinguistic studies. Querying texts for such low-resourced languages raises several questions, issues and challenges. Among them, our work addresses word variation (diacritization, typographic variations) as an optional query expansion mechanism of the search engine. For such processing, taking into account the peculiarities of considered languages is unavoidable. This paper describes the morphology of considered languages, collected linguistic resources, implemented modules (regular expressions, stemming, word clusters) and some preliminary evaluations. Our work will be an opportunity to release resources for those languages. We plan to deepen this work in the near future and hopefully expect it to be useful for other researchers interested in the matter.

Keywords: Under resourced languages · Query expansion · Historical spelling variations

1 The LANGAS Project

The LANGAS database features 3 corpora of historical texts written in Guarani, Quechua and Tupi, dating from the 16th to the 19th century. They are available online, in open-access, and can be browsed text by text or through a bespoke search engine. The purpose of the project is to provide tools allowing comparative analysis of the texts, in order to collect quantitative data for research in linguistics, history and anthropology. The grouping of these corpora is justified by the strong similarities existing between them as regards their history, their linguistic aspects and the current status of the languages in which they are written.

© Springer Nature Switzerland AG 2019
J. A. Lossio-Ventura et al. (Eds.): SIMBig 2018, CCIS 898, pp. 198–211, 2019.
https://doi.org/10.1007/978-3-030-11680-4_20

1.1   Guarani, Quechua and Tupi Languages

Guarani, Tupi and Quechua are Amerindian languages which were widely spread in South America before colonial times. From the 16th century, Spanish and Portuguese colonial institutions made use of them to convert the populations to catholic belief and to govern their territories. Called "general languages", they were written in Latin script, leading to the creation of vast corpora (meta linguistic, religious and mundane). Linguistically, these languages are more sets of varieties than unified languages; we must understand these terms as encompassing several dialects.

Guarani and Tupi are branches of the same linguistic family and share many characteristics: they are polysynthetic languages with prefixes and suffixes and their alphabet features many diacritics.

Quechua is also an agglutinative, but only suffixating language. It has a relatively simple morphology: each word is built from a basic root followed by suffixes ranked in an almost fixed order, more often without alteration of neither the initial form nor the suffix. If the automatic processing can take advantage of that simplicity, it is counterbalanced by the fact that significant differences exist among the varieties spoken in the different parts of the Quechua linguistic area, making it difficult to create generic tools for this language. [14] distinguishes two language families, which are themselves divided into several branches. The LANGAS Quechua corpus contains mainly texts written in various dialects of Southern Quechua (QII.C according to Torero's classification), and some other in Central Quechua (QI) (Table 1).

Table 1. The general languages

Lang.	Geographic distrib.	# speakers
Guarani	Paraguay, Bol., Arg.	~7 millions
Quechua	Peru, Bol., Ec.	~6 millions
Tupi	Brazil	Extinct

1.2   Description of the Corpora

The texts that constitute each corpus are stored in two versions: palaeography and, for Quechua and Guarani, its transliteration. The palaeographic version is a manual raw transcription of the manuscript, which includes indigenous original texts and the original Spanish translation or gloss. The transliteration version is provided by the researcher who submitted the text in the database: it is an adapted transcription of both the indigenous and Spanish/Portuguese palaeographies, using a more modern spelling and if necessary filling the gaps of the manuscript or supplying a translation when the original source does not include one. The four different versions are stored in a database using distinct fields, and are sentence-aligned. In addition to these multiple versions, each text is associated to a metadata file containing general information about the text (author, release date, place of origin, genre) and details about the manuscript.

Corpus Statistics. The corpus contains data as reported in Table 2.

Table 2. Composition of the corpora

Lang.	Texts	Toks pal.	Toks translit.
Guarani	80	29,583	35,035
Quechua	31	250,593	113,547
Tupi	6	2700	NA

1.3   Search Engine

The LANGAS website provides a search engine[1] able to query either the indigenous or the Spanish part of the corpora, browsing them together or separately. A series of options allows to specify the query anchors in relation to the context (perfect match, beginning of string, beginning of section, etc.). By default, queries apply to both palaeographies and transliterations, but the user can also select which one to query. The search displays a list of the contexts where an occurrence of the searched term was found, and the number of matches. In addition, statistics showing the distribution of this term in corpora according to parameters related to metadata are displayed. These basic functionalities give a glimpse of the collected data for descriptive and historical linguistic research, allowing the study of the evolution of these still poorly documented languages. Nevertheless, to reach the effectiveness of a traditional search engine, many additional features remain to be developed. Our work aims at implementing such features at character and word level, with a special attention on Quechua.

1.4   NLP Tasks

We consider several problems related to the nature of our corpora. First, we did deal with the standard problems of digitized historical texts, as described by [9] and [2]: encoding of the texts and linking texts with metadata. Next, the typical problem of non-standard historical orthography is not completely discarded by the presence of transliteration: our corpora are written in languages for which there is no real consensus on orthographic norms in the current practice. In addition, there are variations in both morphology and spelling from one dialect to another. Addressing this issue for search and frequency profiling requires to be able to link lexical equivalents across dialects. To do so, we implemented query expansion, so that search results and displayed statistics can overcome those variations, while preserving the linguistic reality of the corpus.

---

[1] http://www.langas.cnrs.fr/#/recherche_corpus.

2   Related Work and Resources

2.1   Variants Retrieval in Historical Corpus

If the question of spelling variation detection in historical corpora has been widely covered for European language [1,8], the same work for low-resourced languages is less frequent. A corpus very similar to our corpora is that of the *Historical Dictionary of Brazilian Portuguese*, a database of texts from the same period (16th to 19th century) and of comparable genres. From this corpus, [6] extracted spelling variants by grouping forms. Their approach is based on applying generative transformation rules. Sets of potential spellings are created for each word, which are merged if they contain a common variant. A check for an effective occurrence of the variants in the corpus leads to build spelling variants clusters.

2.2   Quechua Resources

One of the most consistent studies for Quechua processing was conducted as part of the SQUOIA project at the University of Zurich. [10] details in her thesis the development of:

- normalization pipeline to transform texts in various dialects to the Cuzco variety and a morphological analyser for Southern varieties [13]
- spell checker plug-in for LibreOffice [12]
- hybrid machine translation system from Spanish to Cuzco Quechua
- dependency treebank of about 2,000 annotated sentences [11], built from Quechua translations of German/Spanish press articles and from one literary text.

A remarkable compilation work has been done by [7] about the question of the different Quechua families and subgroups. A multidialectal dictionary is freely available on his website, grouping 25,200 entries and 255 suffixes with their correspondences in 15 dialects, and, for the more widespread ones, their spelling in several graphic system, along with a translation in 5 European languages. A lighter version featuring only the Southern Quechua to Spanish part is available in the AULEX platform for online search in low-resourced languages dictionaries, and can be freely downloaded. This dictionary is, with [3]'s Southern Quechua-Spanish dictionary (1,753 entries), a very useful resource for NLP work on Southern Quechua varieties. Work has also be done to develop morphological analysers for Quechua [4, 5].

3   Query Expansion for Lexical Variants

3.1   Preliminary Work

Since we wish to develop tools dedicated to our corpora, an extraction of resources drawn from these corpora themselves was straightforward.

For Quechua, we can leverage our parallel data to discover variants across versions. To this end, we first need to align palaeography and transliteration at word level. This work was greatly facilitated by the fact that the two sub-corpora are already sentence-aligned. As there are few segmentation variations between them, once the subcorpora were extracted and punctuation removed, most tokens were actually already aligned. Shifts induced by segmentation differences have been detected by systematic measurement of words lengths between potentially aligned pairs within sentences, and automatically corrected by finding the alignment minimizing the edit distance. For instance, in Table 3, alignment of "ima tapis" was forced with "imatapis". The resource obtained allowed us to gather many lexical variants across palaeographic and transliterated version of the corpora.

Table 3. Alignment example

Pal.	Ni ima tapis ricuicuchu cunancama [...]
Tr.	Ni imatapis rikuykuchu kunankama [...]

We also compiled a lexicon of named entities from the Quechua corpus, including words in Spanish, Latin, proper names and location names. Most occurrences were extracted using an external lexicon[2]. The Spanish words with non canonical spelling missed by the lexicon have been detected thanks to some patterns specific to Spanish: gerund forms, infinitives, or even by matching simple letters such as 'e', 'o', 'b', 'v', 'd', non-existent among Quechua words (at least in the transliterated version). We also relied on the typographical annotations. Each entry extracted by means other than the external lexicon has been manually validated in order to constitute a reliable resource.

3.2 Case of the Guarani

In our corpora, many spelling variations for Guarani are due to variations in diacritics. Each vowel admits several diacritics, as described in Table 4.

The modern Guarani spelling has been standardised and only two diacritizations are now distinguished: the tilde for nasal vowels and consonants, and the grave accent to mark an accented syllable. The careful observation of diacritical variations is a way to better understand the historical practices of the language and to establish hypotheses about its evolution.

A simple yet effective way to retrieve all the variants is to transform the query submitted by the user into a regular expression listing for each character the possible diacritizations, before querying the database and without other condition. This approach allows to retrieve variants without much increasing request time processing. It has been implemented as a fuzzy search option that is automatically activated when querying the Guarani corpus.

---

[2] Unix system's Spanish dictionary.

Table 4. Vowels diacritization

Diacritic	Diacritizations
acute	á é í ó ú ý
grave accent	è ì ò ù ỳ
circumflex	â ê î ô û ŷ
tilde	ã ẽ ĩ õ ũ ỹ
breve	ĕ ĭ ў
inverted breve	ȃ ȇ ȋ ȏ ȗ ŷ

3.3 Case of the Quechua

A Two Step Expansion Mechanism. The variants that we propose to retrieve in the Quechua corpus have more complex characteristics. In previous versions of the search engine, a complex system of transformation rules was implemented as many complex regular expressions. However, the processing had to be performed at each request and significantly decreased the response time. Moreover, it wasn't possible to evaluate the accuracy of the returned results. We consider here a solution allowing to better identify the problems of spelling variations: we first try to evaluate the number and quality of variations (at first for the transliteration corpus), then to constitute an index of the most frequent lexical words. Our presupposition is the following: the corpus contains on the one hand a large quantity of minor regular variations easily detectable by applying known rules (case of the typically dialectal variations which do not change the physiognomy of the word); on the other hand, one also finds complex variants, specific to some words, or hapax not foreseen by the rules. For the latter, an upstream detection is necessary and our solution was to constitute a set of spelling variants clusters covering the corpus.

Quechua Stemmer. As previously mentioned, Quechua words are composed of a root to which suffixes are added. Thus, the orthographical variations that the search engine has to take into account mostly concern the root. We developed a stemmer to extract the roots appearing in our corpus. We then applied a set of transformation rules to each of them in order to generate their possible variations and build clusters of variants.

A naive implementation of a Quechua stemmer could be achieved by sequentially removing the suffixes using an exhaustive list until recovering a basic root. Though, in many cases, this method won't provide as good results as expected since a lot of ambiguities surface during the process, due to a high degree of similarity between some suffixes. Moreover, the stop condition of the stemming is difficult to set without a comprehensive list of roots. To avoid in most cases the ambiguities and provide better stemming for unreferenced words, we have to take into account the Quechua morphosyntactic rules in our stemming strategy.

An extended analysis of Quechua morphology has been performed by [10], and the implementation method of our stemmer is widely inspired by her work about Quechua POS-tagging. However, since we have to deal with a different type of corpus, our approach tends to be more specific to match our data at best.

For our stemming purpose, Quechua basic roots can be classified in two categories. The first category consists in root-words that can be used as it is or with additional suffixes. Since all these roots are nouns or adjectives, we call them **nominal roots**. The second category is composed of **verbal roots**, that need at least one suffix to appear as a word.

Table 5. Roots examples

	Root	Suffixated example
N	llaqta	llaqtapi *in the town*
V	*muna-	munay *to want*

Each category admits specific suffixes in a given order, according to their grammatical roles. In order to build the stemmer, we designed two deterministic automata able to parse respectively nominal and verbal suffixes chains, and to stem them using a suffix-stripping algorithm. The suffixes sequences supported by the automata are detailed in Appendix A. The stemming is completed when the parsing ends or when the remaining string to be parsed reaches a predefined minimal length. When an input word is submitted to the stemmer, both nominal and verbal parsing are performed, and the suffix chains deleted. The minimal stem is considered to be the accurate one. This process is enhanced with the Spanish and named entities lexicon we previously extracted from the corpus: after each strip, the stemmer checks if the resulting string exists in the lexicon. If so, the root is validated with a special tag: named entities spelling indeed varies according to distinct rules, and have to be processed separately (Table 5).

The stemming is performed on the transliteration corpus only, which contains 32,567 single words. To avoid overstemming, words with less than 6 letters are removed from the list; indeed, for the variants retrieval, understemming does not generate errors as long as words with a same suffixation are processed in the same way. Unsuffixed Spanish words are also discarded. Finally, the input file for the stemming consists in 21,581 suffixated words. After the stemming process, 9,300 unique roots are retrieved. The Spanish loan words are filtered for a separated processing. The remaining Quechua roots added to the presupposed unsuffixated words finally constitute the input for the clustering process, with a total of 8,450 entries.

Clustering. From the collected roots, we want to build clusters of spelling variants. One way to do this is to combine the entries by minimal edit distance. However, since Quechua roots often have similar shapes, the results obtained both with the two methods are very noisy. For example, same edit distance is computed from "suyay" (*to wait*) to "yuyay" (*to remember*), as from "llamk'ay" (*to work, Cuzco spelling*) to "llamkay" (*to work, Ayacucho spelling*), but only the second pair of terms is a valid cluster. The amount of false positives being too consequent using edit distance, we choose to apply the method described by [6] (see Sect. 2.1): on every root, a series of transformation rules is applied to generate a set of likely variations. We made a first test of all the rules previously defined by a Quechua specialist to check their existence in the transliteration corpus. Finally, the rules displayed in Table 6 are selected for the clustering implementation. These rules shouldn't generate wrong spelling variants, except for the glottalic consonants spelled with a single quote (t', p', k', ch'): in some dialects, the glottalization is not realized and consequently not transcribed, whilst in others, the spelling difference is semantically relevant (for example: chaki *foot*, ch'aki *dry*); this is a limitation of our system.

Table 6. Equivalence rules

Before [a i u]	p' ~ p ~ pp ~ ph		
	ch' ~ ch ~ chh		
	t' ~ t ~ tt ~ th		
	k' ~k ~ kh		
	q' ~q ~ qh		
	ŝ ~ š		
Global var.	á ~ a	í ~ i	ñ ~ n
	ch ~ ĉ ~ č	s ~ ç	q ~ g
	wa ~ hua ~ gua	wi ~ hui ~ vi	
Before cons./end of word	ch ~ s ~ ŝ ~ š		
	k ~ c ~ q		
	m ~ n		
At end of word	aw ~ ay		

For each set of rules, we obtain a list of clusters whose words differ only in accordance with said rules. As depicted in Table 7, we merge clusters containing common forms. The merging results in the following cluster: {p'unchaw, p'unchay, punchaw, ppunchaw, punĉaw, punčaw}. The effective existence of the generated variants is checked in the corpus for validation. In our example, only {p'unchaw, p'unchay, punchaw, punĉaw} are attested in the corpus.

In so far, as only one transformation set is applied at once, it can happen that no common form is found between two clusters that actually should be merged. Here the edit distance becomes useful: we compute the distance with

Table 7. Cluster example

Rule	Clusters
aw ∼ ay	{p'unchaw, p'unchay}
p' ∼ p ∼ pp	{p'unchaw, punchaw, ppunchaw}
ch ∼ ĉ ∼ č	{punchaw, punĉaw, punčaw}

a random word of each cluster and the other cluster entries (the random choice hardly influences the process, the average distance being next to 1 inside of the clusters and much bigger between two clusters: see Table 8). The pairs of clusters with a low distance are examined and manually merged if relevant.

Table 8. Clusters statistics

Before/after merg.	570/500 clusters
Avg. clusters size	2.16
Avg. inner distance	1.23
Most freq. equival.	ŝ ∼ s (#164)
	ch ∼ ĉ (#71)
	ch' ∼ ch ∼ chh (#67)
	p' ∼ p ∼ pp ∼ ph (#56)
	k' ∼ kh (#53)

Implementation for the Search Engine. The clustering shows that the large majority of the spelling variations attested in the corpus can be detected with simple rules. In consequence, the system of search with regular expression can be maintained, but for a smaller number of rules. For the more complex variations (with more than three alternative spellings), we store these clusters in the database, along with their possible variants, in order to increase recall. Before triggering search, the engine checks if the string, as typed by the user, exists in the clusters and, if so, performs the search for each of the variants. If no variant is found, the sequence of regular expression is applied to the input string and a classic search is performed.

4   Evaluation

4.1   Stemmer Evaluation

In order to evaluate the stemmer, we conducted a first manual validation by an expert of Quechua on the output of a test corpus composed of extracts in several dialects and from different periods (96 single words). At this stage,

the stemmer made 18 errors, which gives an accuracy of 81.2%. An analysis of
the incorrect outputs allowed us to identify some missing suffixes belonging to
a dialect hardly represented in the corpus (Central Quechua), and to fix some
mishandled ambiguities. We improved the stemmer according to results, and a
second evaluation consisting of 100 single words randomly picked in the tokenized
transliteration, gives an accuracy of 82%.

Incorrect roots include:

- Understemmed words extracted before the stemming process. Accuracy could
  be increased by foreseeing a special processing for the very short verbs, especially *ka-* (to be), which is always understemmed. The other cases concern
  the short suffixes such as the agentive -q or the evidential -n/m. Though, the
  presence of the last in the clustering step was useful to study the usage in
  the variations of this suffix. Moreover, we found that if the stemming of these
  short words regulates the issue of understemming for the case of agentive
  nouns, overstemming appears faster and significantly decreases accuracy.
- Words whose suffix order varies according to dialectal peculiarities yet not
  supported
- Words whose last syllable is a suffix homograph, when the suffix chain is not
  long enough to remove the ambiguity
- Compounds

As a complementary experiment, we implemented the model of Quechua suffix system depicted in the appendices, using the corpus processing tool Unitex[3],
in order to have an idea of the coverage of the roots extracted with the stemmer. Only noun and verbal derivations were implemented (deverbal nouns and
denominalized verbs are not supported yet). Among 528 tokens from an extract
of the database, 350 have been recognized, which provides us with a recognition
ratio of 66.3%, a very encouraging result.

4.2 Search Engine Evaluation

In order to quantify query expansion, we randomly selected words (8 for
Quechua, 5 for Guarani) among the most frequent ones of the transliteration
corpora and for which spelling variants exist. For Quechua, we performed a
search for these words in an official spelling[4]. For Guarani, we performed the
search for the non diacritized version of the words. The number of results without and with the query expansion are detailed in Table 9, as well as the ratio
and the number of variants encountered.

The average ratio of expansion is **47.5%** for Quechua. This score decreases
if the spelling used for the query is more represented in the corpus, but remains
significant (32% for the Cuzquenian spelling). For Guarani, the automatic diacritization restores the correct spelling and provides an average expansion of
43.3%. In these cases, no false positive is generated; however, as some diacritics

---

[3] http://unitexgramlab.org.
[4] Dictionary of the Ayacucho dialect provided by the Peruvian Ministry of Education.

Table 9. Query expansion ratio

Word	W/o exp.	With exp.	Ratio	Var
Quechua				
ñuqa	321	333	3.6%	4
kawsa	296	299	1%	2
yacha	230	303	33%	2
sunqu	158	415	8.8%	2
simi	110	371	70.3%	2
mikuy	17	66	74.2%	2
llamka	4	58	93.1%	3
ñisqa	36	485	96.5%	4
Guarani				
tupa	2	321	99.4%	2
nande	2	311	96.2%	2
guasu	135	156	6.5%	2
teko	138	139	0.3%	2
rera	11	57	14.3%	2

are semantically significant, the queries might include word with different meanings. The query expansion for Guarani is still at an experimental stage and we plan to improve it in the future.

5   Conclusion

The LANGAS project, led by anthropologists and socio-linguists, corroborates it is not an easy task to implement a search engine for low resourced languages. In addition to technical issues, it is unavoidable to take into account morphological peculiarities of considered languages, here Quechua and Guarani, as many variants exist in texts, which leads to troublesome silence of the search engine. To address this issue, we implemented query expansion on both character and word level, as a search engine option. Our evaluation already shows the efficiency and benefits of the currently implemented query expansion. As an induced benefit, this also led us to study the morphology of considered language and develop gainful resources, especially for Quechua: lexicons, stemmer, word clusters. We plan to release these resources soon and hope that our work will raise interest of researchers working on these languages or similar ones. For our part, we will keep on working on this database in the near future.

Acknowledgments. The LANGAS project was funded by the French National Research Agency (ANR). This work benefited from the support of Université Sorbonne Paris Cité (USPC) and National Institute for Oriental Languages and Civilizations (INALCO). Many thanks to Joséphine Castaing and Elégant Mateus who developed the site and database.

A   Quechua Suffixes Chains

Root	derivators		poss.	plurals	relators		coord.	
0	1	2	3	4	5	6	7	8
N	cha su niraq ni	sapa yuq niq nti	y yki n nchik yku ykichik nku	kuna pura	kama man manta n pa/p paq pi piwan rayku ta	wan	pas	chá chu chuch chus chusinam má mi/m qa ri si/s yá

Root	Verbal modifiers								p. obj.	aspect	tense	poss. subj.	plur.	cond.	incl.		
0	1	2	3	4	5	6	7	8	9	10	11	12	13	14	15	16	17
V	tya pya	qya raya ykacha ysi kacha rari na naya pa paya	yku rqu	paku tamu rpari	ri	chi	pu ku mu	lla	su wa	chka	rqa sqa	chun ni nki n nchik y sunki yki	chik ku	man	hina puni pas ña raq	taq	chá chu chuch chus chusinam má mi/m qa ri si/s yá
												yman sun waq sqayki saq nqa y					
												chwan					
V ->N									pti spa sqa na stin q y			y yki n nchik ykichik					
N ->V	lli ya ymana																

B   Unitex Graphs

See Figs. 1 and 2.

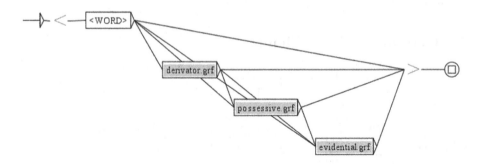

Fig. 1. Noun graph

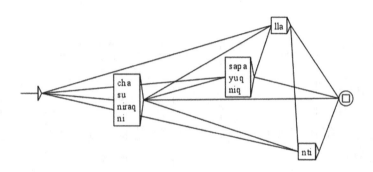

Fig. 2. Derivators subgraph

References

1. Baron, A., Rayson, P.: VARD2: a tool for dealing with spelling variation in historical corpora. In: Postgraduate Conference in Corpus Linguistics (2008)
2. Barteld, F.: Detecting spelling variants in non-standard texts. In: Proceedings of the Student Research Workshop at the 15th Conference of the European Chapter of the Association for Computational Linguistics, pp. 11–22 (2017)
3. Cerrón-Palomino, D.R.: Quechua sureño: diccionario unificado (1994). http://www.illa-a.org/wp/diccionarios/quechua-cerron-palomino/. Accessed 17 Apr 2018
4. Duran, M.: Morphological and syntactic grammars for recognition of verbal lemmas in Quechua. In: Formalising Natural Languages with Nooj 2014, p. 28 (2015)
5. Gasser, M.: Antimorfo 1.0 user's guide (2009)
6. Giusti, R., Candido, A., Muniz, M., Cucatto, L., Aluísio, S.: Automatic detection of spelling variation in historical corpus. In: Proceedings of the Corpus Linguistics Conference (CL) (2007)
7. Jacobs, P.: Vocabulary (2006). http://www.runasimi.de/runaengl.htm

8. Koolen, M., Adriaans, F., Kamps, J., de Rijke, M.: A cross-language approach to historic document retrieval. In: Lalmas, M., MacFarlane, A., Rüger, S., Tombros, A., Tsikrika, T., Yavlinsky, A. (eds.) ECIR 2006. LNCS, vol. 3936, pp. 407–419. Springer, Heidelberg (2006). https://doi.org/10.1007/11735106_36
9. Piotrowski, M.: Natural Language Processing for Historical Texts. Morgan & Claypool, San Rafael (2012)
10. Rios, A.: A basic language technology toolkit for Quechua. Ph.D. thesis, Faculty of Arts, University of Zurich (2015)
11. Rios, A., Göhring, A., Volk, M.: A Quechua-Spanish parallel treebank (12 2008)
12. Rios, A., Mamani, R.: Allin Qillqay! a free online web spell checking service for Quechua (11 2014)
13. Rios, A., Mamani, R.: Morphological disambiguation and text normalization for southern Quechua varieties. In: Proceedings of the First Workshop on Applying NLP Tools to Similar Languages, Varieties and Dialects, pp. 39–47 (2014)
14. Torero, A.: Idiomas de los Andes. Lingüística e historia. Editorial horizonte (2002)

Topic Modeling Applied to Business Research: A Latent Dirichlet Allocation (LDA)-Based Classification for Organization Studies

Carlos Vílchez-Román[1,2]([envelope]) [ORCID], Farita Huamán-Delgado[3] [ORCID], and Sol Sanguinetti-Cordero[4,5] [ORCID]

[1] CENTRUM Católica Graduate Business School (CCGBS), Lima, Peru
[2] Pontificia Universidad Católica del Perú (PUCP), Lima, Peru
cvilchez@pucp.edu.pe
[3] Universidad Nacional Mayor de San Marcos (UNMSM), Lima, Peru
farita.huaman@unmsm.edu.pe
[4] School of Communications, Universidad de Lima (UL), Lima, Peru
[5] Environment Technology Institute (ETI), Jirón Bach BH2, Lima, Peru
solsanguinetti@eti-ngo.org

Abstract. More than 1.5 million academic documents are published each year, and this trend shows an incremental tendency for the following years. One of the main challenges for the academic community is how to organize this huge volume of documentation to have a sense of the knowledge frontier. In this study we applied Latent Dirichlet Allocation (LDA) techniques to identify primary topics in organization studies, and analyzed the relationships between academic impact and belonging to the topics detected by LDA.

Keywords: Text mining · Organization studies · Topic modeling

1 Introduction

In order to make decisions we perceive the environment, select elements from it and find structures that guide our insights, following these implicit patterns for our decision making processes (Dean and Sharfman 1993; Papadakis et al. 1998). Before the exponential growth of information the internet brought with it, decision makers used to extract information at a slower pace; but current rates of information production are very fast, to the degree that some scholars have named this phenomenon as "information glut" (Voss 2001).

In this context, decision makers can be easily overwhelmed by the increasing flow of data, hence, we need new means to find structures that can foster our understanding of it. Text mining techniques provide a good approach to understand hierarchies, and thus structures within documents (Al-Augby and Nermend 2015).

© Springer Nature Switzerland AG 2019
J. A. Lossio-Ventura et al. (Eds.): SIMBig 2018, CCIS 898, pp. 212–219, 2019.
https://doi.org/10.1007/978-3-030-11680-4_21

Among the tools provided by text mining, Latent Dirichlet Allocation (LDA) has a great potential especially for the management of textual data.

Within natural language processing (NLP), algorithms for topic modeling make it possible to analyze huge volumes of textual data and find latent patterns for organizing it. Until 2015 the latent semantic analysis (LSA) text mining technique was one of the most popular NLP techniques; after 2015 LDA became one of the most widely adopted NLP algorithms, according to the number of studies with LDA in the title of the document, which was the result we found after carrying out a simple search in the Scopus database as we had also done with LSA.

LDA is a three-level hierarchical modeling that generates probabilistic topic models, based on the co-occurrence of similar stems (Blei et al. 2003) that can be used to detect structures or features within collections of documents; it is a flexible algorithm for text mining that reveals hidden patterns and trends (Baskara et al. 2017). It has the advantage over LSA that it is highly modular and can be easily escalated. Given that LDA works with the stem of words, and that we can define LDA mathematical models over terms and documents, where terms may be words that were normalized (stemmed or lemmatized) or not, then LDA works more independently of the language, so it can be adjusted to be used efficiently with texts written across languages, e.g. English or Spanish.

Once a corpus is analyzed with LDA the algorithm produces a new set of variables, one for every topic specified in the algorithm settings, computing the probability for each unit of analysis to be classified under the topic with the highest probability. Further to this, the algorithm ranks the words found in the text corpora according to the probability to be assigned to the proper topic.

2   Related Work

From an applied point of view, postgraduate programs in business and management (i.e., business schools) receive students with different academic backgrounds who mostly do not have studies in business-related areas (Martell 2007). They would benefit from having a structure to organize by primary dimensions or topics of all the information they extract from multidisciplinary databases, and could make them more efficient when organizing the contents for their academic papers.

Additionally most of the studies we found in our search that applied LDA as a text-mining technique analyzed online user comments or opinions, from changes in consumer opinions on topics ranging from health (Puranam et al. 2017), to spatial patterns of topical engagement (Brandt et al. 2017). Product managers could to get a deeper understanding of these, using topic models for their decisions (Bendle and Wang 2016).

Another set of studies used LDA for political analysis and marketing, corporate performance and research results in business and management. These ranged from the topic focus of each candidate in the U.S. presidential election

(Ryoo and Bendle 2017), to information requirements of stock market regulators (Dyer et al. 2017), and experiments to quantify risk types from textual risk disclosures (Bao and Datta 2014).

Most relevant for this study are two business and management research papers. In the first one the researcher explored results published in a management journal from 1984 to 2014. The topic modeling algorithm confirmed the existence of trends (e.g., Six Sigma, innovation, etc.), fads and fashion in management research (Carnerud 2017). In the second, the investigator analyzed 858 abstracts published in Annals of Tourism to determine language change in the last 40 years, dividing volumes of the journal in two: 1975 to 1994 for "old" volumes, and 2009 to 2015 for "new" volumes. LDA extracted topics that made it possible to differentiate between old and new abstracts. This result was observed with five topics ($\chi^2 = 30.06, p < 0.001$)), as well as with models with 20, 30, 40 and 50 topics. In all of them, the tests for differences in frequencies for old and new abstracts were statistically significant according to the Mann-Whitney results (Mazanec 2017).

LDA analysis requires a minimum critical mass of documents to be applied, as for example, abstracts available in the records of multidisciplinary databases like Scopus or Web of Science (WoS). Given that this detailed information is available from the 1970s onwards in both databases, but not before, we decided this timeframe was an appropriate starting point to carry out a LDA-based study. We identified organization theory as part of the core knowledge in business and management, since it provides the concepts and models for understanding organizational structure and behavior, hence our research question is the following: Is it possible to detect a topic structure within organization studies published and indexed in Scopus in the last four decades?

3   Methods

3.1   Text Corpora

Information Sources: We chose to use Scopus, as it is a database with a wider coverage for business and management research than that of WoS. According to Scimago Journal and Country Rank, in 2016 there were 1,394 journals indexed in Scopus and classified in the subject area "Business, Management and Accounting"; while on the other hand according to WoS Journal Citation Reports (JCR), in 2016 there were only 356 journals assigned to the subject areas "Business, Business and Finance, and Management". An advanced search on Scopus, carried out in May 24th 2018, retrieved 1,411,684 academic documents under the subject area "Business, Management and Accounting".

Inclusion/Exclusion Criteria: Since we examined the topics structure in the last 46 years, our interest was to detect the longest trends within organizational studies, so we divided that time interval into four 12-year periods: 1970–1981, 1982–1993, 1994–2003, and 2004–2015.

Because we were interested in theoretical development in business and management, as well as in related areas, we only considered for the topic modeling studies classified in five subject areas: (a) Business, Management and Accounting; (b) Decision Sciences; (c) Economics, Econometrics and Finance; (d) Psychology; (e) Social Sciences. We did not filter out by access type (e.g., open-access), language, country or document type (e.g., article, conference paper).

3.2   Textual Data Collection

Textual Datasets: In May 25th 2018, we carried out an advanced search on Scopus using this command: TITLE-ABS-KEY("manag* framework*" OR "manag* model*" OR "manag* structur*" OR "manag* theor*" OR "organizat* behav*" OR "organizat* framework*" OR "organizat* model*" OR "organizat* structur*" OR "organizat* theor") AND SUBJAREA(BUSI OR DECI OR ECON OR PSYC OR SOCI) AND (PUBYEAR > [1969|1981|1993|2003] AND PUBYEAR < [1982|1994|2004|2016]). For each dataset we sorted results by the year of publication. At the end of this step we had four Excel sheets amounting to 28,180 records: 1970–1981 (n = 814), 1982–1993 (n = 2,320), 1994–2003 (n = 5,771), and 2004–2015 (n = 19,275).

Stopwords: Before the topic modeling, we edited the four datasets and through the iterations of the LDA algorithm we identified the stopwords to exclude as non-relevant terms. We built a stopword list for organization studies progressively, and each time we ran the LDA algorithm, we obtained a ranking of the 10 terms with the highest probability to be extracted from the topic. When we identified non meaningful words (e.g., articles, pronouns, etc.), they were manually included in the stopword list. Then the LDA algorithm was ran once more to identify candidate terms for the stopword file. We repeated this process until we did not find any non meaningful word in the ranking of the 10 terms already mentioned above.

3.3   Textual Processing and Algorithm Settings

Textual Data Editing and Pre-processing: Given the fact that the title does not always have a structured and normalized language, and several times the keywords assigned to an article include terms that are too general, we used the content of the abstract for the topic modeling. For this reason we filtered out those records that did not have an abstract, which reduced our initial dataset to 27,226 records: 1970–1981 (n = 674), 1982–1993 (n = 2,150), 1994–2003 (n = 5,601), 2004–2015 (n = 18,801).

Then we deleted all the labels which defined the sections of a structured abstract (e.g., abstract, objective, purpose, summary, and similar terms) and the diacritic characters that do not contribute to modeling. To avoid having duplicated tokens for the same words all the words of the abstracts were changed to lowercase.

When conducting research using Scopus it is common to use programs like JabRef, Mendeley or Zotero, however these are optimized for the management of bibliographic references not paragraph analysis, and as our work was centered on text analysis of the abstracts, where the whole research papers are condensed, therefore LDA analysis was our best option.

LDA Settings: For topic modeling, we used ldagibbs (Schwarz 2018), a Stata package for LDA. We ran the algorithm, with the following settings: $\alpha = 0.25, \beta = 0.10$, seed = 3, samples = 10, minimum characters per word = 5, 6, 7, 8. We incremented the word length to detect any possible variation in the modeling that was associated with the increasing complexity of words. To detect changes among iterations we computed the log-likelihood. Probabilistic estimations for each word were normalized and saved into a matrix.

For each period we ran three topic models, with five, four and three topics, respectively. We applied this increment in topics as in academic literature has not been found a predefined topic structure for organization studies; we adopted a trial and error approach taking advantage of the flexibility of a LDA-based topic modeling. At the end, we set three as the minimum threshold, because in a previous study that used text-mining techniques researchers analyzed the stability and change in organizational theory during 1980–2016, and in the last analyzed period they found that results converged when the clustering algorithm grouped the keywords in three clusters of terms (Vílchez-Román and Huamán-Delgado 2017).

4   Results

4.1   Topic Modeling

Topics Extracted: With five topics we observed a low level of stability since there was a lot of variation in the labels assigned to each topic, therefore it was very difficult to identify stable patterns to develop a topic structure for the discipline; for example, between 1970–1981 and 2004–2015 we identified 14 different structures, and we assigned a label to each one. In contrast, from the same period, and using three topics as default setting for each period, we identified six different structures, and we assigned a label to each one, for an example see Table 1. These results showed more stability with three topics, which makes it suitable for proposing a topic structure that relatively maintains itself stable across time, as a main feature of any structure or taxonomy is its stability.

Model Estimation: The log-likelihood in the three topic models per analyzed period decreased over time, confirming the appropriate adjustment of the three-topics-based model. Regarding the minimum number of characters for selecting the words included in topic modeling, we found that five was an appropriate threshold, as the topics extracted using longer words (characters > 5, 6, 7) were not as intuitive as those extracted with words that had at least five characters.

Table 1. Topic modeling for 2004–2015

Number of topics	Five	Four	Three
Topic 1	Public policy	Organizational management	Management model
Topic 2	Management model	Management model	Organizational management
Topic 3	Organizational management	Public policy	Public management
Topic 4	Organizational learning	Organizational performance	
Topic 5	Organizational behavior		
Log-likelihood	−00000133	−00000134	−00000135

4.2 Classification of Documents

Given the high level of stability of the three-topics-based modeling, for each organization study the LDA algorithm computed the probability to be classified into on the three topics extracted. We created tables for each analyzed period, where studies were sorted out by two criteria: the publication year, then, alphabetically by the first letter of the abstract. For each 12-year period, most of the documents had a higher probability to be classified in one topic, and a very low one to belong to another topic. For example, in the period 1970–1981 (see Table 2), the first listed document had a probability of 68% to belong to the topic "Organizational model", but a 25% probability to be classified under "Organizational behavior", and a very low probability (7%) to belong to the topic "Service management". The same approach was applied to understand the results for the three remaining periods.

Table 2. Probability to be classified in each topic for the first 10 documents in the dataset (1970–1981)

Study	Organizational behavior	Organizational management	Service management
1	.24684685	**.67927928**	.07387387
2	.05101215	**.91740891**	.03157895
3	.17633803	**.62816901**	.19549296
4	.20159363	**.79123506**	.00717131
5	.04581006	**.88379888**	.07039106
6	.02245989	**.78395722**	.19358289
7	.42047244	**.51181102**	.06771654
8	**.54213836**	.34088050	.11698113
9	.36038647	**.52463768**	.11497585
10	.01309255	**.51616766**	.02754491

5   Limitations of the Study

We identified three limitations on the replicability of this study. The first one refers to time intervals used for generating the topic models. We selected 12 years for each analyzed period, however, if the study is replicated with different time intervals, it is possible that results are not comparable.

The second limitation deals with the information source. This LDA-based study can only be replicated using Scopus because it has a free Application Programming Interface (API) and has big topic coverage with broad areas.

The third limitation refers to the validation of the proposed topic structure. As Peruvian researchers, we were able to identify and contact only 18 Peruvian researchers in business and management who came from three leading business schools in our country, and we received only one response from a member of a PUCP-based research group.

6   Discussion and Conclusions

From all we have learned in this study, LDA-based topic modeling is a very promising approach to bringing order and stability to a discipline so important for understanding and improving the performance of organizations as business and management; improving our understanding of topics and thus facilitating decision making processes.

Business and Management is a discipline with a tendency to academic fads and pseudo theories where practitioners prefer to inform themselves with the so called management gurus or best-sellers writers of the moment, rather than to look for scientific research results when making strategic decisions. This can be explained by many reasons, but the main one being the lengthy time required for carrying out a research project or solve an organizational problem from a classic academic perspective.

The application of the probabilistic topic modeling algorithm shows it is feasible to bring some order and stability to a dynamic field like business and management, where theories (academic and pseudo academic) come and go. Even though the tools and knowledge required for applying LDA approach for analyzing high volumes of textual data are not so widespread, the good news is that there is lots of documentation and free tools available for interested practitioners with basic knowledge on descriptive statistics.

References

Al-Augby, S., Nermend, K.: Using rule text mining based algorithm to support the stock market investment decision. Transform. Bus. Econ. **14**, 448–469 (2015)

Bao, Y., Datta, A.: Simultaneuosly discovering and quantifying risk types from textual risk disclosures. Manag. Sci. **60**, 1371–1391 (2014). https://doi.org/10.1287/mnsc.2014.1930

Baskara, A.R., Sarno, R., Solichah, A.: Discovering traceability between business process and software component using Latent Dirichlet Allocation. In: International Conference on Informatics and Computing, ICIC 2016, pp. 251–256 (2017). https://doi.org/10.1109/IAC.2016.7905724

Bendle, N.T., Wang, X.: Uncovering the message from the mess of Big Data. Bus. Horizons **59**, 115–124 (2016). https://doi.org/10.1016/j.bushor.2015.10.001

Blei, D.M., Ng, A.Y., Jordan, M.I.: Latent dirichlet allocation. J. Mach. Learn. Res. **3**, 993–1022 (2003)

Brandt, T., Bendler, J., Neumann, D.: Social media analytics and value creation in urban smart tourism ecosystems. Inf. Manag. **54**(6), 703–713 (2017). https://doi.org/10.1016/j.im.2017.01.004

Carnerud, D.: Exploring research on quality and reliability through text mining methodology. Int. J. Qual. Reliab. Manag. **34**, 975–1014 (2017). https://doi.org/10.1108/IJQRM-03-2015-0033

Dean, J.W., Sharfman, M.P.: Procedural rationality in the strategic decision-making process. J. Manag. Stud. **30**, 587–610 (1993). https://doi.org/10.1111/j.1467-6486.1993.tb00317.x

Dyer, T., Lang, M.H., Stice-Lawrence, L.: The evolution of 10-K textual disclousure: evidence from Latent Direchlet Allocation. J. Account. Econ. **64**, 221–245 (2017). https://doi.org/10.1016/j.jacceco.2017.07.002

Martell, K.: Assessing student learning: are business schools making the grade? J. Educ. Bus. **82**, 189–195 (2007). https://doi.org/10.3200/JOEB.82.4.189-195

Mazanec, J.A.: Determining long-term change in tourism research language with text-mining methods. Tourism Anal. **22**, 75–83 (2017). https://doi.org/10.3727/108354217X14828625279771

Papadakis, V.M., Lioukas, S.K., Chambers, D.J.: Strategic decision-making processes: the role of management and context. Strategic Manage. J. **19**, 115–147 (1998). https://doi.org/10.1002/(SICI)1097-0266(199802)19:2115::AID-SMJ9413.0.CO;2-5

Puranam, P., Narayan, V., Kadiyali, V.: The effect of calorie posting regulation on consumer opinion: a flexible Latent Dirichlet Allocation model with informative priors. Market. Sci. **36**, 726–746 (2017). https://doi.org/10.1287/mksc.2017.1048

Ryoo, J.J., Bendle, N.: Understanding the social media strategies of U.S. primary candidates. J. Polit. Mark. **16**, 244–266 (2017). https://doi.org/10.1080/15377857.2017.1338207

Schwarz, C.: ldagibbs: a command for topic modeling in Stata using Latent Dirichlet Allocation. Stata J. **18**, 101–117 (2018) https://www.stata-journal.com/article.html?article=st0515

Vílchez-Román, C., Huamán-Delgado, F.: Estabilidad y cambio en la teoría organizacional: un estudio basado en la minería de texto y el análisis de co-citación. In: II Congreso Internacional de Ciencias de la Gestión, Lima, Peru (2017)

Voss, D.: Scientists weave new-style webs to tame the information glut. Science **289**, 2250–2251 (2001)

Using Neural Network for Identifying Clickbaits in Online News Media

Amin Omidvar[(✉)], Hui Jiang, and Aijun An

Department of Electrical Engineering and Computer Science, York University,
Toronto, Canada
{omidvar,hj,ann}@cse.yorku.ca

Abstract. Online news media sometimes use misleading headlines to lure users to open the news article. These catchy headlines that attract users but disappointed them at the end, are called clickbaits. Because of the importance of automatic clickbait detection in online medias, lots of machine learning methods were proposed and employed to find the clickbait headlines. In this research, a model using deep learning methods is proposed to find the clickbaits in Clickbait Challenge 2017's dataset. The proposed model gained the first rank in the Clickbait Challenge 2017 in terms of Mean Squared Error. Also, data analytics and visualization techniques are employed to explore and discover the provided dataset to get more insight from the data.

Keywords: Clickbait detection · Text classification · Deep learning

1 Introduction

Today's, headers of news articles are often written in a way to attract attentions from readers. Most of the time, they look far more interesting than the real article in order to entice clicks from the readers or motivate them to subscribe. Online news media publishers rely seriously on the incomes generated from the clicks made by their users, that's why they often come up with likable headlines to lure the readers to click on the headers. Another reason is that there exists numerous online news media on the web, so they need to compete with each other to gain more clicks from readers or subscription. That's why most of the online news media have started following this practice.

These misleading titles, that exaggerate the content of the news articles to create misleading expectations for users, are called clickbaits [1]. While these clickbaits may motivate the users to open the news articles, most of the time they do not satisfy the expectations of the readers and leave them completely disappointed. Since in the clickbaits, the actual article is of low quality and significantly under-delivers the content promised in the headline, it leads to a frustrating user experience. Moreover, clickbaits damage the publishers' reputation, as it violates the general codes of ethics of journalism.

In machine learning and related fields, there have been extensive studies on identifying bad quality content on the web, such as spam and fake web pages. However, clickbaits are not necessary spam or fake pages. They can be genuine pages delivering low-quality content with exaggerating titles.

© Springer Nature Switzerland AG 2019
J. A. Lossio-Ventura et al. (Eds.): SIMBig 2018, CCIS 898, pp. 220–232, 2019.
https://doi.org/10.1007/978-3-030-11680-4_22

Recently, lots of research used state of the art machine learning methods to detect clickbaits automatically. Also, some data science competitions for clickbait detection were announced, such as "Clickbait Challenge 2017", to attract scientists to conduct their researches in this area [2]. In the Clickbait Challenge 2017 competition, different machine learning algorithms were proposed to find the clickbait news headlines. For this particular competition, the goal was to propose a regressor model which can provide a probability of how much clickbait a post is.

In this research, first the provided datasets are explored and analyzed in order to get more insight from data and to understand the problem better. Then, a deep learning model is proposed which gained first ranked in terms of mean squared error on "Clickbait Challenge 2017"'s dataset [3].

2 Related Works

In [4], four online sections of the Spanish newspaper El Paris were examined manually in order to find clickbait features that are important to capture readers' attention. The dataset consists of 151 news articles which were published in June 2015. Some linguistic techniques such as vocabulary and words, direct appeal to the reader, informal language, simple structures were analyzed in order to find their impacts on the attention of the readers.

Two content marketing platforms and millions of headlines were studied to find features that contribute to increasing users' engagement and change of unsubscribed readers into subscribers. This study suggested that clickbait techniques may increase the users' engagement temporarily [5].

In [6], social sharing patterns of clickbait and non-clickbait tweets to determine the organic reach of the tweets were analyzed. To reach this goal, several tweets from newspapers, which are known to publish a high ratio of clickbait and non-clickbait content, was gathered. Then, the differences between these two groups in terms of customer demographics, follower graph structure, and type of text content were examined.

Natural Language processing and machine learning techniques were conducted in order to find clickbait headlines. Logistic regression was employed to create supervised clickbait detection system over 10000 headlines [7]. They tried to detect clickbait in Twitter using common words occurring in the Tweets through mining of some other tweets' specific features.

In [1], a novel clickbait detection model was proposed using word embeddings and Recurrent Neural Network (RNN). Even though they just considered the headings, their results were satisfactory. Their results gained F1 score of 98% in classifying online content as clickbaits or not. Furthermore, a browser add-on was developed to inform the readers of diverse media sites regarding the likelihood of being baited via such headlines.

Interesting differences between clickbait and non-clickbait categories which include -but not limited to- sentence structure, word patterns etc. are highlighted in [8]. They depend on an amusing set of 14 hand-crafted features to distinguish clickbait headlines.

Linguistically-infused neural network model was used in [9] to effectively classify twitter posts into trusted versus clickbait categories. They used word embeddings and a set of linguistic features in their model. The separation between the trusted and clickbait classes is done by contrasting several trusted accounts to various prejudiced, ironic, or propaganda accounts. At the end, their approach could classify the writing styles of the two different kinds of account.

An interesting model was proposed by Zhou for Clickbait Challenge 2017 [10]. He employed automatic approach to find clickbait in the tweet stream. Self-attentive neural network was employed for the first time in this article to examine each tweet's probability of click baiting.

Another successful method [11], which was proposed in Clickbait Challenge 2017, used ensemble of Linear SVM models. They showed that how the clickbait can be detected using a small ensemble of linear models. Since the competitors were allowed to use external data sources, they were used in their research in order to find the pattern of non-clickbait headlines and expand the size of their training set.

In [12], they developed linguistically-infused network model for the Clickbait Challenge 2017 that is able to learn strength of clickbait content from not only the text of the tweets but also the passage of the articles and the linked images. They believed using the passage of the articles and the linked images can lead to a substantial boost in the model's performance. They trained two neural network architectures which are Long Short-Term Memory (LSTM) [13] and Convolutional Neural Network (CNN). Their text sequence sub-network was constructed using embedding layer and two 1-dimensional convolution layers followed by a max-pooling layer. They initialize their embedding layer with pre-trained Glove embeddings [14] using 200-dimensional embeddings.

In [15], another model was proposed using neural networks for the Clickbait Challenge 2017. In the text processing phase, they used whitespace tokenizer with lower casing and without using any domain specific processing such as Unicode normalization or any lexical text normalization. Then all the tokens were converted to the word embeddings which were then fed into LSTM units. The embedding vectors were initialized randomly. They employed batch normalization to normalize inputs to reduce internal covariate shift. Also, the risk of over-fitting was reduced through using dropout between individual neural network layers. At the end, individual networks are fused by concatenating the dense output layers of the individual networks which then were fed into a fully connected neural network.

A machine learning based clickbait detection system was designed in [16]. They extracted six novel features for clickbait detection and they showed in their results that these novel features are the most effective ones for detecting clickbait news headlines. Totally, they extracted 331 features but to prevent overfitting, they just kept 180 features among them. They used all the fields in the dataset such as titles, passages, key words in their model for extracting these features.

In [17], they introduced a novel model using doc2vec [18], recurrent neural networks, attention layers, and image embeddings. Their model utilized a combination of distributed word embeddings and character embeddings using Convolutional Neural Networks. Bi-directional LSTM was employed with an attention layer. Another CNN was employed for learning the embeddings for the images.

3 Data Analytics

The clickbait challenge's dataset includes posts from Twitter. Online news media usually use Twitter to publish their links to attract users to their news website. Each post, which is called a tweet, is a short message up to 140 characters that can be accompanied with an image and a hyperlink. Each post is stored in the dataset using JSON object which its structure is described in the Table 1.

Human evaluators were employed to assign a clickbait score to each tweet. They had four following options for each tweet:

- Score 0: not click baiting (option 1)
- Score 0.33: slightly click baiting (option 2)
- Score 0.66: considerably click baiting (option 3)
- Score 1: clickbait (option 4)

Table 1. Structure of the JSON object

Name of object	Description
ID	The unique ID of the JSON object
postTimestamp	The publish date and time
postText	The text of the tweet
postMedia	The picture that was published with the tweet
targetTitle	The title of the linked article
targetDescription	Description of the article
targetKeywords	The keywords of the actual article
targetParagraphs	The content of the actual article
targetCaptions	All the captions that exist in the article
truthJudgments	Contains 5 scores which were given by human evaluators
truthMean	Mean of human evaluators' scores
truthMedian	Median of human evaluators' scores
truthClass	A binary field that indicates the post is clickbait or not

Each tweet was evaluated by 5 evaluators and all the given scores are saved. They provided three datasets for the contesters which one of them does not have labels. Also, they had test dataset for final evaluation of the models which has not been released yet. The information regarding the size of the provided datasets for the participants are shown in the Table 2. As we can see in the Table 2, both datasets 1 and 2 are imbalanced since the number of non-clickbait tweets in datasets 1 and 2 are 2.1 and 3.1 times bigger than the number of clickbait tweets respectively.

Table 2. Statistical information of the datasets

Datasets	Number of tweets	Clickbaits	Not clickbaits
Dataset 1	2495	762	1697
Dataset 2	19538	4761	14777
Dataset 3	80012	?	?

So, the target variable that competitors should predict is the mean clickbait score of each post. They did not mention how the binary labels are assigned. It is not based on conventional 0.5 threshold on the mean score since the minimum mean score for the clickbait label is 0.39, and the maximum mean score for non-clickbait label is 0.59. However, the median judgment score is completely in line with the clickbait and non-clickbait labels which is shown in the Fig. 1.

As we can see in the Fig. 1, all the tweets that their median judgment score is 1 or 0.66667 are in the clickbait category. In contrast, those their median judgement score is 0 or 0.33333 are in the non-clickbait category. So, we can conclude that if the sum of selected "slightly click baiting" and "not click baiting" options is bigger than the sum of two other options, the tweet will be labeled as non-clickbait. Otherwise, it would be considered as a clickbait. So, for determining the label of the tweets, there is no difference between option 1 and option 2 (i.e. "not click baiting" and "slightly click baiting"). Also, there is no difference between option 3 and option 4 (i.e. "considerably click baiting" and "click bait") as well.

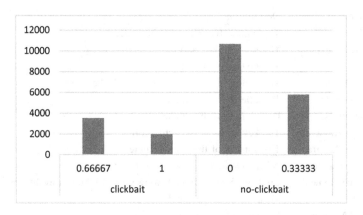

Fig. 1. Total count of tweets based on the median of the tweets' scores for each binary label

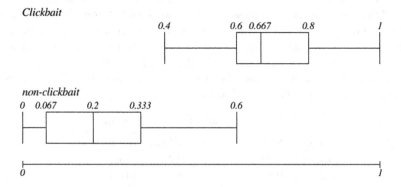

Fig. 2. Boxplot of the tweets' mean judgement score in each binary class

Also, the min, quartile1, median, quartile3, and max of scores for bot clickbait and non-clickbait classes are depicted in the Fig. 2. As we can see in the Fig. 2, the maximum value for the mean judgement score of the tweets in non-clickbait category is equal to 0.6. Also, there are some tweets in clickbait category which their mean judgement score is below 0.5.

Figure 3 shows the distribution of mean judgement score for the tweets in both clickbait and non-clickbait categories. It can be seen how clickbait and non-clickbait tweets have overlap between 0.4 and 0.6 values in terms of mean judgement score.

Figure 4 shows the distribution of tweets based on their post length with respect to the number of tweets in each class. The vertical axis shows the percentage of tweets in each class while the horizontal one represents the post length. As we can see in the Fig. 4, the percentage of short tweets in the clickbait class is higher than the one for the non-clickbait class.

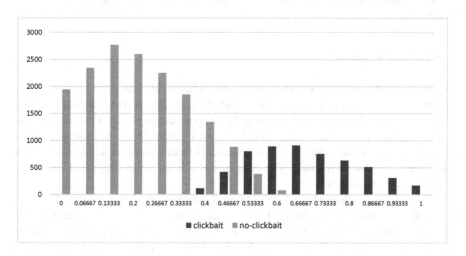

Fig. 3. Distribution of clickbait and non-clickbait tweets based on mean judgement score

One of the issue regarding the data set is that while the evaluators are provided with both the post text and a link to the target article, they were not obliged to read the actual article. So, we can consider the judgements are just based on the post texts.

Also, the scores for the tweets are dependent to the evaluator's background, knowledge, and topics of interest. So, there should be some noises in the dataset because of the existing differences between evaluators. It was found there are 408 post texts that existed in more than one samples. For example, there are nine samples with the post text "10 things you need to know before the opening bell" which 8 of them were labeled as clickbait and one of them was labeled as non-clickbait. Or there are 14 samples with the post text "Quote of the day:" which two of them are labeled as clickbait.

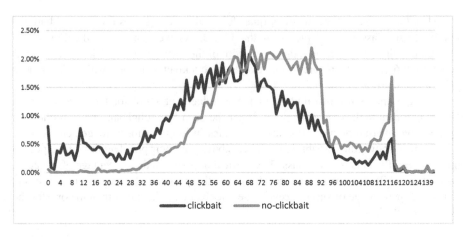

Fig. 4. The distribution of tweets based on their post length with respect to the number of tweets in each class

4 Proposed Model

To find the best model for clickbait detection, different kind of deep learning architectures were implemented and trained, and their results on our test dataset were compared with each other in order to find the best one among them. For example, one of the models was similar to the model that was proposed by Taghipour [19] for automatic essay scoring which used Convolutional Neural Network along with LSTM to find the scores for each article. The other model was similar to [20] which employed CNNs for text classification task.

The model that achieved the lowest Mean Squared Error is shown in the Fig. 5. In this model, we used bi-directional GRU for clickbait detection. Since the test data which was used to compare the contestants' models has not been published yet, we created our test data set in this research using 30% of the Dataset 2 using stratified sampling technique. The models were trained on the training dataset and then they were evaluated using our test data set.

As we can see in Fig. 5, the first layer is an embedding layer which transforms one-hot representation of the input words to their dense representation. We initialized embedding vectors using 50, 100, 200, 300 dimensions using GloVe word embeddings [14].

The next layer is a combination of forward GRU and backward GRU. We evaluated bidirectional simple recurrent units [21], bi-directional GRU, and bi-directional LSTM in order to find the best architecture for the clickbait detection task. The result showed that bi-directional GRU outperformed the two other structures.

The GRU employs a gating approach to trail the input sequences without utilizing separate memory cells. In GRU, there exists two gates which are called update gate z_t

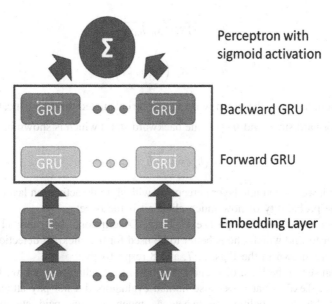

Fig. 5. The proposed model for clickbait detection

and reset gate r_t. These two gates are used together in order to handle how to update information for each state. The reset gate and update gate are calculated for each state based on the formula (1) and formula (2) respectively.

$$r_t = \sigma(W_r x_t + U_r h_{t-1} + b_r) \tag{1}$$

$$z_t = \sigma(W_z x_t + U_z h_{t-1} + b_z) \tag{2}$$

W_r, U_r, b_r, W_z, U_z, b_z are the parameters of GRU that should be trained during the training phase. The candidate state will be calculated at time t using the formula 3.

$$\tilde{h_t} = tanh(W_h x_t + r_t \odot (U_h h_{t-1}) + b_h) \tag{3}$$

$\odot$ denotes an elementwise multiplication between the reset gate and the past state. So, it determines which part of the previous state should be forgotten. Finally, formula (4) is responsible to calculate the new state.

$$h_t = (1 - z_t) \odot h_{t-1} + z_t \odot \tilde{h_t} \tag{4}$$

Update gate in formula (5) (i.e. z_t) determines which information from past should be kept and which new calculated information should be added. The forward way reads the post text from x_1 to x_N and the backward way reads the post text from x_N to x_1. This process is shown through following formulas where $\overrightarrow{h_n}$ and $\overleftarrow{h_n} \in R^{d1}$ and d_1 is a hyperparameter.

$$\overrightarrow{h_n} = \overrightarrow{GRU}\left(x_n, \overrightarrow{h_{n-1}}\right) \tag{5}$$

$$\overleftarrow{h_n} = \overleftarrow{GRU}\left(x_n, \overleftarrow{h_{n+1}}\right) \tag{6}$$

So, the annotation of the given word x_n can be calculated through concatenation of $\overrightarrow{h_n}$ (i.e. the forward state) and $\overleftarrow{h_n}$ (i.e. the backward state) which is shown in formula 7.

$$h_n = [\overrightarrow{h_n}, \overleftarrow{h_n}] \tag{7}$$

At the end, we used single layer perceptron with sigmoid activation layer in order to figure out the probability of how much clickbait is the tweet.

We trained our model on "postText", "targetDescription", and "targetTitle" separately in order to find which one is better to be used for the clickbait detection task and their results are shown in the Figs. 6, 7, and 8 respectively.

As we can see in the Figs. 6, 7, and 8, the best result achieved when we trained our model on "postText". That is because human evaluators did not pay attention to the other data fields for labeling the tweets as much as they paid attention to the "postText".

For the training part, mini batched gradient descent with the size of 64 was selected. Mean squared error was selected for the loss function. Drop out technique was employed for Embedding, forward GRU, backward GRU layers. The model was run with different sets of hyperparameters (i.e. hidden layer sizes, depth, learning rate, embedding vector size, and drop out) in order to find the best tuning for the model. Also, the embedding layer was initialized using the GloVe embedding vectors with different dimensions.

Performance of the models over different number of dimensions was tested, and the result shows 100-dimensional GloVe embeddings have lower mean squared error in comparison with other embedding vectors. After tuning the hyper parameters using our own test dataset, we found out the best value for the dropout of the embedding layer is 0.2, and for the input and output of the bi-directional GRU is 0.2, and 0.5 respectively. For the optimizer, we used RMSprob and the size of both forward GRU and backward GRU is 128 which makes the final representation of the tweets a vector with the length of 256.

Then we used all the available labeled datasets (i.e. test, validation, and training datasets) to train our model. Then the model was run on the Clickbait Challenge's test dataset using TIRA environment [22]. The proposed model gained the first rank among other models in terms of Mean Squared Error. Mean Squared Error was used as a main measure to rank the proposed models. Also, the proposed model gained the lowest runtime among all others as well. The result of the proposed model on Clickbait Challenge's test dataset is shown in Table 3. Moreover, the result of the proposed model is compared with the other models in terms of Mean Squared Error in Fig. 9. The result of the proposed model along with the full list of other models are accessible through the competition's result webpage [23] which the name of the proposed model is **Albacore**.

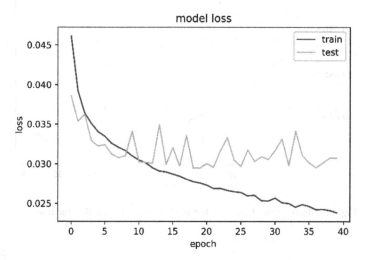

Fig. 6. Mean squared error of the model using "postText" for training.

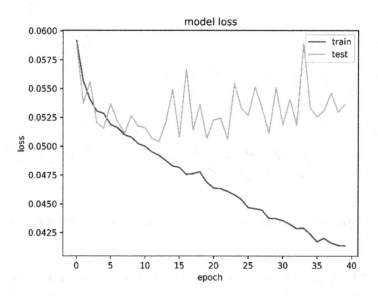

Fig. 7. Mean squared error of the model using "targetDescription" for training.

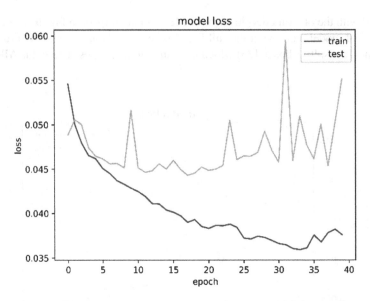

Fig. 8. Mean squared error of the model using "targettitle" for training.

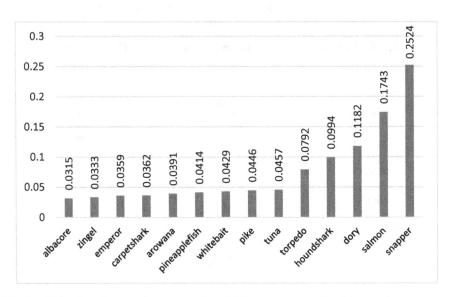

Fig. 9. Comparison between the proposed model (i.e. albacore) with other models in terms of MSE

Table 3. Result of the proposed model on clickbait challenge's test dataset

MSE	Median absolute error	F1 score	Precision	Recall	Accuracy	R2 score	Runtime
0.0315	0.122	0.67	0.732	0.619	0.855	0.571	00:01:10

5 Conclusion

In this research, the state of the art machine learning algorithms which were proposed for clickbait detection, are introduced. Then a recurrent neural network model was proposed which beat the first ranked model in the clickbait challenge 2017 in terms of the mean squared error measurement. We used mean squared error for model comparison since Clickbait challenge 2017 used this measurement to rank the models.

The proposed model does not rely on any feature engineering tasks which means they are able to learn the representation automatically in order to classify tweets into clickbait and non-clickbait categories. There exist some very complex models in clickbait challenge 2017 that they did not achieve good result. They tried to utilize all the provided information in the dataset such as images, external linked articles, keywords, etc. to decide whether the headlines are clickbaits or not. In contrast, the proposed model only uses "postText" field. Also, the proposed model does not calculate the distribution of the annotations' probability. Instead of it, just the probability of the clickbait will be calculated which made the proposed model much simpler by converting multi classification task to the binary classification.

References

1. Anand, A., Chakraborty, T., Park, N.: We used neural networks to detect clickbaits: you won't believe what happened next! In: Jose, J., et al. (eds.) ECIR 2017. LNCS, vol. 10193, pp. 541–547. Springer, Cham (2017). https://doi.org/10.1007/978-3-319-56608-5_46
2. Potthast, M., Gollub, T., Hagen, M., Stein, B.: The clickbait challenge 2017: towards a regression model for clickbait strength. In: Proceedings of the Clickbait Challenge (2017)
3. Potthast, M., et al.: Crowdsourcing a large corpus of clickbait on Twitter (2017, to appear)
4. Palau-Sampio, D.: Reference press metamorphosis in the digital context: clickbait and tabloid strategies in Elpais. com. vol. 29 (2016)
5. Rony, M., Hassan, N., Yousuf, M.: Diving deep into clickbaits: who use them to what extents in which topics with what effects? In: Proceedings of the 2017 IEEE/ACM International Conference on Advances in Social Networks Analysis and Mining, pp. 232–239. ACM (2017)
6. Chakraborty, A., Sarkar, R., Mrigen, A., Ganguly, N.: Tabloids in the era of social media? Understanding the production and consumption of clickbaits in Twitter (2017)
7. Potthast, M., Köpsel, S., Stein, B., Hagen, M.: Clickbait detection. In: Ferro, N., et al. (eds.) ECIR 2016. LNCS, vol. 9626, pp. 810–817. Springer, Cham (2016). https://doi.org/10.1007/978-3-319-30671-1_72
8. Chakraborty, A., Paranjape, B., Kakarla, S., Ganguly, N.: Stop clickbait: detecting and preventing clickbaits in online news media. In: IEEE/ACM International Conference Advances in Social Networks Analysis and Mining (ASONAM), pp. 9–16. IEEE (2016)
9. Volkova, S., Shaffer, K., Jang, J., Hodas, N.: Separating facts from fiction: linguistic models to classify suspicious and trusted news posts on twitter. In: Proceedings of the 55th Annual Meeting of the Association for Computational Linguistics (Volume 2: Short Papers), vol. 2, pp. 647–653 (2017)
10. Zhou, Y.: Clickbait detection in tweets using self-attentive network. arXiv preprint arXiv: 1710.05364 (2017)

11. Grigorev, A.: Identifying clickbait posts on social media with an ensemble of linear models. arXiv preprint arXiv:1710.00399 (2017)
12. Glenski, M., Ayton, E., Arendt, D., Volkova, S.: Fishing for clickbaits in social images and texts with linguistically-infused neural network models. arXiv preprint arXiv:1710.06390 (2017)
13. Hochreiter, S., Schmidhuber, J.: Long short-term memory. Neural Comput. 9(8), 1735–1780 (1997)
14. Pennington, J., Socher, R., Manning, C.: Glove: global vectors for word representation. In: Proceedings of the 2014 Conference on Empirical Methods in Natural Language Processing (EMNLP), pp. 1532–1543 (2014)
15. Thomas, P.: Clickbait identification using neural networks. arXiv preprint arXiv:1710.08721 (2017)
16. Cao, X., Le, T., et al.: Machine learning based detection of clickbait posts in social media. arXiv preprint arXiv:1710.01977 (2017)
17. Gairola, S., Lal, Y., Kumar, V., Khattar, D.: A neural clickbait detection engine. arXiv preprint arXiv:1710.01507 (2017)
18. Le, Q., Mikolov, T.: Distributed representations of sentences and documents. In: International Conference on Machine Learning, pp. 1188–1196 (2014)
19. Taghipour, K., Ng, H.: A neural approach to automated essay scoring. In: Proceedings of the 2016 Conference on Empirical Methods in Natural Language Processing, pp. 1882–1891 (2016)
20. Zhang, Y., Wallace, B.: A sensitivity analysis of (and practitioners' guide to) convolutional neural networks for sentence classification. arXiv preprint arXiv:1510.03820 (2015)
21. Elman, J.: Finding structure in time. Cogn. Sci. 14(2), 179–211 (1990)
22. Potthast, M., Gollub, T., Rangel, F., Rosso, P., Stamatatos, E., Stein, B.: Improving the reproducibility of PAN's shared tasks: In: Kanoulas, E., et al. (eds.) CLEF 2014. LNCS, vol. 8685, pp. 268–299. Springer, Cham (2014). https://doi.org/10.1007/978-3-319-11382-1_22
23. Result of Clickbait Challenge 2017. http://www.tira.io/task/clickbait-detection/dataset/clickbait17-test-170720/

Spanish Named Entity Recognition in the Biomedical Domain

Viviana Cotik[1]([✉]), Horacio Rodríguez[2], and Jorge Vivaldi[3]

[1] Department of Computer Science, FCEyN, Universidad de Buenos Aires, Buenos Aires, Argentina
vcotik@dc.uba.ar
[2] Polytechnical University of Catalonia, Barcelona, Spain
horacio@lsi.upc.edu
[3] Universitat Pompeu Fabra, Barcelona, Spain
jorge.vivaldi@upf.edu

Abstract. Named Entity Recognition in the clinical domain and in languages different from English has the difficulty of the absence of complete dictionaries, the informality of texts, the polysemy of terms, the lack of accordance in the boundaries of an entity, the scarcity of corpora and of other resources available. We present a Named Entity Recognition method for poorly resourced languages. The method was tested with Spanish radiology reports and compared with a conditional random fields system.

Keywords: Named entity recognition · Spanish · Radiology reports · BioNLP

1 Introduction

Named entity recognition (NER) is an information extraction task, whose goal is to identify instances of specific kind of information units in text and assign them a class. It was originally applied to carefully-written text, such as newswire. Afterwards, it began being applied to other domains, such as the biomedical, for identifying genes, proteins, drug names and diseases, among others.

The approaches to solve the NER problem include: dictionary-based, rule-based, statistical-based, machine learning (ML) and combined approaches [10,27].

The biomedical domain is specially challenging due to (1) its highly specialized terminology including a lot of often polysemous abbreviations and acronyms, (2) the use of non-standardized naming conventions and the lack of standards, even among specialists, regarding to which is the boundary of an entity, and (3) the variety of genres and author profiles, owning specific jargon and sublanguages. In addition, following situations, that highlight the challenges of NER task in the biomedical domain are described in [10,18]:

- the absence of complete dictionaries for some biological or medical named entities (NEs) and the fact that new entities are added frequently,

© Springer Nature Switzerland AG 2019
J. A. Lossio-Ventura et al. (Eds.): SIMBig 2018, CCIS 898, pp. 233–248, 2019.
https://doi.org/10.1007/978-3-030-11680-4_23

- abbreviations and other medical terms are often polysemous,
- there might be different ways of referring to the same entity, and
- frequently, medical terms are multi-word units, so there is a need for determining name boundaries and resolving overlap of candidate names. As mentioned in [18], it is easier for a system and for a human to determine if an entity is present or not in a text than to determine its boundaries.

Additionally, there is no standard criteria in the evaluation of biomedical NER systems. Not only the boundary of named entities, but also their class is often ambiguous, due to criteria differences among specialists. Therefore, different matching criteria have been used for Bio-NER[1] system evaluations. Furthermore, datasets are usually not published due to confidentiality issues. Accordingly, usually gold standards have to be generated. The lack of standard metrics, of publicly available datasets, and of standard annotation criteria makes the comparison of different implementations difficult. The processing of medical reports in languages other than English, such as Spanish adds a further difficulty, since there are less resources available.

In this paper we describe different approaches we have followed in order to detect anatomical entities (AEs) and clinical findings (FIs) in a set of Spanish radiology reports. The recognition of these entities is useful because: (a) it enables the possibility to structure and normalize the information, (b) it offers the opportunity to detect relations among findings and anatomical sites where they occurred, (c) if negation is taken into account, identifying which reports contain clinical findings could allow the indexing of only relevant documents and discard those which are not relevant (do not contain clinical findings). This is as a classification task and can serve for the purposes of identifying later on, which are the specific occurrence of clinical findings in the relevant reports, and (d) it could serve to notify physicians about the findings, some of which could require immediate action (alert generation). The obtention of timely information is critical in case of urgent or important findings [6]. Its automatic detection and communication is being studied [12,17].

Most of the work in biomedical NER has focused in the recognition of gene and protein names in formal texts and for English.

To detect entities, we propose and evaluate two different approaches: (1) SiM-REDA, a Simple Entity Detection Algorithm for Medium Resources languages, that is based on a lookup of terms from a specialized vocabulary, on morphological knowledge and on knowledge of PoS tag patterns of AEs and FIs, and that was conceived by us as a method for BioNER in poor and medium resource languages, and (2) a ML approach, based on conditional random fields (CRF). The rest of the paper is organized as follows. Section 2 presents previous work in the NER domain. Section 3 presents the data used and the methods developed. Section 4 shows the results obtained, which are discussed after, in Sect. 5. Finally, Sect. 6 presents conclusions and future work.

---

[1] Bio-NER refers to biomedical named entity recognition systems.

2   Previous Work

A number of surveys have been carried out on the NER task. Various address the biological domain [3,8,28,30].

Spanish has been introduced in CoNLL-2002 and MET-1 events. Usually efforts in NER are dedicated to a specific genre and domain. To port a system to a new domain or textual genre constitutes a major challenge [22]. An overview of works dedicated to different genres and domains and also reference to previous studies can be seen in [20]. The initial approaches for NER were dictionary and rule-based. The first ones look for the appearance of terms belonging to terminologies in the texts (with exact or exact string matching). Rule-based techniques use domain knowledge or information obtained through analysis of a subset of the data. They usually have good results [31], but its construction is time consuming and often not reusable in other datasets.

Statistical methods are also used for NER. They are sometimes combined with dictionary or rule-based techniques [4]. ML methods can be supervised, for which a considerable amount of training data is needed, semi-supervised, as bootstrapping, or unsupervised. Among the supervised methods, there are classification-based and sequence-based approaches. Examples of the first are Naive Bayes (NB) and Support Vector Machines (SVM) [29]. Sequence-based approaches consider sequences of words instead of individual words or phrases considered in the classification-based approaches. Some examples include Hidden Markov Models (HMM) [26] and Conditional Random fields (CRF).[2] CRFs were the best performing systems in various challenges and have been highly ranked in others [27]. Some implementations can be seen in [5,25]. Different features used for these methods are described in [28]. See [27] for more HMM and CRF approaches descriptions. Nowadays, models developed using deep neural networks architectures provide very competitive results. As a drawback, a big amount of training data has to be available. Many semi-supervised methods for NER in the general domain are reviewed in [20]. Unsupervised learning methods are typically based on clustering. Methods are usually based on lexical resources and on large corpus of statistics taken from unannotated texts. See [20] for a review.

The impact of feature engineering in order to improve the performance of different models, such as CRF, SVM or neural networks in the clinical NER task for Spanish, English and Swedish is reported in [33].

A NER system for Spanish electronic health reports with the goal to access their factuality with a NegEx[3] implementation has been presented in [24]. Different techniques are evaluated. Their best result consists in a CRF implementation, tested with 75 electronic health reports annotated with an IAA of 90.53%. As features they use four characters prefixes and suffixes and transform terms to lower case. They consider entities that overlap as partial match. Freeling-Med [21] is used in another study as a way to automatically tag named entities. They test it

---

[2] CRF, are defined in Sect. 3.3.

[3] Negex is the most popular system for detecting negations and their scope.

with 20 clinical reports looking for diseases, drugs and substances. They achieve high F1s, but use following extremely loose matching criteria: "two elements are considered to be equivalent if an element given by the system is entirely contained within an extension of a manually tagged element by six positions both to the left and to the right". A tool similar to UMLS MetaMap Transfer (MMTx)[4] has been presented in [7] for the identification of Spanish SNOMED CT[5] terms corresponding to the *procedures* and *disruptions* hierarchies in Spanish clinical notes. The tool is tested with 100 clinical notes. An inverted index is used and a score is assigned to the retrieved terms, depending on the length of the query with respect to the retrieved terms. It is integrated with MOSTAS [13], a tool that normalizes abbreviations and acronyms, anonymizes reports and corrects spelling errors. Table 1 shows the results for Spanish NER in the medical domain.

Table 1. NER results for Spanish in the medical domain. References, type of documents: ClR: clinical reports, CN: clinical notes, EHR: electronic health reports. Entity types: DS: diseases, DRP: disruptions, DR: drugs, PR: procedures, SB: substances, and SN: SNOMED CT. Other references: doc.: documents, ent.: entities. First results correspond to exact matches. Results marked with (*) correspond to lenient matches and (**) to extremely loose lenient matches.

Paper	# reports	IAA	P	R	F1	doc. types	ent. types
[7]	100	66%	0.43 (0.72*)	0.06 (0.09*)	0.11 (0.16*)	CN	DRP (SN)
			0.35 (0.70*)	0.07 (0.55*)	0.06 (0.10*)		PR (SN)
[24]	75	90.53%	0.36 (0.70*)	0.45 (0.83*)	0.40 (0.76*)	EHR	DS
[21]	20	-	(0.97**)	(0.80**)	(0.88**)	ClR	DS
			(1.00**)	(0.96**)	(0.98**)		DR
			(0.84**)	(0.92**)	(0.88**)		SB

3   Material and Methods

In this section we will explain the data used for training (when it applied) and for testing purposes, the preprocessing applied to reports, and the lexicons used. SiMREDA algorithm and the CRF algorithm and its feature selection are presented next. As previously mentioned, SiMREDA was thought as a solution for cases were there are no low or medium resources available (lexicons, corpora, software tools). CRF was thought as a relatively easy to implement solution for NER when annotated datasets are available. Then, we explain the exact match and a lenient matching evaluation metric used and how they work.

---

[4] https://mmtx.nlm.nih.gov/MMTx/.
[5] https://www.snomed.org/snomed-ct.

3.1 Data

We worked with 513 radiology reports of an Argentinian hospital that were anonymized and annotated by us. Reports are short, approximately 6% of AEs and FIs are written in an abbreviated way, it contains some non-sentences and lack of punctuation signs. Furthermore, many texts lack diacritics. For the sake of uniformity we decided to remove all of them. We also normalized our reports transforming every word to lowercase. Table 2 describes the composition of the dataset and Table 3 shows the number of AEs and FIs and the number of abbreviations and acronyms found in the annotated dataset. For details about the process followed, schema and elaborated guidelines to annotate the dataset refer to [11].

Table 2. Composition of the dataset.

Concept	Number
Number of radiology reports	513
Total amount of words	36,211
Total amount of sentences	4,175
Avg. sentences per report	8
Avg. words per sentence	9

Table 3. Type and amount of entities, modifiers and other characteristics in the annotated reports.

Type	Total	Different
Anatomical entities	4,398	405
Finding	2,637	745
Abbreviations	880	105

Both NER algorithms are going to be evaluated with the same dataset. We split the annotated dataset into the *development dataset* (80%) and the *testing dataset* (20%). For CRF we tested different features with 5-fold cross-validation in the development dataset. Both algorithms were tested with the testing dataset.

3.2 SiMREDA Algorithm

We proposed and implemented SiMREDA, a Simple Entity Detection Algorithm for Medium Resources languages. The algorithm takes as input radiology reports and gives as output the same reports with the anatomical entities and clinical findings automatically tagged. It has three modules and some variants, as shown in Fig. 1. The first module is its basic module. Modules 2 and 3 are additions. Next, we describe the three modules.

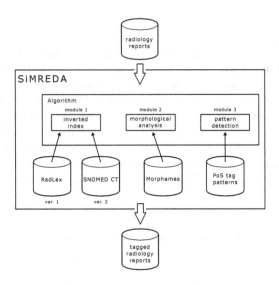

Fig. 1. Schema of SiMREDA algorithm. Its modules and variants.

Module 1: Inverted Index. The module consists in a lookup of terms that come from a specialized vocabulary through the use of an inverted index. As specialized vocabulary we try two alternatives: RadLex[6] a vocabulary specific of the radiology domain, but that had to be translated into Spanish (variant 1) and SNOMED CT, that is not specific of the radiology domain, but exists in Spanish (variant 2). In variant 1, we translate to Spanish all RadLex AEs and FIs. Therefore, we use Google Translate (GT), enhanced through mappings with UMLS and Wikipedia. Also a subset of the translated terms were corrected by a physician of the radiology domain. Variant 2 takes SNOMED CORE Problem list subset as FIs and a subset of its *Body Structure* and *Substance categories* as AEs. No translation is needed. In variant 1 each word appearing in the translated terms is added to an inverted index. Stopwords are excluded. Each entry of the inverted index points to the RadLex terms where it appears and gets the most frequent class assigned (anatomical entity or clinical finding). The process with variant 2 is similar using SNOMED CT instead of RadLex. See Table 4 for an example of an inverted index of RadLex terms translated into Spanish.

Those words that appear in the reports and that also belong to the inverted index are tagged as anatomical entities or as findings, according to the class assigned in the inverted index. Adjacent sequence of words belonging to the same class are tagged together with their corresponding class. For example, lets assume we have following text: *se visualiza prolapso de la válvula mitral* (*a mitral valve prolapse has been noticed*). After running the algorithm that tags terms according to their presence in RadLex we would get: "se visualiza <FI>prolapso</FI>de la<AE>válvula</AE><AE>mitral </AE>" if we

---

[6] https://www.rsna.org/RadLex.aspx.

Table 4. Example of inverted index for RadLex terms *heart, heart valve, ischemic heart disease, basal zone of the heart, aortic valve, mitral valve, mitral valve insufficiency, insufficiency fracture* and *heart failure* translated to Spanish. The first column has the indexed words. The second column has the RadLex terms, where the words occur, and the third column has the class assigned to the word, that depends on the class of the RadLex terms, where the word appears. The table should also have entries for the words *ischemic, disease, basal, zone, aortic, mitral, fracture* and *failure* (we do not add them because of space constraints).

Word	RadLex terms	Class assigned
corazón (heart)	"corazon" (AE), "válvula del corazon" -heart valve- (AE), "enfermedad isquémica del corazon" -ischemic heart disease-(FI), "zona basal del corazón" -basal zone of the heart- (AE),...	AE
válvula (valve)	"válvula del corazon" -heart valve- (AE), "válvula aórtica" -aortic valve- (AE), "válvula mitral" -mitral valve- (AE), "insuficiencia de la valvula mitral" -mitral valve insufficiency- (FI),...	AE
insuficiencia (insufficiency)	"insuficiencia de fractura" (FI) -insufficiency fracture-, "insuficiencia de la valvula mitral" -mitral valve insufficiency- (FI), "insuficiencia cardiaca" -heart failure- (FI)...	FI

assume that *prolapso* appears in RadLex more times in terms referring to FIs than in terms referring to AEs and if we consider the class assigned to *válvula* in Table 4. Then, if there are contiguous words of the same class (in this case we have *válvula* and *mitral*, both tagged as anatomical entities) we tag them together with their corresponding class. In this case we would get: "Se visualiza <FI>prolapso</FI> de la <AE>válvula mitral</AE>". As a result, as the algorithm output, we have a set of radiology reports with terms referring to AEs and to FIs automatically tagged according to the translation to Spanish of RadLex anatomical and clinical finding terms.

Module 2: Morphological Analysis. Graeco-Latin morphemes are used in medical terms of many languages, including Spanish. Even a small number of morphemes of Greek and Latin origin can generate a large amount of terms [15,32]. Therefore, their lookup can help discover clinical findings that do not appear in the lexicons, that are not correctly translated to Spanish or that are not well written in reports. Thus, the second module considers the appearance of those morphemes.

We implemented a simple module to detect Graeco-Latin morphemes. Therefore, we compiled a dictionary of morphemes, that includes their type -prefix or suffix- and meaning. The dictionary was built based on a reduced subset of [2]. Those words, that include morphemes corresponding to findings, in the correct position (as suffix or as prefix) are tagged as FIs replacing the tag assigned based on RadLex terms (Module 1). For example, *ascitis -ascites-* is not tagged as a finding based on RadLex, but our morpheme detection module detects the suffix *-itis*, so it assumes that *ascitis* is a FI and tags it as such.

The detection of morphemes related to the medical domain might also help us improve the dictionary-based approach by detecting terms that are misspelled. For example, *epatitis* for *hepatitis*. Nevertheless, not all the words that contain the previously described morphemes are medical terms (consider, for example, *homologo* -homologous- for suffix *logo*). Furthermore, there are words that contain more than one morpheme related with the medical domain (***peritonitis** -peritonitis-*).

Module 3: Pattern Detection. Usually AEs and FIs satisfy certain PoS tagging patterns. For example, from 20% of our development dataset that we used to analyze the PoS tag sequences of the annotated anatomical entities and clinical findings, we discovered that many of the anatomical terms beginning with a noun continue with an adjective, that is also considered part of the AE (e.g. *testículo izquierdo -left testicle-* and *pared abdominal -abdominal wall-*, both nouns followed by adjectives). In many cases only the noun is tagged by our Module 1. We analyze the PoS tag patterns present in the previously mentioned subset of our development dataset and look for these patterns in the radiology reports in order to improve SiMREDA results, expanding the named entities to the adjectives (in this example). So, "*[testículo]*(AE) *izquierdo*" is expanded to "*[tesículo izquierdo]*(AE)". This constitutes module 3. Tables 5 and 6 show the most frequent PoS tagging sequences of anatomical entities and clinical findings appearing in the selected subset of the development dataset. In these tables, columns 3 and 4 show the percentage of annotated entities that have the pattern listed in column 1 and the accumulated percentage. The last column shows the probability that a sequence that has the PoS tags analyzed in the row and whose first word is tagged as an AE (Table 5) or a FI (Table 6) is an AE or a FI respectively.

Table 5. Detected anatomical entity patterns.

PoS tag sequence	Example	Perc. (%)	Acum. perc.(%)	Prob. of being AE
NC	bazo (spleen)	75.88	75.88	1.00
NC-AQ	músculo pilórico (pyloric muscle)	17.31	93.18	0.76
NC-NC	venas porta (portal veins)	1.66	94.84	0.63

3.3 Conditional Random Fields (CRF)

Conditional random fields are probabilistic models used to predict sequences of labels based on sequences of input samples. A text can be seen as a sequence of tokens. We can say that each token has an associated vector of features, such

Table 6. Detected finding patterns. *liquido* is tagged as a verb, while it should be a noun (this happens because the accent is missing, *líquido* is the correct word).

PoS tag sequence	Examples	Perc. (%)	Acum. percentage (%)	Prob. of being FI
NC	ovariocele (ovariocele)	35.65	35.65	1.00
VMP	dilatadas (dilated)	14.57	50.22	1.00
VMI-AQ	liquido libre (free fluid)	12.61	62.83	1.00
NC-AQ	hipertrofia pilrica (pyloric stenosis)	9.57	72.4	0.81
VMP-SP-NC	aumentada de tamao (increased in size)	4.57	76.97	0.92
AQ	bffida (bifid)	2.39	79.36	1.00
NC-SP-DA-NC	incremento de la vascularizacion (increase in vascularization)	1.96	81.32	0.60

as the word's part of speech tag, the word's suffix of a given length and an indication as to whether the word is capitalized or not. The input of CRF is the sequence of tokens of the text. The features of a token and the pattern of labels assigned to previous words are used to determine the most likely label for the current token. In linear chain CRF only the label of the previous token is used. As mentioned in a previous section, CRF have been successfully used for NER and also for some other natural language processing tasks, such as PoS tagging.

We tried different set of features, some provided in different NER tasks and a set of features proposed by ourselves. We used our development dataset in order to decide the best set of features. Once we decided which to use, we used the whole development dataset as training set, and we tested the results with our testing dataset. The best set of feature is one proposed in [14] for the solution of CLEF eHealth 2015 task 1b (NER for French). Nevertheless, the relative difference among its F1 and those of our proposed feature set is very low (0.57 % relative improvement). We selected this set of features, that includes: lexical (lower case), morphological (four characters prefix and four characters suffix), reduced PoS tags, orthographic features and shape-related features (length of the token, whether the token begins with a capital letter, whether all its characters are capital letters, whether it contains only digits, only letters or letters and digits). It also takes into account context for morphology features and for PoS tags.

3.4 Evaluation of Results

We will measure our algorithms with the classical exact match metrics (precision (P), recall (R) and F1) and with a lenient (or approximate) match metric, based on the MUC challenge evaluation metric and that scores partial matches (matches with wrong boundary and same entity type) as half of an exact match. Metrics are explained in detail in [9] and in the *Scoring Software User's Manual.*[7]

---

[7] The Message Understanding Conference Scoring Software User's Manual. https://www-nlpir.nist.gov/related_projects/muc/muc_sw/muc_sw_manual.html, accessed June 2017.

4   Results

In this section we present SiMREDA and CRF algorithms results, that can be seen in Table 7.

Precision, recall and F1 measure were calculated against every entity type (AE and FI) and a final overall score, that considers both entity types is also given for all the measurements. Similarly, precision, recall and F1 for partial boundary matching is calculated for every entity type (AEPM and FIPM) and a final overall score (totalPM) is calculated. In both cases we used the testing dataset composed by 20% of the annotated 513 reports (103 reports).

We are interested in a solution that retrieves a high rate of relevant entities and that the entities retrieved by the solution are actually positive (high recall and high precision). Hence, we will choose F1 metric, that balances precision and recall, as the metric in order to compare results.

Table 7. SiMREDA implementation compared to CRF implementation. Exact and partial match results are shown for each entity type and the overall measure (total) is also shown.

NE	SiMREDA compared to CRF					
	SiMREDA			CRF		
	P (%)	R (%)	F1 (%)	P (%)	R (%)	F1 (%)
AE	58.74	73.25	65.20	92.09	91.56	91.82
FI	56.21	48.68	52.17	85.78	74.95	80.00
total	58.00	64.10	60.90	89.68	84.70	87.12
AEPM	65.46	77.23	70.86	95.00	92.61	93.79
FIPM	59.86	54.49	57.05	88.46	80.06	84.05
totalPM	63.73	68.9	66.21	92.42	87.45	89.87

Table 8 presents the Graeco-Latin morphemes related to findings, that were discovered in the testing dataset.

Table 8. Morphemes related with medical terms appearing in the test set.

Morpheme	Category	Number of appearances (distinct)
-itis	Finding	5 (2)
-megalia	Finding	18 (3)
-osis	Finding	8 (4)

5 Analysis of Results

Regarding SiMREDA, results are lower than CRF. Even though, they are better than results for NER detection in Spanish clinical texts presented in Table 1. SiMREDAs results are similar to best CLEF 2015 and 2016 results (for MEDLINE articles written in French). Nevertheless, results are difficult to compare since the definition of named entities differ, in some of the cases the languages are different, and also lenient match definitions differ.

Other results, not shown in Table 7 is described next. The use of translated RadLex AEs and FIs had better results than the use of SNOMED CT clinical findings and anatomical entities. This is probably because the second vocabulary has many terms that do not belong to the radiology domain, which decreases SiMREDA's precision. Furthermore, it does not contain terms specific of the radiology domain (as RadLex does), which decreases recall.

The improvement of only 10% of RadLex translations derived in a relative F1 increase of anatomical entities and findings of ~7% and ~4% respectively. Therefore, we conclude that it makes sense to invest effort in improving the translations.

Only 31 findings with Graeco-Latin suffixes appear in our testing dataset. Therefore, the addition of Module 2 does not improve the results in a very noticeable way (overall F1 increase of less than 1%). However, the detection of morphemes related to the medical domain helped us to detect terms that are misspelled. For example, *etenosis* for *estenosis -stenosis-* were found in reports and detected as findings by Module 2.

As expected, partial match results are always higher than exact match results. For example, as reported in Table 7, the overall SiMREDAs F1 is 60.90 with exact match. Partial match achieves a relative increase in F1 of 8.72%.

Also, findings show a greater increase in partial match F1s than AEs. We believe this is motivated, because it is much more complex to determine the boundaries of a FI than those of an AE. This issue was also reported during annotation. Furthermore, in our dataset findings have longer terms (in amount of words composing them) than anatomical entities, which makes its boundary detection a harder problem.

Some errors are due to following causes:

- tokenization problems: the text *(...)ascitis-* appeared in one of the reports. The tokenizer did not separate the word *ascitis* from the symbol -, so ascitis was not recognized by our algorithm as a finding.
- annotation criteria (a decision was taken to annotate implants, such as *kidney implant* as an AE. The algorithm does not annotate implant as part of an anatomical entity. Also, for example, *ovarian cyst* should be annotated as [ovarian cyst](FI), while the algorithm detects [ovarian](AE) [cyst](FI)).

244 V. Cotik et al.

- annotation inconsistencies: there is a number of errors and inconsistencies in the annotations. Some of them, like the omission of annotation of entities (such as *bile duct* and *dilated*), the incorrect classification of entities (such as *gallbladder* as FI) erroneously worsens the results. The annotation of entities with wrong boundaries explains, in part, the difference of performance among the exact match and the partial match.

Our CRF results outperform others obtained with the same feature set for French [14] (the original proposal of the feature set) and for German [23].[8] Since all results are tested with different genre of data and in different languages it is not easy to draw a conclusion about the differences in the results. In Spanish and in French anatomical entities have a higher F1 than findings. That is what usually happens. It can be also noticed that results with our Spanish dataset are better in both entity types than in the original French implementation. This might have to do with the fact that our corpus is of a restricted domain -only radiology reports, while the French implementation has EMEA and MEDLINE articles-, that in our case we had two entity types, while the other case had to select among 10 entity types, and that we trained with 410 reports and tested with 103, while in the French case, 836 MEDLINE titles and 4 EMEA documents were used for training and 832 MEDLINE titles and 12 EMEA documents were used for testing. Besides, the definition of AE and FI among both systems does not necessarily coincide.

As can be seen in Table 7, as expected, CRF outperforms SiMREDA for exact as well as for partial match. Both methods require manually created resources: SiMREDA a lexicon and the elaboration of rules and CRF an annotated corpus. The CRF algorithm is much better, but SiMREDA is adequate when there are few resources available for annotation.

Concluding, the development of a dictionary-based algorithm enhanced with rules is more laborious than a ML approach such as CRF. In cases as ours, where there do not exist specific resources for the radiology domain in Spanish it is even more difficult. Nevertheless, this method has the advantage of needing few annotated data. Based on the good perspective of CRFs results, feature engineering can be carried out in order to improve results.

6 Conclusions and Future Work

In this paper we presented SiMREDA, a dictionary-based entity recognition algorithm, enhanced with morphology analysis and with a post-processing based on the analysis of PoS patterns of the entities of interest, and an algorithm based on CRF. SiMREDA approach can be used when there are no datasets annotated for implementing ML techniques and when there are no lexicons in the language

---

[8] We consider that AE and FIs in the French dataset are *anatomy* and *disorders* hierarchies of UMLS. In the case of for German, what we consider AEs corresponds to *organs* and what we consider FI corresponds to *symptoms*, *diagnoses* and *observations*.

of the reports. From the results obtained and the analysis carried out we can draw following conclusions.

Despite the conclusion about the coverage of SNOMED CT terms in the radiology domain obtained in [1],[9] we obtained better results with SiMREDA using a translated version of RadLex -although it is not a high-quality translation- than with SNOMED CT terms that are already in Spanish.

Based on results obtained comparing the original GT translation and a correction of a portion of it by a physician of the radiology domain, we can conclude that our algorithm is sensitive to a poor translation. The improvement of only 10% of RadLex translations improves our results. Therefore, we conclude that it makes sense to invest effort in improving the translation.

The rules added to SiMREDA in Module 3, based on the analysis of its PoS tagging patterns improved the results. It also could be noticed that the morphological processing improvement is almost imperceptible, but we can appreciate that it recognizes more AEs and FIs and that the limited increase in performance is probably due to the reduced size of the test set. We could also see that the morphological module helped in recognizing misspelled entities.

In this paper we only show the final results. But, lenient match draws better results than the exact matching for every entity type across all settings of both algorithms tested. Besides, in this use case it is more important to determine if an entity is present than to correctly determine its boundaries. Therefore, we conclude that it is important to report a precisely defined partial metric accompanying the exact match results.

We can also conclude that despite having a small annotated dataset (513 reports -see Tables 2 and 3-), we could successfully apply ML.

There are many studies than can be carried out as future work. There are some phrases, we call *prefix terms*,[10] such as "could suggest", "is visualized", that usually determine that the following noun phrase corresponds to a clinical finding. Detecting those phrases and the noun phrases that come after them, could help improve the recall of retrieved findings.

The construction of abbreviation databases for Spanish radiology reports, would be probably less useful than others existing for English [19,34], since many of the abbreviations used in these kinds of reports do not follow naming conventions and would, therefore, be difficult to generalize to other texts. However, the subject could be studied and an abbreviation database could be constructed. Therefore, previous efforts could be studied [16].[11]

It would be interesting to detect of all the morphemes composing a word, as [32] carried out. This can help to a better understanding of the words. For

---

[9] The paper is not available online. Results were discussed in a personal communication.

[10] In Spanish they usually occur before the terms of interest.

[11] Acronyms and abbreviations provided by the National Academy of Medicine of Colombia http://dic.idiomamedico.net/Siglas_y_abreviaturas and by the Spanish Ministry of Health http://www.redsamid.net/archivos/201612/diccionario-de-siglas-medicas.pdf?0.

instance, words that have more than one morpheme related with the medical domain (e.g. **peritonitis** -peritonitis-) can be found, and their semantics can be better comprehended. Consider also *cardiopatía* -cardiopathy- and *linfoadenopatía* -lymphadenopathy-, whose decomposition into morphemes (cardio-patía and linf-o-adeno-patía) explains in which anatomical entity the findings have occurred.

There are some patterns that would also probably help to improve finding retrievals. Consider:

- AE FI, as in [ovarian](AE) [cyst](FI),
- FI AE,
- and FI (en (el |la(s?) |los|λ) |de (la(s?) |los |λ) |del) AE,[12,13] as in "[luxación](FI) de la [cadera](AE)" (hip dislocation).

With the current version of SiMREDA, these patterns are not considered as findings, but they were annotated as findings. An additional SiMREDA module that detects those patterns as entities could be constructed. It is also important to notice that detecting [ovarian](AE) [cyst](FI) as a first step, has as advantage, that it can be determined where the finding is located. If [ovarian cyst](FI) would have been detected, then this understanding would be lost.

Finally, a deep learning architecture could be implemented to improve CRFs results. Character based convolutional neural networks (CNN), recurrent neural networks (probably biLSTM), and CRF could be considered as layers. The non-existence of sufficient data to train word embeddings in this particular domain and language, might make them not very beneficial in this particular case.

References

1. Aleksovski, Z.: Testing RadLex for completeness using large database of radiology reports. In: Society for Imaging Informatics in Medicine, Annual Meeting (2014)
2. Ambulódegui, E.S.: Manual de Terminología Médica N 2 (2012)
3. Ananiadou, S., Friedman, C., Tsujii, J.: Introduction: named entity recognition in biomedicine. J. Biomed. Inform. **37**(6), 393–395 (2004)
4. Basaldella, M., Furrer, L., Tasso, C., Rinaldi, F.: Entity recognition in the biomedical domain using a hybrid approach. J. Biomed. Semant. **8**(1), 51 (2017)
5. Batista-Navarro, R.T., Rak, R., Ananiadou, S.: Chemistry-specific features and heuristics for developing a CRF-based chemical named entity recogniser. In: Proceedings of the Fourth BioCreative Challenge Evaluation Workshop, vol. 2, pp. 55–59. Citeseer (2013)
6. Cascade, P.N., Berlin, L.: Malpractice issues in radiology. AJR Am. J. Roentgenol. **173**(6), 1439–1442 (1999)
7. Castro, E., Iglesias, A., Martínez, P., Castaño, L.: Automatic identification of biomedical concepts in Spanish-language unstructured clinical texts. In: Proceedings of the 1st ACM International Health Informatics Symposium, pp. 751–757. ACM (2010)

---

[12] in |in the |from.
[13] Written as a regular expression.

8. Chapman, W.W., Cohen, K.B.: Current issues in biomedical text mining and natural language processing. J. Biomed. Inform. **42**(5), 757–759 (2009)
9. Chinchor, N., Hirschman, L., Lewis, D.D.: Evaluating message understanding systems: an analysis of the third message understanding conference (MUC-3). Assoc. Comput. Linguist. **19**(3), 409–449 (1993)
10. Cohen, A.M., Hersh, W.R.: A survey of current work in biomedical text mining. Brief. Bioinform. **6**(1), 57–71 (2005)
11. Cotik, V., Filippo, D., Roller, R., Uszkoreit, H., Xu, F.: Annotation of entities and relations in Spanish radiology reports. In: Proceedings of the International Conference Recent Advances in Natural Language Processing, RANLP 2017, pp. 177–184 (2017)
12. Do, B., Wu, A., Maley, J., Biswal, S.: Automatic retrieval of bone fracture knowledge using natural language processing. J. Digit. Imaging **26**(4), 709–713 (2013)
13. Iglesias, A., et al.: Mostas: Un etiquetador morfo-semántico, anonimizador y corrector de historiales clínicos. Procesamiento del lenguaje Nat. **41**, 299–300 (2008)
14. Jiang, J., Guan, Y., Zhao, C.: WI-ENRE in CLEF eHealth evaluation lab 2015: clinical named entity recognition based on CRF. In: Working Notes of CLEF 2015 - Conference and Labs of the Evaluation Forum, Toulouse, France (2015)
15. López Piñero, J.M., Terrada Ferrandis, M.L.: Introducción a la terminología médica. Masson S.A. (2005)
16. Laguna, J.Y.: Diccionario de siglas médicas y otras abreviaturas, epónimos y términos médicos relacionados con la codificación de las altas hospitalarias
17. Lakhani, P., Langlotz, C.P.: Automated detection of radiology reports that document non-routine communication of critical or significant results. J. Digit. Imaging **23**(6), 647–57 (2009)
18. Leaman, R., Gonzalez, G.: BANNER: an executable survey of advances in biomedical named entity recognition. In: Proceedings of the Pacific Symposium on Biocomputing, vol. 13, pp. 652–663 (2008)
19. Moon, S., Pakhomov, S.V.S., Liu, N., Ryan, J.O., Melton, G.B.: A sense inventory for clinical abbreviations and acronyms created using clinical notes and medical dictionary resources. JAMIA **21**(2), 299–307 (2014)
20. Nadeau, D., Sekine, S.: A survey of named entity recognition and classification. Linguist. Investig. **1**(30), 3–26 (2007). https://doi.org/10.1075/li.30.1.03nad
21. Oronoz, M., Casillas, A., Gojenola, K., Perez, A.: Automatic annotation of medical records in Spanish with disease, drug and substance names. In: Ruiz-Shulcloper, J., Sanniti di Baja, G. (eds.) CIARP 2013. LNCS, vol. 8259, pp. 536–543. Springer, Heidelberg (2013). https://doi.org/10.1007/978-3-642-41827-3_67
22. Poibeau, T., Kosseim, L.: Proper name extraction from non-journalistic texts. In: Computational Linguistics in the Netherlands 2000, Selected Papers from the Eleventh CLIN Meeting, Tilburg, 3 November 2000, pp. 144–157 (2000)
23. Roller, R., et al.: Detecting named entities and relations in German clinical reports. In: Rehm, G., Declerck, T. (eds.) GSCL 2017. LNCS (LNAI), vol. 10713, pp. 146–154. Springer, Cham (2018). https://doi.org/10.1007/978-3-319-73706-5_12
24. Santiso, S., Casillas, A., Pérez, A., Oronoz, M.: Medical entity recognition and negation extraction: assessment of NegEx on health records in Spanish. In: Rojas, I., Ortuño, F. (eds.) IWBBIO 2017. LNCS, vol. 10208, pp. 177–188. Springer, Cham (2017). https://doi.org/10.1007/978-3-319-56148-6_15

25. Settles, B.: Biomedical named entity recognition using conditional random fields and rich feature sets. In: Proceedings of the COLING 2004 International Joint Workshop on Natural Language Processing in Biomedicine and Its Applications (NLPBA/BioNLP), COLING 2004. Association for Computational Linguistics, Stroudsburg (2004)

26. Shen, D., Zhang, J., Zhou, G., Su, J., Tan, C.L.: Effective adaptation of a hidden Markov model-based named entity recognizer for biomedical domain. In: Proceedings of the ACL 2003 Workshop on Natural Language Processing in Biomedicine, vol. 13, pp. 49–56. Association for Computational Linguistics (2003)

27. Simpson, M.S., Demner-Fushman, D.: Biomedical text mining: a survey of recent progress. In: Aggarwal, C., Zhai, C. (eds.) Mining Text Data, pp. 465–517. Springer, Boston (2012). https://doi.org/10.1007/978-1-4614-3223-4_14

28. Sondhi, P.: A survey on named entity extraction in the biomedical domain (2008)

29. Takeuchi, K., Collier, N.: Bio-medical entity extraction using support vector machines. Artif. Intell. Med. **33**(2), 125–137 (2005)

30. Tasneem, A., Archana, B.: A survey on biomedical named entity extraction. Asian J. Eng. Technol. Innov. **4**(7), 25–28 (2016)

31. Uzuner, Ö., Solti, I., Cadag, E.: Extracting medication information from clinical text. J. Am. Med. Inform. Assoc. **17**(5), 514–518 (2010)

32. Estopà, R., Vivaldi, J., Cabré, M.T.: Use of Greek and Latin forms for term detection. In: Proceedings of the Ninth International Conference on Language Resources and Evaluation (LREC 2000), vol. 78, pp. 855–859 (2000)

33. Weegar, R., Casillas, A., de Ilarraza, A.D., Oronoz, M., Prez, A., Gojenola, K.: The impact of simple feature engineering in multilingual medical NER. In: Proceedings of the Clinical Natural Language Processing Workshop, pp. 1–6 (2016)

34. Xu, H., Stetson, P.D., Friedman, C.: A study of abbreviations in clinical notes. In: AMIA 2007, American Medical Informatics Association Annual Symposium, Chicago, IL, USA (2007)

Ontology Modeling of the Estonian Traffic Act for Self-driving Buses

Alberto Nogales[1(✉)], Ermo Täks[2], and Kuldar Taveter[2]

[1] CEIEC, Research Institute, Universidad Francisco de Vitoria (UFV),
Carretera Pozuelo-Majadahonda km. 1,800, 28223 Pozuelo de Alarcón, Spain
alberto.nogales@ceiec.es
[2] Department of Informatics, Tallinn University of Technology, 15 Raja Street,
12618 Tallinn, Estonia
ermo.taks@ttuu.ee, kuldar.taveter@ttu.ee

Abstract. The development of self-driving cars is a major research area that has led to several still unresolved issues. One of them is the need to abide by the legal stipulations fixed by a traffic act concerning the territory of operation. An appropriate solution to make text understandable by machines is the use of ontologies. This paper presents a first approach where the Estonian Traffic Act is transformed from text into populated ontologies, so it can be understood by machines. The proposal is a (semi)-automatic ontology learning process that combines natural language processing (NLP) and ontology matching techniques with a deep learning model. The results show that 78% of the norms that have been considered valid can be modelled with the method described in the paper.

Keywords: Ontology learning · Ontology matching · Deep learning

1 Introduction

Self-driving vehicles (SDV), also known as autonomous, automated or driverless vehicles, have become a technological trend in recent years. SDVs are defined in [1] as a new era of vehicle systems where part or all of the driver's actions may be removed or limited, and where cars involve a combination of new technologies including sensors, computing power, and short-range communications, effectively creating a new human-automobile hybrid.

By the end of 2016, it was announced that self-driving buses would be used in Tallinn (Estonia). These buses need to abide by the Estonian Traffic Act to move around the city. In other words, the system responsible for driving the bus must be able to understand the text describing the traffic norms.

Legal text has its own specific limitations in their organization and formulation because of the very function of such texts. The smallest meaningful representation of a norm in legal text is a clause. By definition, a clause is a group of words containing a subject and predicate and functioning as a member of a complex or compound sentence.

Software run by SDVs has to interpret very large and complex information. Thus, when there is a modification in the traffic act, changes are an expensive and cumbersome task. A good solution to this problem is to make a representation of legal texts

J. A. Lossio-Ventura et al. (Eds.): SIMBig 2018, CCIS 898, pp. 249–256, 2019.
https://doi.org/10.1007/978-3-030-11680-4_24

that is understandable by machines, as the ontologies. These are defined as a specification of a conceptualization [2]; they can describe actors, relations and situations of a particular field.

In this paper, the first approach to modeling norms from the Estonian Traffic Act as ontologies is described. The process of constructing ontologies by the integration of a multitude of disciplines is referred to as ontology learning [3].

The rest of this paper is structured as follows: Sect. 2 is a discussion of the state of the art of previous studies made in the fields of this paper. Section 3, "Materials and methods," gives a deeper idea of how the approach was achieved and what resources were used for that purpose. In Sect. 4, the results are analyzed, and a use case is presented. Finally, Sect. 5 offers some conclusions about the research and proposes future lines of work.

2 Background

There are previous studies benefitting from using ontologies to represent legal texts for traffic. An intelligent transportation system using ontologies is proposed in [4]; it was tested in a situation where two cars are travelling along four roads with four intersections. An Advanced Driver Assistance System (ADAS) ontology for autonomous driving tasks and an ADAS ontology-based map data was described in [5]. Finally, a conceptual description of all road entities with their interactions is defined in [6].

Ontology learning can be found in [7], where a system called Concept-Relation-Concept Tuple-based Ontology Learning (CRCTOL) is described. This system uses statistical algorithms for word sense disambiguation. Then, Web folksonomies were used for ontology learning in [8]. Another ontology learning tool called Text2Onto was applied to Spanish legal texts using language-specific algorithms [9].

However, the present research differs from the studies cited above. This work describes, in particular, norms from the Estonian Traffic Act using ontologies. Also, a (semi-)automatic ontology learning process has not been previously applied in the field of texts describing the traffic laws of a country.

3 Materials and Methods

As has been said before, the Estonian Traffic Act will be annotated with ontologies so machines can understand them. An ontology learning method is directly joined to the process of ontology development defined in [10]. This process is divided into a six layers cake that can be seen in Fig. 1.

In the following subsections, the proposed (semi)automatic process of the paper, which is divided into three stages, is explained. Section 3.1 describes the preprocessing stage of the Traffic Act, which goes from the raw XML document to the extraction of the terms for each norm. Then, Part of Speech (PoS) is applied to obtain the grammatical function of these terms. These two tasks correspond to the first level of the layer cake presented in the figure above.

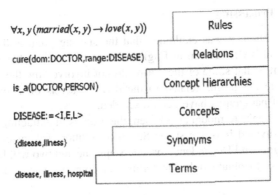

$\forall x, y\,(married(x, y) \rightarrow love(x, y))$ — Rules

cure(dom:DOCTOR,range:DISEASE) — Relations

is_a(DOCTOR,PERSON) — Concept Hierarchies

DISEASE:=<I,E,L> — Concepts

{disease,illness} — Synonyms

disease, illness, hospital — Terms

Fig. 1. Ontology learning cake described in [10]

Section 3.2 consists of obtaining synonyms for these terms and finding mapping between terms and synonyms with ontologies, which is a similar process to the "Synonyms", "Taxonomy" and "Relations" layers. The annotation will be performed by using ontology matching, which is defined as the technique used to find the relationships among entities [11]. First of all, synonyms for all the terms of previous step are obtained. For that purpose, WordNet, which is a large database of nouns, verbs, adjectives and adverbs grouped as cognitive synonyms [12], has been used. The terms plus its synonyms have mappings with vocabularies from Linked Open Vocabularies (LOV) which is a catalogue of the vocabularies used in the Web of Data [13]. Mappings of two types are established in order to find relations between different classes and properties. The mappings will add extra knowledge understandable by machines to the terms. Finally, a hierarchy between concepts has to be established. At this point, the PoS tags from Sect. 3.1 are mapped with OLiA Annotation Model [14]. In particular, it will be used the version for Penn Treebank PoS annotation. The information from the mappings will be used to establish a relation of subclasses between the PoS tags mapping and the ones provided by the vocabularies. This is used, for example, to clarify if a term is a noun or adjective in a sentence.

Finally, Sect. 3.3 builds an instance upon one or more ontologies depending on the results obtained in the previous steps. This final stage has a manual task where a user decides which of the mappings fit better and an automatic task where, depending on the type of norm, the ontology is built with a different structure. To simplify the latter task, a deep learning classifier is depicted to categorize each norm depending on its level of restriction. Deep learning is defined by [15] as a technique allowing computational models that are composed of multiple processing layers to learn representations of data with multiple levels of abstraction. Finally, depending on the type of norm, Semantic Web Rule Language (SWRL) will be applied. SWRL is a language that can be used to express rules as logical conditions.

3.1   Raw Text Processing

In the previous section, it was explained that the starting point is the Traffic Act of Estonia. An official XML version in English from which norms have to be extracted [16]. The problem is that some of the norms do not involve how the vehicle behaves. For example, there are norms related to the driver's levels of alcohol intoxication. Thus, this kind of paragraphs have been discarded.

Once the paragraphs to model are chosen, the process of extracting terms will start. This is typically divided into five steps: Sentence Splitter, Tokenizer, Morphological analyzer, PoS Tagger and Dependency parser. Steps one and two will be applied to that part of the process to obtain only the terms that compose each norm without stop words.

Paragraphs have a set of sentences; each sentence will be considered a norm. Thus, the interest of this research lies in modeling each norm with a set of ontologies. This process starts with a "Sentence Splitter", which consists of dividing a paragraph into its sentences to process each one separately. Then, each sentence is divided into a set of terms by removing its stop words; for that purpose, a "Tokenizer" is used. Both the "Sentence Splitter" and the "Tokenizer" have been developed in a Python script with the help of the NLTK package version 3.3 which is used to work with human language data [17]. At the end of this step, a list of the terms without stop words will be obtained, and the order of the terms will be the same as in the sentence corresponding each list to a norm.

At this step, there is a need to know how the terms are related to understand the sentence as a whole. According to [18], a rule or norm in a legal text has the structure of a case, condition, subcondition, legal subject and legal action. This means that a subject will perform an action when a condition or some conditions are achieved. PoS tagging is the process of giving a grammatical category to every word in a sentence. This information will show which words are part of the subject, which are part of the action and which are part of the conditions.

To obtain this, another script has been developed using a tool called Stanford Parser [19] that is included in the NLTK package. By using the parser with a sentence, each word of a sentence will be related with a tag that defines its function in it. Tags correspond to the Penn Treebank tag set [20], which at the time of its publication contained 4.5 million words in American English. By knowing which words correspond to the conditions, which to the subject and which to the action, the norm can be built following structure: if the "condition" is accomplished, the "subject" has to perform an "action". Table 1 summarizes how each kind of norm can be structured.

Table 1. Norm structures.

Type of norm	Structure
Permission	if (Condition==True) then (Subject → (Action OR NOT Action))
Obligation	if (Condition==True) then (Subject → Action)
Prohibition	if (Condition==True) then (Subject → NOT Action)

3.2 Mapping Words with Vocabularies

At this step, there is a need to give extra knowledge to each term extracted in the previous section. To achieve this, mappings between these terms and the vocabularies are needed. The mappings are performed with two different aims: the first set of mappings will take into account the tags given by the PoS techniques and will map them with the OLiA annotation model which is an ontology describing PoS and the syntactic tags of Penn Treebank.

The second set of mappings relates to the terms with all the vocabularies in LOV. It was decided to use this set of vocabularies, as not all the terms in the Traffic Act are covered by legal or traffic ontologies. Here, the mappings are made on two levels. The morphological level, which considers that two words are the same if they are written in the same way, and the semantic level, which considers that two words are the same if they have the same meaning. At the time of the experimentation, LOV was composed of 601 vocabularies in different fields.

As previously mentioned, these mappings are made in two steps. The first corresponds to the morphological level and will take the set of words of a sentence obtained in the previous section and try to map it with all the vocabularies in LOV. To obtain a mapping, a word needs to be exactly the same as a class or property of a vocabulary. To find the mappings, a Python script has been developed using the RDFLib package [21], which allows users to work with the resource description framework (RDF) representations. The process will consist of taking a word and comparing it to all the classes and properties of a vocabulary and then going through all the vocabularies in LOV. For the second type of mappings, there is a need to work with synonyms. In this case, two words are considered the same depending on their meaning. Again, a Python script has been developed using RDFLib as in the previous mapping approach. To find the mappings, the synonyms of a word will be obtained from WordNet, and then these synonyms will be compared with the terms provided by the LOV vocabularies. A mapping will be found if a synonym and a term are equal by comparing them string by string. In Table 2, there are examples of the three types of mappings described above.

Table 2. Examples of different mappings sentences.

Word	Mapped with	Type of mapping
(NN driver)	http://purl.org/olia/penn.owl#NN	PoS
Light	https://w3id.org/saref#Light	Morphological
Cycle	http://linkedgeodata.org/ontology/Motorcycle	Semantic

3.3 Populating the Ontologies

The final step consists of constructing an instance of ontologies for each norm, which is a total of 420 at this point. It uses the vocabularies extracted with the vocabulary mappings to give some knowledge to the words and the information given by the PoS tagging mappings. This information in combination with SWRL, will build some logic rules that could be interpreted by machines.

First, there is the information obtained in the step of mapping the text with the vocabularies. As more than one mapping can be obtained for a word, there is a need to choose one vocabulary to represent the word. It has been decided to do this manually.

Then, the rule needs to be built based on the PoS tagging and applying SWRL. Taking into account [22], it is known that norms in traffic can be classified into permissions, obligations and prohibitions, each having a different representation. A permission denotes that the action could be done or not, an obligation denotes that the action must be done and the prohibition that the action cannot be done.

To make the classification of different norms easier and more precise, a deep learning classifier has been developed. Based on [23], a convolutional neural network (CNN) was used. After the training stage, the model had a loss of 9% and an accuracy of 96%. The classifier was built with the Keras library, an API written in Python for managing neural network models [24].

At this point, there was a set of words with their corresponding tags that denoted the grammatical function of each one. By using these tags and the type of norm depending on the category, SWRL can be applied so that machines can understand them.

4 Analysis and Discussion

This section presents an analysis of the different results obtained during the research. The research used a set of 420 norms. After classifying them as permissions, obligations and prohibitions, there were 97 permissions, 164 obligations and 33 prohibitions. The remaining 126 are what has been called "environment", which include properties such as the speed limit of a road.

Each sentence was also mapped with the catalog of vocabularies provided by LOV. Taking into account Table 3, a sentence is considered fully mapped when all the terms at the time of obtaining the mappings have a correspondence with a term in LOV. It has been established 65% as the minimum of words needed to understand a text, [25]. Then, in 328 norms it has been possible to model the information. That was approximately 78% of the possible norms. It should be noted that stop words were removed previously.

Table 3. Percentage of mapped sentences.

Percentage of mapped words	Number of norms
100%	5
90–99%	18
80–89%	97
70–79%	127
65–69%	81

5 Conclusions and Future Work

During this research, several issues were addressed. A tool able to extract traffic norms from an XML document was developed, mapping these norms to a catalog of vocabularies called LOV and to an ontology describing PoS tags. LOV mappings were built with two perspectives. From a morphological point of view and from a semantic view.

Also, a classification of norms was developed, distinguishing between permissions, obligations, prohibitions and "environment" norms. This classification was automatized with the help of a deep learning model.

In future works, some improvements can be made. N-grams can be used to make mappings. Also, word sense disambiguation can be applied at this step. The process of choosing the most accurate vocabulary can be performed automatically. Deep learning techniques can also be applied at the stage of building the rules with SWRL, in order to obtain more accurate results. Once the process is as automatic as possible, created ontologies can be integrated with a multiagent system, thus allowing for experiments that could prove its application in the field.

Acknowledgements. The work providing these results has received funding with Dora Plus Action scholarship from Tallinn University of Technology in Estonia.

References

1. Blyth, P., Mladenović, M., Nardi, B., Su, N., Ekbia, H.: Driving the self-driving vehicle: expanding the technological design horizon. In: Proceedings of the International Symposium on Technology and Society (ISTAS), pp. 1–6 (2015)
2. Gruber, T.: A translation approach to portable ontologies. Knowl. Acquis. **5**(2), 199–220 (1993)
3. Maedche, A., Staab, S.: Ontology learning for the semantic web. IEEE Intell. Syst. **16**(2), 72–79 (2001)
4. Fernandez, S., Ito, T., Hadfi, R.: Architecture for intelligent transportation system based in a general traffic ontology. In: Lee, R. (ed.) Computer and Information Science 2015. SCI, vol. 614, pp. 43–55. Springer, Cham (2016). https://doi.org/10.1007/978-3-319-23467-0_4
5. Zhao, L., Ichise, R., Mita, S., Sasaki, Y.: Ontologies for advanced driver assistance systems. J. Jpn. Soc. Artif. Intell. (2015)
6. Armand, A., Filliat, D., Guzman, J.I.: Ontology-based context awareness for driving assistance systems. In: IEEE Intelligent Vehicles Symposium Proceedings, pp. 227–233 (2014)
7. Jiang, X., Tan, A.: CRCTOL: a semantic-based domain ontology learning system. J. Am. Soc. Inform. Sci. Technol. **61**(1), 150–168 (2009)
8. Tang, J., Leung, H., Luo, Q., Chen, D., Gong, J.: Towards ontology learning from folksonomies. In: Proceedings of the 21st International Conference on Artificial Intelligence (JCAI 2009), pp. 2089–2094 (2009)
9. Völker, J., Fernandez Langa, S., Sure, Y.: Supporting the construction of Spanish legal ontologies with Text2Onto. In: Casanovas, P., Sartor, G., Casellas, N., Rubino, R. (eds.) Computable Models of the Law. LNCS (LNAI), vol. 4884, pp. 105–112. Springer, Heidelberg (2008). https://doi.org/10.1007/978-3-540-85569-9_7

10. Buitelaar, P., Cimiano, P., Magnini, B.: Ontology learning from text: an overview. In: Buitelaar, P., Cimiano, P., Magnini, B. (eds.) Frontiers in Artificial Intelligence and Applications, Ontology Learning from Text: Methods, Evaluation and Applications, vol. 123, pp. 3–12 (2005)

11. Shvaiko, P., Euzenat, J.: Ontology matching: state of the art and future challenges. IEEE Trans. Knowl. Data Eng. **25**(1), 158–176 (2013)

12. Kilgarriff, A., Fellbaum, C.: WordNet: an electronic lexical database. Lang. speak Commun. **76**(3), 706 (2000)

13. Linked Open Vocabularies. https://lov.linkeddata.es/dataset/lov/

14. Chiarcos, C., Sukhareva, M.: OLiA – ontologies of linguistic annotation. Semant. Web **6**(4), 379–386 (2015)

15. Bengio, Y., Courville, A.C., Goodfellow, I.J., Hinton, G.E.: Deep learning. Nature **521** (7553), 436–444 (2015)

16. English Traffic Estonian Act in XML. https://www.riigiteataja.ee/en/tolge/xml/5070120-14005

17. NLTK Python package. https://www.nltk.org/

18. Fung, S.Y.C., Watson-Brown, A.: The Template: A Guide for the Analysis of Complex Legislation. Institute of Advanced Legal Studies Location, London (1994)

19. Klein, D., Manning, C.D.: Accurate unlexicalized parsing. In: Proceedings of the 41st Annual Meeting of the Association for Computational Linguistics (ACL-03), pp. 423–430 (2003)

20. Santorini, B.: Part-of-speech tagging guidelines for the Penn Treebank Project. Department of Computer and Information Science, University of Pennsylvania (1990)

21. RDFlLib Python package. https://rdflib.readthedocs.io/

22. Baumfalk, J., Dastani, M., Poot, B., Testerink, B.: A SUMO extension for norm-based traffic control systems. In: Behrisch, M., Weber, M. (eds.) Simulating Urban Traffic Scenarios. LNM, pp. 55–82. Springer, Cham (2019). https://doi.org/10.1007/978-3-319-33616-9_5

23. Kim, Y.: Convolutional neural networks for sentence classification. In: Proceedings of the Conference on Empirical Methods in Natural Language Processing (EMNLP), pp. 1746–1751 (2014)

24. Keras Python package. https://keras.io/

25. Laufer, B., Sim, D.D.: Taking the easy way out: non-use and misuse of contextual clues in EFL reading comprehension. Engl. Teach. Forum **23**, 7–10 (1985)

Thought Off-line Sanitization Methods
for Bank Transactions

Isaias Hoyos and Miguel Nunez-del-Prado[✉]

Universidad del Pacífico, Av. Salaverry 2020, Lima, Peru
{isaias.hoyos,m.nunezdelpradoc}@up.edu.pe

Abstract. In the digital era, people generate a lot of digital traces ranging from posts on social networks, call detail records and credit or debit
banks transactions among others. These data could help society to understand different urban phenomena such as what citizens are talking about,
how they commute or what are their spending behaviors. Therefore, the
use of such data trigger privacy issues. In the present effort, we study four
different Statistical Disclosure Control filters to sanitize off-line credit
or debit bank transactions. Consequently, we analyze Noise Addition,
Microaggregation, Rank Swapping and Differential Privacy filters concerning Disclosure Risk, Information Loss, and utility. We observed that
Microaggregation and Different Privacy perform very well for minimizing
Disclosure Risk while providing a good utility for statistics of spending
amounts per industry type.

Keywords: Privacy filters · Statistical Disclosure Control (SDC) ·
Microaggregation · Differential Privacy

1 Introduction

Nowadays in the digital society, all the information produced by the citizens
could be used for common social wealth. By releasing these datasets, researchers
and policymakers could study and understand different social phenomena such
as, how people commute, the economic wealth of cities, even the detection of
anomalous events. The problem making these datasets publicly available is the
possible privacy breach, which is also regulated by the GDPR[1].

In the present effort, we take into account three different SDC mechanisms
of the perturbative family namely, Noise Addition, Microaggreggation, and Rank
Swapping. Besides, we apply a Differential Private to complete a benchmark.
Hence, by applying the aforementioned SDC filters, we aim to find the most
suitable filter to have a good trade-off between utility and privacy. Thus, in our
scenario, we release transactional bank data to allow policymakers to perform
statistics on the amount of spent money in a given sector or industry.

---

[1] General Data Protection Regulation: www.eugdpr.org.

Supported by the research fund projects of the Vicerrectorate of the Universidad del
Pacífico PY-ESP-0210013216.

© Springer Nature Switzerland AG 2019
J. A. Lossio-Ventura et al. (Eds.): SIMBig 2018, CCIS 898, pp. 257–264, 2019.
https://doi.org/10.1007/978-3-030-11680-4_25

The rest of the paper is organized as follows. Section 2 summarizes the related works. Section 3 presents the adversary model and the privacy mechanisms to reduce re-identification risk. Section 4 describes the experiments and results of the privacy mechanisms applied to raw bank transaction data, respectively. Finally, Sect. 5 concludes our work and proposes new research avenues.

2   Related Works

In the literature, there are *non-perturbative* and *perturbative* statistical disclosure control (SDC) methods [8]. The former SDC method suppresses samples partially or reduces their details. For instance, *Sampling* [8] takes some records instead of all the information. Thus, indirectly it suppresses records to publish a partial version of the dataset; *Global recoding* [8] generalizes variables, it replaces the continuous value for their discrete counterpart; *Top and bottom coding* [24] is a particular case of *Global recoding* that treats variables that can be ranked. The basic idea of this technique is to replace values with respect to certain thresholds at the top and the bottom of a rang; and *Local suppression* [2] replaces quasi-identifiers variable for missing values when they are combined to re-identify a subject.

The latter SDC method adds some noise to samples to make re-identification more difficult. For example, *Additive noise masking* [1,5,9,22] adds correlated or not correlated Gaussian noise with a mean zero. *Multiplicative noise masking* [16,17] takes into account the value of the variable for adding small noise to small values and important noise to high values. *Microaggregation* [10,16,21] replaces original values by the average value of a certain variable belonging to a group of at least k members. *Rank swapping* [18,19] exchange values of a sensitive variable between users; *Data shuffling* [14] is a particular case of *Rank swapping*, where the mechanism conserves marginal distributions of the sanitized variable. Finally, *Rounding* [25] changes the value of the sensitive value for its rounded equivalent.

3   Background

In this section, the adversary model and the Statistical Disclosure Control (SDC) filters used in order to obfuscate the data and achieve privacy will be described. It is worth noting that we have selected only perturbative filters because re-identification is easier over non-perturbed values. In addition, the Noise Addition, Microaggregation, Rang Swapping and Differential Privacy filters were chosen for their use in the literature [7,20,23]. Besides, the metric to quantify the impact of the SDC filters on Information Loss (utility) and Disclosure Risk (privacy) will be introduced.

3.1   Adversary Model

From the user point of view, there is a risk of discrimination by releasing credit and debit card transaction. More precisely, if an adversary is able to re-identify users, he will be capable of estimating the clients spending and associate a socioeconomic category [11]. Thus a possible loan denial due to the socioeconomic category. The dataset use in the present effort contains three variables, such as *client ID*, *type of commerce* and the *spending* on the commerce; where the *client ID* is a pseudonymous identifier, the *type of commerce* is a quasi-identifier but non-confidential, and the *spending* is a private variable, we sanitize. In the present work, we test the worst case scenario. Therefore, the adversary has original records with identity and try to link identities in both sanitized and raw dataset.

3.2 Noise Addition

The Noise Addition filter [8] adds uncorrelated noise to the values of the attributes of a certain user. This filter takes a as a parameter, which is a raw value between the [0,1] range. We denote by x_i, the *i-th* value of the attribute x, and by x_i' its obfuscated counterpart. In this way, the noisy values are calculated as follows: $x_i' = x_i + a \times \sigma \times c$. Where σ, is the standard deviation from the attribute that will be obfuscated; and ϵ is a Gaussian random variable in the range [0,1].

3.3 Microaggregation

The Microaggregation filter [6] relies on grouping similar users with a clustering algorithm. First, the algorithm computes the similarity, using the Euclidean distance, between a target user and the set of other users. Therefore, the k most similar users become neighbors of the target user. Thus, a cluster with $k+1$ elements is obtained.

Then, the centroid of the cluster is computed, the values of the clusters are replaced with it. The elements replaced are then removed from the dataset and a new target user is chosen in order to create a new cluster. This filter receives as parameter the minimum number of neighbors elements k that the cluster must have.

3.4 Rank Swapping

The Rank Swapping filter [15] is based on the idea of data swapping, which consists in transforming a dataset by exchanging the values of confidential variables. First, the values of the target variable are ranked in ascending order. After that, for every ranked value, another ranked value is select for swapping within a p range. An input parameter p controls this range, thus a value x will be swapped with any of the higher or lower p values.

3.5 Differential Privacy

This filter relies on the concept of Differential Privacy and the use of the Laplace Mechanism in order to provide privacy guarantees over numeric variables [12,13]. To understand this method, we review the global sensitivity concept of a function. Global sensitivity is defined as the upper bound on the difference between the sums of any two data sets D_1 and D_2 differing on a single record (neighbor datasets), and is calculated as follows: $\Delta(f) = max||f(D_1) - f(D_2)||$.

In this way, given a database D, a mechanism M release a function f achieving ϵ-Differential Privacy if $M(D) = f(D) + L$. Where L is a vector of random variables drawn form Laplace distribution with parameters location and scale equals to 0 and $(\frac{\Delta(f)}{\epsilon})$, respectively; and M is the function that applies the Microaggregation to the dataset.

3.6   Information Loss (IL)

This measure provides a way to measure the detriment of information produced by different SDC methods. The estimation is based on the Sum of Squared Errors (SSE) value between the original x and anonymized x' samples, evaluating in this way the amount of distortion introduced by each SDC method [3]. To compute IL, we rely on the following equation: $IL = \sum_{x \in X} \sum_{x' \in X'} (dist(x, x'))^2$.

3.7   Disclosure Risk (DR)

This metric uses a record linkage approach [4] to estimate the risk of records re-identification. In our case, this metric is computed as the ratio of correctly identified individual and the total number of individuals in the data set.

In the next section, we describe the performed experiments using the SDC filters and metrics presented in the current section.

4   Experiments and Results

In the present section, we perform some experiments to find the best SDC filter with an optimal parameter configuration. The SDC filters take as input transactional historical data belonging to users from two districts in Lima, Peru as detailed in Table 1. The present datasets contains three variables, such as *client ID*, *type of commerce* and the *spending* on the commerce; where the client ID is a pseudonymous identifier, the type of commerce is a quasi-identifier but non-confidential, and the spending is a private variable, we try to protect.

Table 1. Datasets characteristics

Number of	Barranca	Independencia
Transactions	465	449 709
Commerce	88	1 673
Clients	148	32 665

Table 2. Parameter values

Privacy filter	Parameters	
	Name	Selected values
Noise addition	c	0.1, 0.25, 0.5, 0.75, 1.0
Microaggregtion	k	3, 5, 10, 15, 20, 25, 50, 100
Rank swapping	p	10, 25, 50, 75, 80
Differential	k	3
Privacy	ϵ	0.01, 0.1, 1, 10, 100

The before described filters are applied to sanitize the spending variable. It is worth mentioning that each one of the filter is applied using different parameter settings on the Barranca dataset to find the most suitable configuration. The parameters are empirically chosen to find the trade-off, which minimizes privacy risks and maximizes utility. The selected parameters are presented in Table 2. The IL and DR values of these metrics are computed for every configuration setup introduced in Table 2. Due to the lack of space, we could not present all the analysis done.

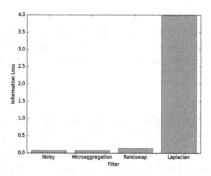 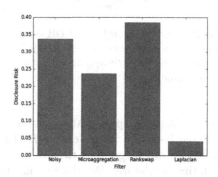

Fig. 1. In green IL and in red DR comparison of the four SDC filters' best configurations in Barranca dataset, where Noise Addition with $c = 0.75$, Microaggregation with $k = 100$, Rank Swapping with $p = 25$, and Differential Privacy with $\epsilon = 1$. (Color figure online)

Once the fine-tuning of parameters is done, we compare the four filters with their best configuration. Figure 1 depicts the Laplacian Differential Privacy filter as the one that introduces more noise. Thus, it has the highest IL. Besides, it is the most reliable filter in terms of DR. The second most performing filter is Microaggregation with small IL and the second lowest DR.

After computing different experiments with the Barranca dataset, we test the performance of the four filters in bigger datasets *i.e.*, Independencia dataset. We have used the best setups for the four filters to build Fig. 2. In these figures, we observe the same pattern. The Laplacian Differential Privacy filter has the biggest IL and the smallest DR follows by the Microagreggation filter. Using both Barranca and Independencia datasets, we notice that Noise Addition and Rank Swapping do not perform well with the credit and debit card transactions.

In order to measure the utility of the sanitized data. We analyze the use case of spending statistics per business category. From Fig. 4, one can observe that categories *RESTBAR* and *ROPMOD* are the most frequent categories with the major number of transactions. Hence, we have computed the spending statistics per business category by taking as input the sanitized datasets produced in the precedent section. It is worth noting that for lack of space only some figures are included.

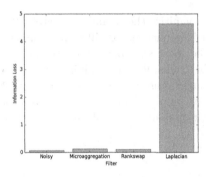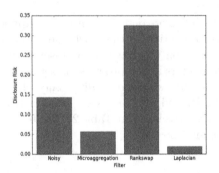

Fig. 2. In green IL and in red DR comparison of the four SDC filters' best configurations in Independencia dataset, where Noise Addition with $c = 0.75$, Microaggregation with $k = 100$, Rank Swapping with $p = 25$, and Differential Privacy with $\epsilon = 1$. (Color figure online)

The Rank Swapping, Microaggregation, and Differential Privacy filters have a similar behavior. In this case, $RESTBAR$, $ROBMOD$, and $TIENDEPART$ categories are bigger and the other ones are less than the statistics computed with the raw dataset. We observe that Differential Privacy and Microaggregation filters give more privacy guarantees at the cost of some IL and a detriment of the utility. Nevertheless, it conserves the shape of the raw dataset. Regarding the Rank Swapping and Noise Addition filters, they do not introduce much noise but they do not protect well. It is worth mentioning that the high or low differences in the sanitized data could be consider as good from the user point of view, while inaccurate but acceptable from the use case perspective (*c.f.*, Fig. 3). Finally, the categories of *bars & restaurants RESTBAR* and *clothes & fashion ROPMOD* are the most disturbed categories because they are more frequent. Thus, filters affect them more than the other categories as depicted in Fig. 4.

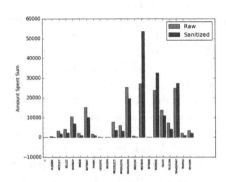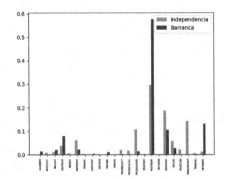

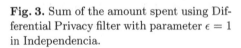

Fig. 3. Sum of the amount spent using Differential Privacy filter with parameter $\epsilon = 1$ in Independencia.

Fig. 4. Amount of transactions in Barranca and Independencia.

5   Conclusion

In the present effort, we tested and benchmarked four different perturbative SDC mechanisms, such as Noise Addition, microagreggation, rank swapping, and Laplacian Differential Privacy filter over raw bank data transaction. We have evaluated their performance using Disclosure Risk and Information Loss measures. As we can observe, Noise Addition does not protect data due to the uncorrelated generated noise added to the protect variable. Concerning Rank Swapping, it needs a high percentage of elements for swapping to achieve a small Disclosure Risk probability. Regarding Microaggregation, it reduces the Disclosure Risk because it replaces close values with the average value of a given variable in the group. Finally, Differential Privacy outperforms the other methods in terms of Disclosure Risk and Information Loss due to the way it adds noise using Laplacian distribution with a small ϵ parameter. As we can observe, the two most important SDC filters are Microaggregation and Laplacian Differential Privacy. Both are able to achieve very low DR with a price of the increment of IL. Consequently, these filters could be applied to make re-identification more difficult as demanded by GDPR regulation.

In the future, we plan to implement more filters to test them, and also apply the filters to a more large dataset. Another research direction is to MapReduce the filters for sanitizing massive datasets.

References

1. Brand, R.: Tests of the applicability of sullivan algorithm to synthetic data and real business data in official statistics. European Project IST-2000-25069 CASC (2002)
2. DeWaal, A., Willenborg, L.: Global recodings and local suppressions in microdata sets. In: Proceedings of Statistics Canada Symposium, vol. 95, pp. 121–132 (1995)
3. Domingo-Ferrer, J., Mateo-Sanz, J.M.: Practical data-oriented microaggregation for statistical disclosure control. IEEE Trans. Knowl. Data Eng. **14**(1), 189–201 (2002)
4. Domingo-Ferrer, J., Torra, V.: Disclosure risk assessment in statistical data protection. J. Comput. Appl. Math. **164**, 285–293 (2004)
5. Domingo-Ferrer, J., Sebé, F., Castellà-Roca, J.: On the security of noise addition for privacy in statistical databases. In: Domingo-Ferrer, J., Torra, V. (eds.) PSD 2004. LNCS, vol. 3050, pp. 149–161. Springer, Heidelberg (2004). https://doi.org/10.1007/978-3-540-25955-8_12
6. Domingo-Ferrer, J., Sebé, F., Solanas, A.: A polynomial-time approximation to optimal multivariate microaggregation. Comput. Math. Appl. **55**(4), 714–732 (2008)
7. Domingo-Ferrer, J., Torra, V.: A quantitative comparison of disclosure control methods for microdata. In: Confidentiality, Disclosure and Data Access: Theory and Practical Applications for Statistical Agencies, pp. 111–134 (2001)
8. Hundepool, A., et al.: Statistical Disclosure Control. Wiley, Hoboken (2012)
9. Kim, J.: A method for limiting disclosure in microdata based on random noise and transformation (2002)

10. Laszlo, M., Mukherjee, S.: Minimum spanning tree partitioning algorithm for microaggregation. IEEE Trans. Knowl. Data Eng. **17**(7), 902–911 (2005)
11. Leo, Y., Karsai, M., Sarraute, C., Fleury, E.: Correlations of consumption patterns in social-economic networks. In: 2016 IEEE/ACM International Conference on Advances in Social Networks Analysis and Mining (ASONAM), pp. 493–500. IEEE (2016)
12. Leoni, D.: Non-interactive differential privacy: a survey. In: Proceedings of the First International Workshop on Open Data, WOD 2012. ACM, New York, pp. 40–52 (2012)
13. Li, N., Lyu, M., Su, D., Yang, W.: Differential privacy: from theory to practice. Synth. Lect. Inf. Secur. Priv. Trust. **8**, 1–138 (2016)
14. Muralidhar, K., Sarathy, R.: Data shufflinga new masking approach for numerical data. Manag. Sci. **52**(5), 658–670 (2006)
15. Nin, J., Herranz, J., Torra, V.: Rethinking rank swapping to decrease disclosure risk. Data Knowl. Eng. **64**(1), 346–364 (2008)
16. Oganian, A.: Multiplicative noise protocols. In: Domingo-Ferrer, J., Magkos, E. (eds.) PSD 2010. LNCS, vol. 6344, pp. 107–117. Springer, Heidelberg (2010). https://doi.org/10.1007/978-3-642-15838-4_10
17. Oganian, A.: Multiplicative noise for masking numerical microdata with constraints. SORT **35**, 99–112 (2011)
18. Reiss, S.P.: Practical data-swapping: the first steps. ACM Trans. Database Syst. (TODS) **9**(1), 20–37 (1984)
19. Reiss, S.P., Post, M.J., Dalenius, T.: Non-reversible privacy transformations. In: Proceedings of the 1st ACM SIGACT-SIGMOD Symposium on Principles of Database Systems. ACM, pp. 139–146 (1982)
20. Rodríguez, D.M., Nin, J., Nuñez-del Prado, M.: Towards the adaptation of SDC methods to stream mining. Comput. Secur. **70**, 702–722 (2017)
21. Sebé, F., Domingo-Ferrer, J., Mateo-Sanz, J.M., Torra, V.: Post-Masking optimization of the tradeoff between information loss and disclosure risk in masked microdata sets. In: Domingo-Ferrer, J. (ed.) Inference Control in Statistical Databases. LNCS, vol. 2316, pp. 163–171. Springer, Heidelberg (2002). https://doi.org/10.1007/3-540-47804-3_13
22. Sullivan, G.R.: The use of added error to avoid disclosure in microdata releases. Iowa State University, Unpublished Ph.D. dissertation (1989)
23. Templ, M.: Statistical Disclosure Control for Microdata: Methods and Applications in R. Springer, Cham (2017). https://doi.org/10.1007/978-3-319-50272-4
24. Templ, M., Meindl, B.: Robust statistics meets SDC: new disclosure risk measures for continuous microdata masking. In: Domingo-Ferrer, J., Saygın, Y. (eds.) PSD 2008. LNCS, vol. 5262, pp. 177–189. Springer, Heidelberg (2008). https://doi.org/10.1007/978-3-540-87471-3_15
25. Willenborg, L., De Waal, T.: Statistical Disclosure Control in Practice, vol. 111. Springer, New York (1996). https://doi.org/10.1007/978-1-4612-4028-0

Big Data for Development: An Approach as a State Government Capacity in the Countries

Marcelino Villaverde Aguilar[✉]

Peruvian Engineers Association, Lima, Peru
mvillaverde@cip.org.pe

Abstract. The implementation of Big Data architectures has proven to be very useful in the management of companies and global corporations, allowing a greater profitability and shorter investment returns, due to the power of processing and the use of analysis algorithms. However, in the public sector, its development still has a long way to go. This paper provides a public governance perspective which is managed by data based on Big Data platforms, achieving a better orientation of public policies with a holistic focus, turning public agencies into creative and innovative entities that provide not only goods and services, but also, provide relevant knowledge of their respective sectors, obtained as a result of their integration and interactions with other public entities, constituting Big Data as a necessary state capacity that governments should strengthen. Thus, the definition of the components of the proposed architecture is shown, whose main orientation is to ensure that the state policies proposed by governments have an impact on quality of life of its citizens.

Keywords: Big Data · Public management · Public policies · Information technology

1 State Capacity and Public Policies

1.1 Conceptualization and Difficulties

State capacity is defined as the ability of government agencies to express, through public policies, the highest possible levels of social value, considering the fundamental public problems of population [1]. Success or failure of any public policy generally depends on capabilities of state institutions. If government institutions have the necessary and sufficient capacities, they will be able to achieve their objectives, making them more reliable before society, otherwise, their functioning will be deficient. Therefore, it is necessary to elevate, improve, build, rebuild or strengthen the levels of State capacity for effective, efficient and sustainable management of public sphere [2].

Intergovernmental articulation is a critical limitation in the management of governments, making their policies and efforts to face urgent problems such as

J. A. Lossio-Ventura et al. (Eds.): SIMBig 2018, CCIS 898, pp. 265–272, 2019.
https://doi.org/10.1007/978-3-030-11680-4_26

poverty or climate change, difficult to implement, costly and ineffective [3]. This lack of articulation reflects the need to propose approaches that allow a better cohesion between different plans and budgetary process to address as a priority problems that afflict its population. To achieve acceptable levels of articulation, information gaps related to implementation and evaluation of public policies must be eliminated, in order to identify a set of indicators to strengthen the state's capacity and make decisions about design and reformulation of policies based on evidence. In this context, adoption of ICT in public management helps define more collaborative and participatory relationships that allow interested parties (citizens, companies and non-governmental organizations) to actively influence the prioritization of public policies and collaborate in design of public services, generating greater cohesion and integration of solutions to complex citizens problems [4]. This expected effect, as a result of technologies adoption, has forced governments to redefine and create capacities through Big Data, as it is one of the tools that allows a better management of information, because it examines large amounts of data, from various data sources and in different formats, allowing decision making at different levels of government in real time or almost real time [5].

1.2 Planning and Complexity

Public policies become instruments of planning and management of the State, which allow it to guide, direct, manage and strategically implement matters of national interest. In this way, the actions of the State entities, around a problem of the population, are under the referent of a public policy [1]. To understand the complexity that this level of articulation requires, we will take the case of Peru. The National Center for Strategic Planning (CEPLAN by its initial in Spanish) has defined several types of plans for each level of government. Table 1, shows the different types of plans, as well as the number of government entities required to prepare them.

Table 1. Types of plans in government of Peru.

Level of government	Plan name	Plans numbers
National	Multianual Sector Strategic Plan (PESEM)	28 Sectors and 19 Ministries
Regional	Regional Concerted Development Plan (PDRC)	26 Regional Governments
Local	Plan of Concerted Local Development. (PDLC)	196 Provincial Mayorships and 1874 District Mayorships

Coordination between plans, as instruments of implementation of public policy, can be considered a complex process due to interaction between various components involved [6], represented by: Government agencies (PO), Public policies of the government (PP), Government Levels (GL) and variables that intervene in each public policy (PV). The 28 national sectors are specialized entities (health, education, agriculture, energy, defense, security, etc.) and multidisciplinary, where public policies are defined within the framework of their competencies, delegating some functions to regional and local governments, in which the following scenarios are presented:

1. Government entities and authorities that converge in public policy with reduced integration and coordination.
2. Variety of criteria and use of autonomy during definition of objectives, goals and interests of population.
3. Dispersion of non-aligned functions and a weak integral approach.
4. Interactions between components of the system and the influence that some components have on behavior of others.

The complexity associated with public policy reflects the need for information that will allow a better articulation among related government entities, interactions between them, and provide information to the affected population, creating a state capacity capable of making decisions and executing actions, concrete actions of governments jointly with interest groups, also providing openness of data and transparency about the policy results.

2   Big Data as State Capacity

2.1   Big Data and Data Analytics in Public Policies

Big Data term is associated with multiple concepts related to data, technologies for processing and techniques, and technologies for the analysis of information. The massive data by themselves, lack value; and this is only obtained after processing and an analysis that allows building valuable knowledge to contribute to governments' decision-making [7] through the application of multiple techniques, such as: association rule learning aimed at discovery of relationships in databases; A/B testing allowing for comparison of control group with a test group, pattern recognition, among others [8]. In this way, Big Data can help governments improve policy design and service delivery by strengthening state capacity and technical considerations that facilitate data integration that fosters greater knowledge exchange to improve the existing government policies. In relation to decision making, the objective is to produce evidence that is relevant, of quality and timely, in order to base and guide decisions that generate solutions that meet population needs in social, demographic and territorial contexts [9].

The data analytic allows to answer the questions and/or hypotheses formulated from modeling and analysis techniques following a certain cycle (see Fig. 1) [7].

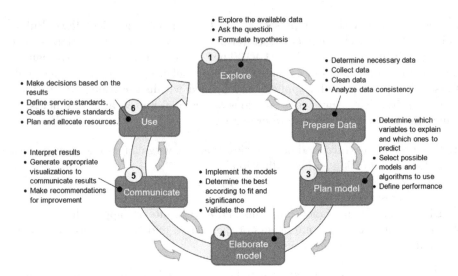

Fig. 1. Life cycle of data analysis as an iterative process. You can go back to previous stages if you need to reformulate the questions based on availability data, or reinterpret the results due to new evidence.

2.2 Critical Factors to Consolidate Big Data as a State Capacity

To develop Big Data and convert it as a strategic state capacity for the interaction of public organisms and decision making in design, implementation and evaluation of public policies, two aspects have been identified:

1. Institutional capacities

   To implement Big Data projects it is necessary to generate a series of institutional capacities within the government [7], at least, in the following three dimensions:

   (a) Development of strategies: A plan that determines problems and which questions are urgent to answer, what data to collect, what techniques will be used to analyze them and which government agencies to articulate.
   (b) Human capital: For the functions of analysis of available information: clean, prepare, format and ensure the reliability of the data and perform specific training in data analysis and solutions based on them.
   (c) Technology: Technological resources for use of large datasets and the software and storage services associated with them. In the same way, it seeks to achieve interoperability between the information systems of public agencies involved in public policy in order to share data.

2. Technical Capabilities

The success of Big Data technologies requires the following requirements [10]:

(a) Simplified use: Development of algorithms and simple analytical applications, easy to access large volumes of data.
(b) Easy Implementation: Implementation schemes for different Big Data applications. In the development stage it can be executed in a single machine and in public or private cloud infrastructure, while the production systems could have to be implemented in a dedicated cluster.
(c) Management of heterogeneity: Big Data applications originate with a large number of heterogeneous, distributed datasets. Only after its aggregation and integration is when Big Data emerges, for that reason, data models, schemes, formats and governance schemes are required.
(d) Improvement of scalability: Especially if several storage and processing tools are used. Scalability is still a problem.

2.3 Types of Data and Need for a Big Data Architecture

Although governments have a large amount of electronic data, the knowledge and trends that may reveal their processing, are usually hidden from public officials, because government information systems process isolated transactions and data formats incompatible, for that reason, knowledge is not available to the government [11]. This scenario is unsustainable considering that citizens are more aware of their rights and demand better access to information and better quality of public services. In this way, the Big Data solution architecture is the basis for discovering, acquiring, refining and providing knowledge to improve public policies of governments, focusing mainly on processing the following types of data [12]:

1. Operational data: information of citizens, suppliers, other government entities and employees, whose source is based on transactions or databases;
2. Dark data: historical information from government archives that cannot be clearly structured. This data includes email messages, contracts, multimedia information;
3. Commercial Data: information obtained from financial history, properties of goods, etc., applicable to any type of data legally available in government entities. Even, data related to a population problem can be shared through the interoperability of the computer systems of public agencies.
4. Public data: data belonging to public institutions, including the national government, ministries, regional governments and local governments;
5. Data of social networks: Activity of citizens in social networks, useful to determine trends, preferences or attitudes.

2.4 Reference Big Data Architecture

The proposed reference architecture is based on that defined by the National Institute of Standards and Technology (NIST) of the United States of America

Table 2. Functional components in the reference Big Data architecture.

Component/Role	Description
System Orchestrator (SO)	Defines and integrates the required data application activities into an operational vertical system
Data Provider (DP)	Introduces new data or information feeds into the ecosystem
Big Data Application Provider (AP)	Executes a life cycle (collection, processing, dissemination) controlled by the system orchestrator to implement specific vertical applications requirements and meet security and privacy requirements
Big Data Framework Provider (FP)	Establishes a computing fabric (computation and storage resources, platforms, and processing frameworks) in which to execute certain transformation applications while protecting the privacy and integrity of data
Data Consumer (DC)	Includes end users or other systems who utilize the results of the Big Data Application Provider

[13] and the ISO working group [14]. Table 2 shows the functions in the Big Data architecture of reference, and the levels of government proposed as responsible for their definition.

In this way, the state agency responsible for the implementation of Public Policy must develop the institutional and technical capacities as activities prior to the implementation of proposed architecture, because the implementation, orchestration and management of the components of the NIST architecture proposal, requires these capabilities to provide the desired results. Similarly, levels of articulation with other public bodies should be clearly defined in service level agreements that establish responsibility for data quality, security levels, rules of use and exploitation of data. Figure 2 shows the interaction between the state body responsible for public policy, the state agencies of different levels of government and the proposed NIST architecture.

The different levels of government intervene as Data Providers, depending on the type of data they contribute to the implementation and monitoring of the public policy of the government, while the person in charge of the implementation is responsible for defining the Big Data Framework Provider and Big data Application Provider, where the techniques of storage, processing, analysis and access to information obtained from the platform are defined. Finally, the Data Consumer are in charge of the study of implementation results of Public Policy.

The proposed architecture is characterized by its scalability, which facilitates the multisectoral articulation to improve the public policies of the governments in areas of citizen security, education, health, infrastructure and citizen

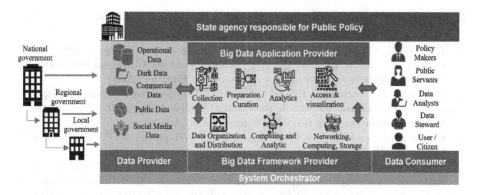

Fig. 2. Interaction between the state entity responsible for public policy, other state agencies and the components of the proposed NIST architecture.

attention, among others. Although no result is causally attributed to the existence of Big Data projects, the results that can be observed in international experiences indicate value that a Big Data architecture can provide to government in compliance with its policies, requiring this, the development of a critical mass in all its components.

3   Conclusions

Countries execute interventions and consider as a fundamental principle, the pursuit of public interest and carry it out through Public Policies, establishing a solid relationship between government and different economic agents that promote the development of its citizens and generate an impact on their life quality. Achieving this efficiency is difficult in a scenario where public administrations are organized in silos, with often overlapping mandates, with fragmented functions in various state entities, generators of large volumes of data that are not processed and a lack of articulation strategies between entities.

To address this problem, many governments have identified as a critical point, the weakness of processing information they have and articulate with other entities to obtain any other information that allows them, a better implementation of public policies in their charge, concluding that, information technologies, and specifically Big Data projects, as the effective tool to obtain knowledge in various sectors of government responsibility, considered in this way, as a necessary state capacity to improve public policies in favor of its citizens, promoting their development and well-being. Regardless of the development and implementation of the technology, it is observed that state entities must develop certain institutional and technical capacities, which will allow Big Data projects to be developed in a sustainable manner over time and guarantee the investment of the assigned resources. In this way, considering Big Data as part of capacities of the states to articulate their institutions and make decisions about problems of their citizens, is fundamental to generate the society development.

References

1. Repetto F.: Capacidad estatal: requisito necesario para una mejor política social en América Latina. In: VIII Congreso Internacional del CLAD sobre la Reforma del Estado y de la Administración Pública, p. 6, Panamá (2003)
2. Completa, E.: Capacidad Estatal: "Qué tipo de capacidades y para qué tipo de estado". POSTData: Revista de Reflexión y Análisis Político **22**, 111–140 (2017). Grupo Interuniversitario Postdata, Buenos Aires, Argentina
3. Organisation for Economic Co-operation and Development - (OECD): Integrated Governance for Inclusive Growth, p. 68, 249, París-Francia (2016)
4. Organisation for Economic Co-operation and Development - (OECD): Recommendation of the Council on Digital Government Strategies, p. 7. Public Governance and Territorial Development Directorate, Editions OECD, París-Francia (2014)
5. Alexandru, A., Alexandru, C. A., Coardos, D., Tudora, E.: Big data: concepts, technologies and applications in the public sector. In: 18th International Conference on Applied Computer Science and Engineering, Conference Proceedings, vol. 18, no. 10 (2016)
6. Mercurea, J., Pollittc, H., Bassid, A., Viñualesb, J., Edwardse, N.: Modelling complex systems of heterogeneous agents to better design sustainability transitions policy. Glob. Environ. Chang. **37**, 102–115 (2016)
7. Rodríguez, P., Palomino, N., Mondaca, J.: El uso de datos masivos y sus técnicas analíticas para el diseño e implementación de políticas públicas en Latinoamérica y el Caribe. Banco Interamericano de Desarrollo, pp. 1–8 (2017)
8. Leszek, A.: The role of big data solutions in the management of organizations. Review of selected practical examples. In: International Conference on Communication, Management and Information Technology, pp. 1008–1010 (2015)
9. Tomar, L., Guicheney, W., Kyarisiima, H., Zimani, T.: Big data in the public sector: selected applications and lessons learned. Inter-American Development Bank, pp. 5–7 (2016)
10. Auer, S., et al.: The BigDataEurope platform – supporting the variety dimension of big data. In: Cabot, J., De Virgilio, R., Torlone, R. (eds.) ICWE 2017. LNCS, vol. 10360, pp. 41–59. Springer, Cham (2017). https://doi.org/10.1007/978-3-319-60131-1_3
11. Goldsmith S., Crawford S., Grohsgal B.: Predictive analytics: driving improvements using data. Inter-American Development Bank, pp. 1–3 (2016)
12. Gartner: Big Data Strategy Components: Business Essentials. http://www.iab.fi/media/tutkimus-matskut/gartner_big_data_strategy_components.pdf. Accessed 21 June 2018
13. National Institute of Standards and Technology (NIST): Big Data Interoperability Framework: Volume 6, Reference Architecture. Special Publication 1500-6, pp. 8–13 (2015)
14. International Organization for Standardization (ISO): ISO/IEC JTC 1 Information technology. Big data preliminary report 2014, pp. 3–11. ISO Published, Switzerland (2015)

Towards Real-Time Automatic Stress Detection for Office Workplaces

Franci Suni Lopez[1,2], Nelly Condori-Fernandez[3,4(✉)], and Alejandro Catala[5]

[1] Universidad Católica San Pablo, Arequipa, Peru
[2] Universidad Nacional de San Agustín de Arequipa, Arequipa, Peru
franci.suni@ucsp.edu.pe, fsunilo@unsa.edu.pe
[3] Universidade da Coruña, A Coruña, Spain
[4] Vrije Universiteit Amsterdam, Amsterdam, The Netherlands
n.condori.fernandez@udc.es, n.condori-fernandez@vu.nl
[5] Centro Singular de Investigacion en Tecnoloxias da Informacion (CiTIUS),
Universidade de Santiago de Compostela, Santiago de Compostela, Spain
alejandro.catala@usc.es

Abstract. In recent years, several stress detection methods have been proposed, usually based on machine learning techniques relying on obstructive sensors, which could be uncomfortable or not suitable in many daily situations. Although studies on emotions are emerging and rising in Software Engineering (SE) research, stress has not been yet well investigated in the SE literature despite its negative impact on user satisfaction and stakeholder performance.

In this paper, we investigate whether we can reliably implement a stress detector in a single pipeline suitable for real-time processing following an arousal-based statistical approach. It works with physiological data gathered by the E4-wristband, which registers electrodermal activity (EDA). We have conducted an experiment to analyze the output of our stress detector with regard to the self-reported stress in similar conditions to a quiet office workplace environment when users are exposed to different emotional triggers.

Keywords: Stress detection · Physiological data · Emotional trigger

1 Introduction

Wearable technology is gaining popularity and the interest to include these devices as useful input for diverse software applications beyond simply gathering data have awakened the interest of software industry too. Moreover, wearable sensing technology for emotion recognition is becoming less obtrusive and inexpensive, what have favored considering biosignals in different sectors such as e-health, e-commerce, wellness, e-learning, and games. Leading organizations recognize that social aspects are just as important to long-term success as economic aspects. Particular focus is given to the labor conditions for reducing risks

J. A. Lossio-Ventura et al. (Eds.): SIMBig 2018, CCIS 898, pp. 273–288, 2019.
https://doi.org/10.1007/978-3-030-11680-4_27

associated with work-related stress. Physiological stress is one of the factors that are most affecting current modern industries. For instance, Graziotin *et al.* [19] proposed a theory about the impact of the effects on programming performance. While studies on emotions are emerging and rising in Software Engineering (SE) research, stress has not been yet well investigated in the SE literature despite its negative impact not only on stakeholder performance but also on user satisfaction and acceptance (*e.g.*, [11,18]).

Currently, most of the stress detection methods are based on Machine Learning (ML) techniques (*e.g.*, support vector machine [35]). The main disadvantage of using these methods is the need for big data sets to carry out the training stage, where the machine learns about the user behavior in relationship with predefined tasks. To address this issue regarding the training of stress detector, we use an arousal-based statistical approach for detecting stress in real-time. This introduces two main advantages for the resulting stress detector: (i) reliability is independent of a training data set, in contrast to the requirement imposed by approaches based on machine learning algorithms; (ii) higher flexibility is provided since the detector can be used in different user conditions. In this paper, we present and evaluate experimentally an automatic stress detector that uses wearable sensors for gathering physiological data. We targeted office employees (programmers and junior researchers) working with computers (*i.e.*, desktop, laptop), who were exposed to several types of emotional triggers.

The paper is organized as follows. Section 2 discusses related works on stress recognition, and Sect. 3 presents some scenarios where a real-time stress detector can be useful in SE. Section 4 presents the theory about emotional triggers. The description of the algorithms used in our stress detector is presented in Sect. 5. Section 6 provides the experiment design and results. Finally, conclusions and future work are discussed in Sect. 7.

2   Related Work

During many years, researchers in several computer science fields have paid attention to developing methods to recognize and understand human emotions. For instance, based on natural language processing (*e.g.*, [27,38,41]); or emotion recognition through facial expression (*e.g.*, [8,25,28]) or using physiological data (*e.g.*, [5,9,15–17,20,21,31,35,36]).

In the strand of works dealing with stress detection relying only on physiological data, Mozos *et al.* [31] proposed to combine machine learning techniques using EDA, photoplethysmogram (PPG) and heart rate variability (HRV) signals to detect stress in social situations using The Trier social stress test (TSST) as stressful. Garcia *et al.* [15] used accelerometer (ACC) data of a mobile phone to recognize stress in real workplace environments of thirteen subjects using two classification models: naive bayes and decision trees. They obtained an accuracy of 71% and their study lasted 8 weeks.

Sanno and Picard [35] implemented different machine learning classifiers to detect stress: Support Vector Machine (SVM) with linear kernel, SVM with

Radial basis function (RBF) kernel, k-nearest neighbors, Principal component analysis (PCA) and SVM with RBF kernel and k-nearest neighbors. Their work is focused on comparing the performance the implemented algorithms using the collected data of the subjects (skin conductance, ACC and mobile phone usage) of five days. Kocielnik *et al.* [23] described a framework to detect stress in the context of a person's activities. They use a min-max algorithm and ACC as source data. Bogomolov *et al.* [5] collected mobile phone activity (i.e. call log, SMS log, Bluetooth interactions) of 117 subjects to recognize stress during common daily activities. They applied different classifiers: SVM, artificial neural networks, ensemble of tree classifiers based on a Breiman's Random Forest (RF) and Friedmans Generalized Boosted Model (GBM). Similarly, using a range of machine learning techniques, some other examples can be found in the existing literature (*e.g.*, [3,10,21,34,36,37]).

Table 1. Comparative chart of the most representative related works of stress recognition.

Author	Classification algorithm	Source data	Evaluation tools	Context
Mozos *et al.* [31]	SVM, AdaBoost, and k-nearest neighbor	EDA, PPG and HRV	Accuracy of 89.75%, precision of 89.5% and recall of 95%	Social situations using the TSST
Garcia-Ceja *et al.* [15]	Naive bayes and decision trees	ACC	Accuracy of 71%	Real working environments
Sano and Picard [35]	SVM, RBF, k-nearest neighbors, PCA, SVM and PCA	SC, ACC and mobile phone usage	Accuracy of 75%	Stress detection that subjects are able to perceive and report
Kocielnik *et al.* [23]	Min-max algorithm	SC and ACC	No reported	Subject's activities
Bogomolov *et al.* [5]	SVM, ANNs, tree classifiers based on RF and GBM	Mobile phone activity	Accuracy of 72.39%	Common daily activities

Table 1 summarizes the most representative related work, illustrating the diversity of used algorithms to recognize stress. Most of these works use a machine learning method to implement the classifier; as we indicated previously, there can be some issues concerning the large training datasets required and their request to gather data and train new classifiers for every single task/context of use. In contrast, we use an arousal-based statistical approach that does not need a big dataset to learn a model for recognizing stress and can work in different tasks. Next we envision prospective scenarios of use in the context of SE in which a real-time stress detector could be useful.

3   Scenarios of Use in the Context of SE

A real-time stress detector could enhance/contribute to the **emotional labor in SE**, which refers to the process of managing feelings and expressions to fulfill the emotional requirements of a software engineer. For instance, in tasks that demand the collaboration of stakeholders with different perspectives, such as reviews-based requirements validation [13], analysts could get awareness on their stress level, which can be helpful not only for regulating their negative emotions but also for having a better performance in validating requirements.

Another potential scenario is in the development of large and complex software projects that require a continuous evolution, and maintenance, where the history of stress could help human resources managers in their decision-making processes. For instance, identifying members of a development team, who could be experiencing long-term stress that might be affecting their productivity. These members suffering stress could become potential deserters from the company. Two general worth exploring scenarios where a real-time stress detector can be useful are:

Usability and Software Testing Based on Emotions for Quality Assurance and User Experience: To detect interaction pitfalls or defects in the user interface and/or software functionality that could cause certain level of stress on end-users, which can be used to enhance the software quality [11]. It has a diagnosis purpose of the software developed, and therefore as an additional input to assure quality in future software versions. The real-time feature is relevant to determine which variations in the stress detection are associated to the use of specific parts of the software (*e.g.*, elements of the user interface [32], awareness elements of a software game [39]) which can facilitate us to discover new quality requirements from actual user needs (*e.g.*, [12]).

Development of Self-adaptive Software Systems Guided by Emotion: As a kind of context-aware system, in which part of the user context is provided by the emotional states over time while interacting with software systems, self-adaptation could be guided by emotions ([29,30]).

4   Emotional Triggers

An emotion is just a response we give to a stimulus or event, whether it is external, or even internal, such as a memory or an idea [14]. Additionally, in experimental settings, researchers can generate emotions on users intentionally, by using specific emotional triggers determined by the emotions to be induced.

In this respect, an emotional trigger is any stimulus that generates a negative or positive emotion (*e.g.*, uncomfortable or comfortable temperature, environmental noise, *etc.*). According to Kanjo *et al.* [22], emotional triggers can be classified into seven types: environment, physical movements, memories, perception, interacting with others, accomplishments and failure.

Nowadays, different kinds of emotional triggers exist. We briefly introduce existing stress triggers that will be used to evaluate our stress detector: Westman and Walters [40] and Passchier-Vermeer and Passchier [33] considered an environmental trigger, where participants are exposed by five minutes to listen fire alarm sounds. The Sing-a-Song Stress Test [7] is a social trigger, where participants are asked to sing a song aloud for 30 s with their arms still. An example of a cognitive trigger is the Stroop Task [24], where participants have to pay attention and react to the color of a word while ignoring the word itself. A reduced version of this trigger works with 4 colors, using the words "green","red", "yellow","blue" written in all four different colors. Words are presented randomly for each participant.

5   Automatic Stress Detector

Figure 1 shows the stress detection process of our approach that has been automated to detect stress of individuals (*e.g.*, programmers, testers) in real time. We use wearable sensors (*i.e.*, E4-wristband[1]) to collect physiological data. In this first version, we focus only on sensing electrodermal activity (EDA) as a main input for the implementation of the stress detector. A transient increase on the EDA signal is proportional to sweat secretion and it is related to stress [2,6]. The main functionality of the stress detector is to determine whether the user is stressed or not. The detector will mark a label of *"stressed"* or *"not stressed"*. We have implemented the preprocessing steps proposed by Bakker *et al.* [2] for arousal detection in an integrated pipeline to enable real-time processing (see Fig. 1 for the involved preprocessing steps). Next, we explain the methods/algorithms that were used in the stress detection process.

Fig. 1. Overview of a stress detection process.

5.1   Noise Filter

We use *EDA* to recognize stress changes, then the first stage in the pipeline of the stress detector is to collect raw signals by the Empatica's E4-wristband, Fig. 2 (part a) presents a common sample of *EDA* signals, which is measured in microSiemens (μS), a unit of electric conductance. Usually for measuring EDA is required two electrodes that need skin contact to produce a reliable signal, therefore the quality of the collected *EDA* signals depends on the continuity of

---

[1] https://www.empatica.com/e4-wristband.

the contact between user skin and the device's sensors. However, the contact is not the same in all users and noise could be introduced in the signal. Hence, noise filtering is needed to mitigate these issues in the input (*e.g.,* in Fig. 2 (part a), we can find some gaps as a consequence of weak skin contact). Before analyzing *EDA* signals, it is important to clean raw data, because noise might be mistaken as genuine peaks. Therefore, the first step of the preprocessing is to apply a median filter over a moving window of size $n = 100$ EDA samples, as suggested in [2]. Figure 2 (b) shows the noise filtering of the collected raw data.

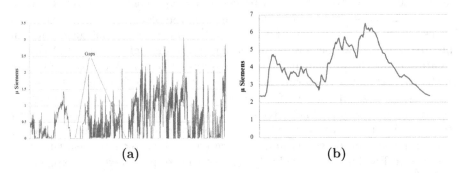

Fig. 2. (a) Gaps occur when the contact between the user skin and the sensors is not tight. (b) Filtered raw *EDA* signals.

5.2   Aggregation

The EDA signal acquired by the E4-wristband is sampled at 4 Hz (*i.e.,* the device provides 4 samples or readings per second, which means 240 samples per minute). Based on [2], we apply an aggregation step of each minute over the filtered input signal: given y' is a moving window of size $m = 240$ (the EDA samples of one minute), where $y_1, ..., y_m$ is aggregated to a single value y'' where $y'' = max(y')$. For instance, Fig. 3 (part a) shows the aggregation of approximately 7200 filtered EDA samples to 30 representative points (collected signals of 30 min).

5.3   Discretization

In this step, the data is discretized using the symbolic aggregate approximation (*SAX*) method [26]. It is a means for very efficient local discretization of time series subsequence from 1 to 5 that can be interpreted as levels of stress variation (1: completely relaxed to 5: maximum arousal). Those levels should not be understood as absolute levels of arousal, but rather as a local relative measure of arousal.

The input for *SAX* is a time-series X of length n and the output is a string of length w, where $w < n$ typically, the output string is normalized to an alphabet of size > 2. The algorithm consists of the following two stages:

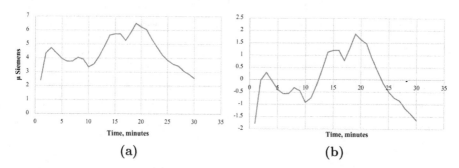

Fig. 3. (a) Aggregated process over previous filtered data. (b) Z-normalization of aggregated data.

- Transformation of original time-series into a Piecewise Aggregate Approximation (PAA) representation. To do this, first it is necessary a Z-normalization
  (see Eq. 1), where the mean is around 0 and the standard deviation is close
  to 1, using the following formula:

$$x_i' = \frac{x_i - \mu}{\rho} \tag{1}$$

$$\overline{x}_i = \frac{M}{n} \sum_{j=\frac{n}{M}(i-1)+1}^{(\frac{n}{M})i} x_j \tag{2}$$

where μ is the mean of the time series and ρ is the standard deviation. After
the Z-normalization, we can apply PPA transform, which approximates the
time-series into vector $\overline{X} = (\overline{x}_1, ..., \overline{x}_M)$ of length $M \leq n$ (See Fig. 3 (b)).
Where each $\overline{x}_i$ is calculated with the Eq. 2. With the objective to reduce the
dimensionality from n to M, first we divide the time-series to n/M equally
sized samples and calculate the mean for each sample (See Fig. 4 (a)).
- Transformation of the PAA data into a string. The method use a breakpoint or
  cuts $B = \beta_1, \beta_2, ..., \beta_{\alpha-1}$ such that $\beta_{i-1} < \beta_i$ and $\beta_0 = -\infty, \beta_\alpha = \infty$ divides
  the total area in equal subareas. Additionally, it assigns a symbol $alpha_j$ to
  each interval $[\beta_{j-1}, \beta_j)$, and the final conversion from PAA coefficients $\overline{C}$ into
  a SAX string $\hat{C}$ is with the Eq. 3. Figure 4 (b) shows SAX transformation of
  the previous preprocessing signals.

$$\hat{c} * i = alpha * j, iif, \overline{c} * i \in [\beta_{j-1}, \beta_j) \tag{3}$$

5.4   Change Detection

We use a change detection algorithm based on ADaptive WINdowing (ADWIN)
method [4]. ADWIN computes the mean for each split of a sequence of signals and
analyzes the statistically significant difference between two consecutive splits.

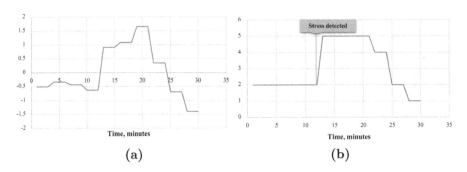

Fig. 4. (a) PAA representation of the preprocessing data. (b) SAX representation and stress detection using ADWIN algorithm.

When a statistically significant difference is detected at point p_i, ADWIN drops the data backwards from p_i, after it repeats the splitting procedure until no significant differences be found in the sequence. For instance, given ϕ_1 and ϕ_2 as the means of two splits of a sequence of EDA signals, then $|\phi_1 - \phi_2| > \epsilon_{cut}$ is the condition for a change detection that is computed with the Eq. 4.

$$\epsilon_{cut} = \sqrt{\frac{2}{m}.\sigma_W^2.ln\frac{2}{\delta'} + \frac{2}{3m}ln\frac{2}{\delta'}} \qquad (4)$$

where σ_W^2 is the variance of the elements of W. δ is the desired confidence and $\delta' = \delta/(ln\ n)$ [2]. Figure 4 (b) shows the output the algorithm detecting a stress change.

6   Experiment

In order to validate the stress detector implemented, we designed an experiment where participants experienced different stressful situations caused by emotional triggers introduced in Sect. 4.

The goal of the experiment is to evaluate the performance of our stress detector in terms of its accuracy. This evaluation was performed from the viewpoint of office workers in the context of performing certain tasks that cause stress (emotion trigger). From this goal, the following research question is derived:

RQ$_1$: *How accurately is the stress detector able to recognize subjects stress under different types of emotional triggers?*

Based on the defined research question, we have the **independent variables** *emotional trigger*, originally with 3 levels (environmental: fire alarm; cognitive: Stroop Task; and social: Sing-a-Song Stress Test). After running a pilot study, we decided to remove the social trigger to reduce the length of the experiment to thirty minutes. This is further explained in the data collection section. Figure 5 (c) shows a screenshot of the instructions for the Stroop Task.

As **dependent variables:** *subject stress status*, which is measured in a nominal scale (stressed or not stressed); and *perceived stress* measured by means of a self-response questionnaire.

Our hypothesis is that *when different types of emotional triggers are delivered, the stress detector is able to recognize stress with a similar accuracy.* Accuracy refers to the closeness of a measured value to a "true value". In our study, the true value of perceived stress was determined by the subjects of the experiment.

6.1 Subjects

Twelve subjects from University of Twente (The Netherlands), involved in research in computing areas (*i.e.*, Master students, PhD candidates), participated voluntarily in the experiment, whose ages ranged between 21 and 32 years old. Seven are women and five men.

6.2 Instrumentation and Procedure

The experiment was carried out in a quiet room equipped with a table and a chair as shown in Fig. 5(a). Subjects interacted with a laptop where the Stroop Task (cognitive trigger implemented with Psychopy[2]) was installed. Also, subjects wore the E4-wristband and headphones to interact with the environmental trigger. Figure 5 (b) shows the correct position of the E4-wristband on the non-dominant hand of the subject.

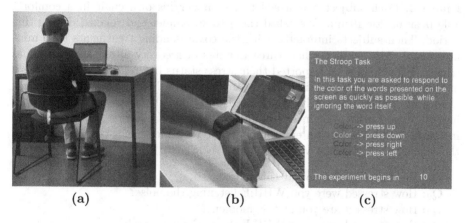

(a)　　　　　　　　　(b)　　　　　　　　　(c)

Fig. 5. (a) A subject in the experiment room interacting with emotional triggers. (b) E4-wristband placed on the non-dominant hand. (c) Instructions for interacting with cognitive trigger (stroop task).

The evaluation followed a within-subjects design, where all subjects were exposed individually to both cognitive and environmental triggers (treatments).

---

[2] http://www.psychopy.org/.

The order in which the subjects interacted with the treatments were assigned randomly. Figure 6 shows the procedure of the experiment that consists of two phases:

Phase 1. Firstly, the subjects were asked to read and sign the informed consent form, which described the purpose and structure of the experiment. Subjects were informed beforehand about the sensing device and the possibility of experiencing some stress during the experiment. Furthermore, they were informed that they could pass on the task at any time if they considered stress unbearable. After signing the consent form, each subject got put on and adjusted the E4-Wristband to enable the gathering of physiological data. Then subjects were asked to complete a demographic questionnaire, and press a button to start the experimental tasks when they were ready. This phase lasts around five minutes.

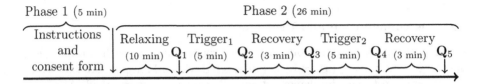

Fig. 6. Experiment procedure and timeline.

Phase 2. Each subject was asked to sit on her/his own chair in a comfortable position for 10 min. We asked them to stay quiet and relaxed during this period. Then subjects interacted with the corresponding treatments (five minutes each). Also, subjects had three minutes of recovery after each emotional trigger. Participants self-reported their stress status before, during the delivery of the corresponding emotion trigger and after the last trigger, the questions were answered progressively. The closed questions that were formulated during the experiment were in a 7-point-ordinal-scale (presented on-screen). For instance, delivering first an environmental trigger and then a cognitive trigger, the sequence of questions were as follows (see Fig. 6):

- Q_1: How stressed are you at this moment?
- Q_2: How stressed were you WHILE listening the noise?
- Q_3: How stressed are you at this moment?
- Q_4: How stressed were you WHILE doing the color task?
- Q_5: How stressed are you at this moment?

6.3 Data Collection

The twelve subjects S01-S12 interacted with two emotional triggers successfully (i.e. environmental and cognitive). The experiment obtained an ethical approval from the Ethics Committee of the Faculty of Electrical Engineering, Mathematics and Computer Science of the University of Twente. Raw data and questionnaires answers were encrypted (WinZip AES encryption: 128-bit AES) and stored in a secured remote location for later analysis.

We validated the experimental design with a pilot study involving two participants (who did not take part in the final evaluation), to ensure the task descriptions were fully understandable, its implementation error-free, and to check the time and any further issue regarding the experimental design. The initial experiment was originally designed with three emotional triggers (cognitive, social and environmental), and approximately it lasted forty minutes. The feedback collected in the pilot suggested that the experiment was going to last too long and that the social trigger was not causing stress as expected given the time available. Hence, we changed our design to exclude the social-emotional trigger in order to prevent issues and reduce the experiment time approximately to thirty minutes (around 6480 EDA samples).

Table 2. Labeled results of the questionnaires and stress detector.

Subject	Trigger	Reported stress	Stress detector	Trigger	Reported stress	Stress detector
S01		Stressed	Stressed		Not stressed	Not stressed
S02		Not stressed	Not stressed		Not stressed	Not stressed
S03		Not stressed	Not stressed		Stressed	Stressed
S04		Not stressed	Not stressed		Not stressed	Not stressed
S05	Cognitive	Stressed	Not stressed	Environmental	Stressed	Not stressed
S06		Stressed	Not stressed		Not stressed	Not stressed
S07		Not stressed	Not stressed		Not stressed	Not stressed
S08		Not stressed	Not stressed		Not stressed	Not stressed
S09		Not stressed	Not stressed		Not stressed	Not stressed
S10		Stressed	Stressed		Not stressed	Not stressed
S11		Not stressed	Not stressed		Not stressed	Stressed
S12		Not stressed	Not stressed		Not stressed	Stressed

6.4   Threats to Validity

Internal Validity. As our main objective is to evaluate the performance of our stress detector, in this study, we decided to use three well-known emotional triggers from the psychology community. However, given that these triggers had not been used previously in Software Engineering, we acknowledge the fact that the selected triggers could not always generate stress on the subjects(programmers and researchers in computer science) due to different other factors (*e.g.*, greater resilience) that were not investigated in this study.

Given this interaction with different emotional triggers (treatments), with the purpose of avoiding that the the first emotional trigger does not affect on the next one, we set out relaxing and recovery periods.

Another possible threat is the effect of the instrumentation used during the experiment (*i.e.,* E4-Wristband), which could also have been causing any stress level. In order to know whether this instrument could be considered as additional potential triggers, we asked participants to complete a post-questionnaire regarding this issue for further investigation.

External Validity. Given the low number of subjects and the fact they were researchers working in computing-related areas but not fully working as software engineers, one potential threat to external validity is regarding the generalization of our results. Moreover, as our controlled experiment was conducted in a lab setting, involving real practitioners would have been harder. We think that having this lab-setting still allowed us to evaluate the stress detector without interruption of external factors (*i.e.,* meetings, calls) as a first necessary step to validate and continue developing the detector.

Construct Validity. The use of a single device to measure physiological stress (a construct) could be considered as main threat to construct validity of this study.

6.5   Results

The self-reported stress was rated in 7-point ordinal-scale questions that were gathered before and during the experiment (1 = "*not stressed*" to 7 = "*extremely*"). We labeled the overall self-reported score as "*stressed*" when the difference led to an increase equal o higher than 3 (threshold) in the perceived value of self-reported stress; otherwise, it was labeled as "*not stressed*" according to [5]. Table 2 summarizes the labels for each subject to assess the accuracy or *trueness* of our detector, by comparing the computed label with the final self-reported label; red cells indicate the cases where the stress detector missed[3].

For answering our research question, we use the following (well-known) metrics regarding: precision (Eq. 5), recall (Eq. 6) and accuracy (Eq. 7):

$$precision = \frac{TP}{TP + FP} \tag{5}$$

$$recall = \frac{TP}{TP + FN} \tag{6}$$

$$accuracy = \frac{TP + TN}{TP + TN + FP + FN} \tag{7}$$

Where TP indicates true positives, TN true negatives, FP false positive and FN false negatives. In our case, we consider true positive cases the examples where reported stress and stress detector are labeled as *stressed* (see Table 2).

---

[3] Raw data and details of each subject can be found at https://goo.gl/eQ4KC2.

We obtained an accuracy of 79.17%, a precision of 60% and a recall of 50%. By comparing our results with other machine-learning based recognition methods, we consider that our method has a good accuracy because it oscillates between 70% and 85% [1, 15], values reported in the literature of stress recognition using physiological data. It is also important to remark that most of these existing recognition methods do not report precision and recall measures. Overall, a method that gets a high recall value it is considered as a good detector. In our study, despite we obtained a 50% of recall, we consider that it can be due to the threats that were identified after the results analysis.

7   Conclusions and Future Work

In this paper, we present the stress detection process of the arousal-based statistical method. The different algorithms and techniques used for supporting such detection process were implemented and integrated as part of our real-time automatic stress detector, in a single processing pipeline. In order to evaluate its accuracy, an experiment was conducted with 12 subjects using the E4-wristband device to gather physiological data. Comparing the outcome of our stress detector with the reported by each subject (perceive stress), the detector obtained an accuracy of 79.17%.

An interesting observation is that although some subjects did not feel stress when an emotional trigger was delivered, the outcome of the detector was consistent with the corresponding perceived stress value. However, from this observation, we can also see that the emotional trigger was not very effective for generating stress in all cases (subjects). A possible explanation for this might be due to the different resilience extent of our participants or the need to exposing them longer to the stimuli. As a future empirical work, beyond checking the accuracy of the detector, more research is needed regarding the role of emotional triggers and resilience of people working in office workplaces (e.g., developer facing stress in unexpected situations). To do this, we plan to conduct a field experiment with practitioners from a Spanish SME involved in software projects with multi-tenancy characteristics.

Acknowledgments. Authors would like to thank to Dirk Heylen, head of HMI Lab of University of Twente, for facilitating us the HMI Lab to conduct the experiments and his early feedback. Also, We thank all the participants who took part in our research. This work has been supported by grant 234-2015-FONDECYT (Master Program) from Cienciactiva of the National Council for Science, Technology and Technological Innovation (CONCYTEC-PERU). Moreover, this work has received financial support from the Spanish Ministry of Economy, Industry and Competitiveness with the Project: TIN2016-78011-C4-1-R; Council of Culture, Education and University Planning with the project ED431G/08, the European Regional Development Fund (ERDF).

References

1. Alberdi, A., Aztiria, A., Basarab, A.: Towards an automatic early stress recognition system for office environments based on multimodal measurements: a review. J. Biomed. Inf. **59**, 49–75 (2016)
2. Bakker, J., Pechenizkiy, M., Sidorova, N.: What's your current stress level? detection of stress patterns from gsr sensor data. In: Proceedings of the 2011 IEEE 11th International Conference on Data Mining Workshops ICDMW 2011, pp. 573–580. IEEE Computer Society, Washington (2011)
3. Bauer, G., Lukowicz, P.: Can smartphones detect stress-related changes in the behaviour of individuals? In: 2012 IEEE International Conference on Pervasive Computing and Communications Workshops, pp. 423–426, March 2012
4. Bifet, A., Gavaldà, R.: Learning from time-changing data with adaptive windowing. In: Proceedings of the 2007 SIAM International Conference on Data Mining, pp. 443–448. Society for Industrial and Applied Mathematics, April 2007
5. Bogomolov, A., Lepri, B., Ferron, M., Pianesi, F., Pentland, A.S.: Pervasive stress recognition for sustainable living. In: 2014 IEEE International Conference on Pervasive Computing and Communication Workshops (PERCOM WORKSHOPS), pp. 345–350, March 2014
6. Boucsein, W.: Electrodermal Activity. Springer, New York (2012)
7. Brouwer, A.M., Hogervorst, M.A.: A new paradigm to induce mental stress: the sing-a-song stress test (SSST). Front. Neurosci. **8**, 224 (2014)
8. Busso, C., et al.: Analysis of emotion recognition using facial expressions, speech and multimodal information. In: Proceedings of the 6th International Conference on Multimodal Interfaces ICMI 2004, pp. 205–211. ACM, New York (2004)
9. Canento, F., Fred, A., Silva, H., Gamboa, H., Lourenço, A.: Multimodal biosignal sensor data handling for emotion recognition. In: 2011 IEEE Sensors Proceedings, pp. 647–650, October 2011
10. Carneiro, D., Castillo, J.C., Novais, P., Fernández-Caballero, A., Neves, J.: Multimodal behavioral analysis for non-invasive stress detection. Expert Syst. Appl. **39**(18), 13376–13389 (2012)
11. Condori-Fernandez, N.: Happyness: an emotion-aware QoS assurance framework for enhancing user experience. In: Proceedings of the 39th International Conference on Software Engineering Companion ICSE-C 2017, pp. 235–237. IEEE Press, Piscataway (2017)
12. Condori-Fernandez, N., Catala, A., Lago, P.: Discovering requirements of behaviour change software systems from negative user experience. In: Proceedings of the 40th International Conference on Software Engineering: Companion Proceeedings ICSE 2018, pp. 222–223. ACM, New York (2018). https://doi.org/10.1145/3183440.3195072
13. Condori-Fernandez, N., España, S., Sikkel, K., Daneva, M., González, A.: Analyzing the effect of the collaborative interactions on performance of requirements validation. In: Salinesi, C., van de Weerd, I. (eds.) REFSQ 2014. LNCS, vol. 8396, pp. 216–231. Springer, Cham (2014). https://doi.org/10.1007/978-3-319-05843-6_16
14. Ekman, P.: Basic emotions. In: Handbook of Cognition and Emotion, pp. 45–60. Wiley, January 1999
15. Garcia-Ceja, E., Osmani, V., Mayora, O.: Automatic stress detection in working environments from smartphones x2019; accelerometer data: a first step. IEEE J. Biomed. Health Inform. **20**(4), 1053–1060 (2016)

16. Girardi, D., Lanubile, F., Novielli, N.: Emotion detection using noninvasive low cost sensors. In: 2017 Seventh International Conference on Affective Computing and Intelligent Interaction (ACII), IEEE, October 2017
17. Gouizi, K., Maaoui, C., Reguig, F.B.: Negative emotion detection using EMG signal. In: 2014 International Conference on Control, Decision and Information Technologies (CoDIT), pp. 690–695, November 2014
18. Graziotin, D., Wang, X., Abrahamsson, P.: Software developers, moods, emotions, and performance. IEEE Softw. **31**(4), 24–27 (2014)
19. Graziotin, D., Wang, X., Abrahamsson, P.: How do you feel, developer? an explanatory theory of the impact of affects on programming performance. PeerJ Comput. Sci. **1**, e18 (2015)
20. Guendil, Z., Lachiri, Z., Maaoui, C., Pruski, A.: Emotion recognition from physiological signals using fusion of wavelet based features. In: 2015 7th International Conference on Modelling, Identification and Control (ICMIC), pp. 1–6, December 2015
21. Healey, J.A., Picard, R.W.: Detecting stress during real-world driving tasks using physiological sensors. Trans. Intell. Transp. Syst. **6**(2), 156–166 (2005)
22. Kanjo, E., Al-Husain, L., Chamberlain, A.: Emotions in context: examining pervasive affective sensing systems, applications, and analyses. Pers. Ubiquit. Comput. **19**(7), 1197–1212 (2015)
23. Kocielnik, R., Sidorova, N., Maggi, F.M., Ouwerkerk, M., Westerink, J.H.D.M.: Smart technologies for long-term stress monitoring at work. In: Proceedings of the 26th IEEE International Symposium on Computer-Based Medical Systems, pp. 53–58, June 2013
24. Lattimore, P.: Stress-induced eating: an alternative method for inducing ego-threatening stress. Appetite **36**(2), 187–188 (2001)
25. Le, H.T., Vea, L.A.: A customer emotion recognition through facial expression using kinect sensors v1 and v2: a comparative analysis. In: Proceedings of the 10th International Conference on Ubiquitous Information Management and Communication IMCOM 2016, pp. 80:1–80:7. ACM, New York (2016)
26. Lin, J., Keogh, E., Lonardi, S., Chiu, B.: A symbolic representation of time series, with implications for streaming algorithms. In: Proceedings of the 8th ACM SIGMOD Workshop on Research Issues in Data Mining and Knowledge Discovery DMKD 2003, pp. 2–11. ACM, New York (2003)
27. Liu, Y., Yu, X., Chen, Z., Liu, B.: Sentiment analysis of sentences with modalities. In: Proceedings of the 2013 International Workshop on Mining Unstructured Big Data Using Natural Language Processing UnstructureNLP 2013, pp. 39–44. ACM, New York (2013)
28. Menne, I.M., Lugrin, B.: In the face of emotion: a behavioral study on emotions towards a robot using the facial action coding system. In: Proceedings of the Companion of the 2017 ACM/IEEE International Conference on Human-Robot Interaction HRI 2017, pp. 205–206. ACM, New York (2017)
29. Mocholi, J., Jaen, J., Krynicki, K., Catala, A., Picón, A., Cadenas, A.: Learning semantically-annotated routes for context-aware recommendations on map navigation systems. Appl. Soft Comput. J. **12**(9), 3088–3098 (2012). https://doi.org/10.1016/j.asoc.2012.05.010
30. Mocholí, J.A., Martínez, J.J., Catalá, A.: Towards affection integration on context-aware recommendation of semantically annotated routes. In: Workshop Proceedings of the 7th International Conference on Intelligent Environments, IE 2011, Nottingham, 25–28 July 2011, pp. 51–62 (2011). https://doi.org/10.3233/978-1-60750-795-6-51

31. Mozos, O.M., et al.: Stress detection using wearable physiological and sociometric sensors. Int. J. Neural Syst. **27**(02), 1650041 (2017)
32. Ormeño, Y.I., Panach, J.I., Condori-Fernandez, N., Pastor, O.: Towards a proposal to capture usability requirements through guidelines. In: IEEE 7th International Conference on Research Challenges in Information Science (RCIS), pp. 1–12, May 2013. https://doi.org/10.1109/RCIS.2013.6577677
33. Passchier-Vermeer, W., Passchier, W.: Noise exposure and public health. Env. Health Perspect. **108**(suppl 1), 123–131 (2000)
34. Sandulescu, V., Andrews, S., Ellis, D., Bellotto, N., Mozos, O.M.: Stress detection using wearable physiological sensors. In: Ferrández Vicente, J.M., Álvarez-Sánchez, J.R., de la Paz López, F., Toledo-Moreo, F.J., Adeli, H. (eds.) IWINAC 2015. LNCS, vol. 9107, pp. 526–532. Springer, Cham (2015). https://doi.org/10.1007/978-3-319-18914-7_55
35. Sano, A., Picard, R.W.: Stress recognition using wearable sensors and mobile phones. In: 2013 Humaine Association Conference on Affective Computing and Intelligent Interaction, pp. 671–676, September 2013
36. Sriramprakash, S., Prasanna, V.D., Murthy, O.R.: Stress detection in working people. Procedia Comput. Sci. **115**, 359–366 (2017). 7th International Conference on Advances in Computing & Communications, ICACC-2017, 22–24 August 2017. Cochin, India (2017)
37. Sun, F.-T., Kuo, C., Cheng, H.-T., Buthpitiya, S., Collins, P., Griss, M.: Activity-aware mental stress detection using physiological sensors. In: Gris, M., Yang, G. (eds.) MobiCASE 2010. LNICST, vol. 76, pp. 282–301. Springer, Heidelberg (2012). https://doi.org/10.1007/978-3-642-29336-8_16
38. Tang, D.: Sentiment-specific representation learning for document-level sentiment analysis. In: Proceedings of the Eighth ACM International Conference on Web Search and Data Mining WSDM 2015, pp. 447–452. ACM, New York (2015)
39. Teruel, M.A., Condori-Fernandez, N., Navarro, E., González, P., Lago, P.: Assessing the impact of the awareness level on a co-operative game. Inf. Softw. Technol. **98**, 89–116 (2018). https://doi.org/10.1016/j.infsof.2018.02.008. http://www.sciencedirect.com/science/article/pii/S0950584918300314
40. Westman, J.C., Walters, J.R.: Noise and stress: a comprehensive approach. Environ. Health Perspect. **41**, 291–309 (1981)
41. You, Q.: Sentiment and emotion analysis for social multimedia: methodologies and applications. In: Proceedings of the 2016 ACM on Multimedia Conference MM 2016, pp. 1445–1449. ACM, New York (2016)

SoTesTeR: Software Testing Techniques' Recommender System Using a Collaborative Approach

Ronald Ibarra[1,2] and Glen Rodriguez[1(✉)]

[1] Universidad Nacional Mayor de San Marcos, Lima, Peru
{ronald.ibarra, grodriguezr}@unmsm.edu.pe
[2] Universidad Nacional Agraria de la Selva, Tingo María, Peru
ronald.ibarra@unas.edu.pe

Abstract. Software testing is a key factor on any software project; testing costs are significant in relation to development costs. Therefore, it is essential to select the most suitable testing techniques for a given project to find defects at the lower cost possible in the different testing levels. However, in several projects, testing practitioners do not have a deep understanding of the full array of techniques available, and they adopt the same techniques that were used in prior projects or any available technique without taking into consideration the attributes of each testing technique. Currently, there are researches oriented to support selection of software testing techniques; nevertheless, they are based on static catalogues, whose adaptation to any niche software application may be slow and expensive. In this work, we introduce a content-based recommender system that offer a ranking of software testing techniques based on a target project characterization and evaluation of testing techniques in similar projects. The repository of projects and techniques was completed through the collaborative effort of a community of practitioners. It has been found that the difference between recommendations of SoTesTeR and recommendations of a human expert are similar to the difference between recommendations of two different human experts.

Keywords: Software testing techniques · Recommender system ·
Content-based reasoning · Collaborative repository · k-Nearest Neighbors

1 Introduction

According to [1], software testing is expensive for the industry and is always limited by time and effort. "Software testing is a series of process which is designed to make sure that the computer code does what it was designed to do. The main purpose of testing can be: quality assurance, reliability estimation, validation or verification" [2].

So, testing is present in all software lifecycle and different techniques are applied in each level of testing: unit, integration, system and acceptance testing. "Software testing techniques are based in an amalgam of methods drawn from graph theory, programming language, reliability assessment, reliable-testing theory, etc." [3].

J. A. Lossio-Ventura et al. (Eds.): SIMBig 2018, CCIS 898, pp. 289–303, 2019.
https://doi.org/10.1007/978-3-030-11680-4_28

Considering the consequences of failures and associated costs, the selection of the most efficient and effective testing methods is very important. Although there are many testing techniques, there are no scientific guidelines for the selection of appropriate techniques in different domains and contexts. Moreover, [4] indicates that there are many testing techniques, which make the selection process complex, because professionals want to know which of those techniques will detect defects that interest them the most in the programs they plan to test.

Both [5] and [1] agree that testing professionals have little or no information about available techniques, their usefulness and generally how appropriate they are for their projects, in order to decide which technique to use.

Several frameworks, protocols and checklists have been proposed, some for primary studies such as [6, 7]; some in the software industry [6–9]; some in academic environments [1, 10]; and others mentioned on secondary studies. In both cases, the aim is to support the selection of testing techniques and tools based on the characterization of the software to be tested, as well as the technique and/or a test tool to be used. In that sense [7] indicates that the aim is to ensure that the information from primary studies can become a basis of evidence for secondary studies that support the selection of techniques and testing tools. However, there are limitations of time and resources to carry out the primary studies.

Recommendation system assists users in the process of identifying items that meet their desires and needs. These systems are successfully applied in different e-commerce configurations, for example in the recommendation of news, movies, music, books and digital cameras [11]. There are some works about recommendation system oriented to the domain of software engineering and even less to the domain of software testing.

We propose a content-based recommendation system for software testing techniques that allows practitioners from different organizations to share their knowledge and experience about the use of software testing techniques, feeding collaboratively a repository from which the system will make recommendations of software testing techniques according to the features of a given target project and the performance of the techniques in the historical projects of the repository.

Section 2 shows some work related to the selection of software testing techniques, some of which have been our inspiration or have influenced this work. In Sect. 3, we describe the proposed method of recommendation and the software tool, which was created as part of the present work to implement the recommendation method; and Sect. 4 shows the validation of the proposed method.

2   Related Work

We have revised work related to software engineering domain with emphasis on selection of software testing techniques.

2.1   Recommender Systems and Software Engineering

According to [12], recommender systems make use of different sources of information, offering predictions and recommendations of items (videos, songs, films, etc.) to the users. Recommender systems are based on how humans make decisions throughout history: In addition to their own experiences, they also base their decisions on the experience and knowledge that come from a relatively large group of known people.

"Recommender systems combine ideas from information retrieval and filtering, user modeling, machine learning, and human–computer interaction. Case-based reasoning has played a key role in the development of an important class of recommender system known as content-based or case-based recommenders." [13].

The most widely used algorithm is k-Nearest Neighbors (kNN). There are researchers that proposed recommendation systems in the field of software engineering: [14] presents a method for estimating effort, evaluating the similarity of a project with other previous projects that can predict the effort of the target project; [15] presents a recommendation system which recommends to the developer, Java library class files that were used in the similar programs but were not used in the developer's program; [16] mention two Recommendation Systems: Hipikat, targeting software evolution in open source projects, specifically aiming at helping project newcomers and ImpRec, supporting safety-critical change impact analysis in a company with rich development processes.

2.2   Software Testing Techniques Selection Methods

[17] proposed Porantim as a selection method and tool of testing techniques for Model Based Testing (MBT); from a repository of techniques characterized, their method establishes the level of adequacy of each technique to the project as well as the performance of each technique or combination of techniques. [18] introduced MENTOR, a method and tool for selecting testing tools using Multicriteria AHP and MAUT decision-making methods. [5] proposed a method for the selection of software testing techniques by comparison using a catalogue of techniques developed based on the characterization scheme. Also, they outline the use of a tool to its application.

This paper presents a method that, unlike the afore mentioned works, has a collaborative approach and consists of a dynamic catalogue of projects and instantiation of techniques. The proposed method is carried out with the help of our own software tool.

2.3   Characterization Schemes for Software Testing Techniques

Although there are several characterization schemes of software testing techniques and tools [1, 5, 7, 17, 19], we must highlight the characterization scheme proposed by [5], due to its completeness and adaptability to different contexts, which is evidenced by its instantiation in several of other works related to evaluation and the selection of software testing techniques.

Slight adjustments to that scheme were presented by [7]. It has been instantiated in numerous cases of study in the industry and is aimed at all kinds of testing techniques, which corresponds with the purpose of the recommendation system presented in this work. In addition, to determine the possible values of each attribute of the characterization scheme, we also have considered the schemes presented in [5, 17].

The characterization schema used in this paper has a total of 21 attributes; 12 of them are grouped as "similarity attributes", which are used to compare projects with each other; nine attributes are grouped as "performance attributes", which serve to assess the techniques instantiated in the testing process of a historical project of the repository or to indicate the preferences of a user regarding the expected performance of the techniques that will be recommended.

2.4   Repository of Software Testing Techniques

Some research [5–7, 10, 17, 20] mention the need for a repository or knowledge base for the evaluation and selection of software testing techniques and/or tools (or try to create such repository): [5] propose a mechanism that suggests a way to keep the knowledge base updated. On one hand, testers (consumers) provide information from their experience in the use of testing techniques. On the other hand, information provided by researchers in the area (producers) comes from the result of their research in the development of new techniques as well as the study of the conditions of applicability of the existing ones. They also indicate the need for a librarian who will participate indirectly by supervising the information in order to maintain the repository.

Two repositories have been found available: one from [20] that has a catalogue with characterizations of 13 techniques and other from [21] which manage to characterize 219 techniques or MBT approaches in total. In both cases the characterization of the techniques or approaches was made by review of technical literature whose evidence varies between: speculation, examples, proof of concept, reports of industrial experience and experimentation.

The proposed repository of this work is based on the proposed [5], with the differences mentioned in Table 1.

Table 1. Differences between repositories

Comparison criterion	Repository by [5]	Our repository
Catalog update	Static: Each time a literature review is carried out or a new research work is produced on one or more techniques	Dynamic: Each time a new project is entered into the repository, the instantiated techniques are characterized, and the repository is updated
Actors who take part in the process	Consumer: Academic or industry Producer: Academic who performs a primary study Librarian: Academic who organizes, validates producer information to feed, maintain and update the repository	Consumer: Academic or industry Producer: Academic or industry practitioner using a testing technique (the same consumer) Librarian: not needed
Number of instantiations that feed the repository	Limited to the number of primary studies	Every time a technique is instantiated in the industry, there is an opportunity to feed it to the repository

3   SOTESTER

This is a collaborative method as shown in Fig. 1, which in base of information provided about the application or instantiation of one or more testing techniques in historical projects, a collaborative repository is constructed. From the characterization of a target project, it is possible to obtain a ranking of testing techniques from this repository based on similar historical projects; such ranking is ordered according to the performance of the techniques. From that, the tester or testing team can choose the technique or techniques to be used in their project. Subsequently, the tester or testing equipment must characterize the testing technique(s) in its project to feed the repository and finally determine the level of certainty of each recommendation, which will establish the reputation of each of the instanced technique involved in the recommendations.

The method consists of four steps, of which two are automated and two are manual.

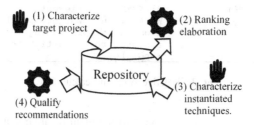

Fig. 1. Recommendation method steps

3.1   Step 1 – Characterize Target Project

The user must record the characteristics of the target project, based on the attributes of similarity of the characterization scheme, which can be seen in Table 3.

3.2   Step 2 – Ranking Elaboration

In order to elaborate the ranking of testing techniques to be recommended, the k Nearest Neighbor algorithm (kNN) is used, as shown in Fig. 2.

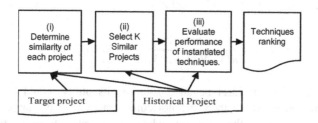

Fig. 2. Elaboration of ranking of testing techniques

In first phase of Fig. 2, it is determined the similarity of each historical project with each target project using the TOPSIS method in analogous way to [22], it places the score of each alternative based on its Euclidean distance to the positive and negative ideal solutions. According to this technique, the best alternative is that with a shorter distance to the ideal positive solution and with the most extensive distance to the ideal negative solution. To calculate the similarity, we should:

Represent as vectors, both the target project and each one of the projects of the repository in a similar way as shown in [17].

Find a relative similarity: To do this, the values of each attribute are converted to numerical values using the Jaccard coefficient of similarity, as shown in the algorithm depicted on Fig. 3, multiplying them by the weight of the attribute, based on MAUT (Multi attribute utility theory), as was done in (Pilar et al.) [18].

```
If (RPA = TPA)              TPA: Target project attribute
                            RPA: Repository project   attribute
Then SR ←1                  SR: Similarity relative to an attribute,
Else, If (RPA ⊂ TPA)        #: Number of elements,
Then SR ← (#RPA /# TPA)*W   W: attribute weight, which is obtained
                            from the simple average of the weights
Else                        assigned to the attribute by users of the
    SR← 0                   repository.
```

Fig. 3. Relative similarity calculation

- Finding the distance from each project to the ideal positive solution

$$d\left(V_j, V^+\right) = \sqrt{\sum_{j=1}^{n} \left(v_j - v_j^+\right)^2} \tag{1}$$

- Finding the distance from each project to the ideal negative solution

$$d\left(V_j, V^-\right) = \sqrt{\sum_{j=1}^{n} \left(v_j - v_j^-\right)^2} \tag{2}$$

- Finding the proximity to the ideal solution.

$$C_j^+ = \frac{d\left(V_j, V^-\right)}{d\left(V_j, V^-\right) + d\left(V_j, V^+\right)}, 0 < C_j^+ < 1 \tag{3}$$

Where V^-: Vector of ideal negative solution, its utility would be zero and it represents situation in which there are not coincidences with the target project; V_j: Vector for each j project in the repository, where j takes values from one to n, being n, the total number of closed projects in the repository; $d\left(V_j, V^+\right)$: distance of ideal positive solution; $d(V_j, V^-)$: distance of ideal negative solution; v_j: value of similarity attribute for j project; v_j^+: positive ideal value for similarity attribute; v_j^+: negative ideal value for similarity attribute.

In the second phase of Fig. 2, the K projects most similar to the target project are selected, and finally, in the third phase, the performance of the techniques instantiated in the selected projects is evaluated, for which:

- The user indicates preference weights for each performance attribute W_l., where $l = 1 \ldots p$, and p is the number of performance attributes of the characterization schema.
- For assure recommendations are based on collaborative knowledge, the method excludes techniques that have not been instantiated at least m times, where m is the average number of instantiations of the techniques registered in the repository; so, we avoid that a technique is recommended based on few and non-representative instantiations.
- Based on MAUT, to each instantiation h extracted, its utility is calculated based on the performance attributes of the characterization scheme and the preferences of the target project.

$$U_h = \sum_{l=1}^{p} X_{hl} W_l \tag{4}$$

- The average utility is calculated for each technique based on the utility and trust of each instantiation h.

$$\bar{U} = \sum_{h=1}^{t} U_h \, C_h' \tag{5}$$

where t: Total selected instantiations of a given technique; and C_h': Average confidence level of the recommendations given by h., being thus elaborated a ranking from which the user is recommended the first Z techniques with greater average utility, where Z is a value entered by the user in base of their need.

3.3   Step 3 – Characterize Instantiated Techniques

As part of the testing process in a software project, one or more recommended techniques are instantiated, so, on this step, the user must characterize them based on the performance attributes, in order to provide feedback. The characterizations of the instantiated techniques will be part of the repository and may participate in the recommendation process. Values were considered in a range of zero to one for each attribute.

3.4 Step 4 – Qualify Recommendations

It consists on determining the error of the recommendations for a technique instantiated in the target project. This error will help determine the level of trust by offering other recommendations.

$$E_r = |(U_h - U_o)| \qquad (6)$$

Where U_h: Performance of the technique according to the instantiation h used in recommendation r and, U_o: Performance of the technique in the target project o.

3.5 SOTESTER – Web Application

In order to implement all steps of the recommendation method, a web application "SOTESTER" was developed. This application has a relational database that constitutes the repository or evidence base that refers to the method.

Among some characteristics of the application, it is possible to emphasize the scalability and adequacy capacity, so that, if it is necessary, the method could be tested with different characterization schemes, without having to make additional programming efforts. It would even be possible to use this application in other domains.

The solution has the follow components:

- "SoTesttersDB": the main database which storages projects data.
- "AccountDB": security database which storage user, roles and login data.
- "SoTestterWebApp": Web application, which serves as interface for enter data to the repository and to obtain recommendations, this web application was built on. NET framework with C# language and a pdf user manual was made for help new users.

On Fig. 4 shows the structure of the relational database of the repository "SoTesttersDB", where:

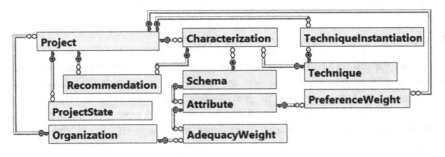

Fig. 4. Repository structure

- "Project": is a table storing general data for project
- "ProjectState": this table, save state flow for project (Non-Characterized, characterizing before instantiation, in testing, selecting instanced techniques, characterizing after instantiation)

- "Technique": software testing technique
- "TechniqueInstantiation": represents each instantiation of testing technique on a project.
- PreferenceWeight: it saves user preferences respect to performance attributes of testing technique instantiated on similar projects to the target project.
- "Characterization": save data of characterization of both, project and technique instantiation.
- "Schema" and "Attribute": save characterization schema and its attributes, so the data structure should support any characterization schema.
- "AdequacyWeight": it saves importance weight assigned by user to similarity attributes. This value is used by method to calculate similarity between projects.
- "Recommendation": saves relation between characterization obtained for offer recommendations and the target project.
- Organization: a person or company who participates giving projects and obtaining recommendations.

4 Validation

The recommendation method was validated by contrasting the recommendations made by the Recommender system with three software testing experts, for a total of 11 target projects. We had to: (1) Initialize the repository with testing techniques and historical projects, (2) Register target projects in the repository and get recommendations for testing techniques for each one, (3) Request recommendations of 3 testing techniques from a group of experts for a total of 11 target projects and (4) Contrast the recommendations made by experts with those of the system in base to the agreement between experts and recommendation system taking in account that we do not want to prove only that system recommends good techniques, we are also interested in not recommending bad techniques, because a bad recommendation is expensive for the tester.

4.1 Results

Initial Feeding of the Repository. 23 Software testing techniques compiled from [23, 24] were recorded in the repository. 12 professionals who occupy positions in the software testing field and who have an average experience of six years contributed with 26 historical projects in total. A total of 71 instantiations of testing techniques were recorded, 18 techniques were instantiated at least once and five were not instantiated. The most instantiated technique is "Decision tables" with 18 instances and the mean of instantiations per technique is 3.08.

Target Projects. Eleven target projects were obtained from real projects of software companies, which are shown in Table 2. The person in charge of each target project, performed the characterization and expressed preferences regarding the testing techniques. On Table 3, the characterization of the target project 4 is shown as an example. Table 4 shows preferences for the same project.

Table 2. Target projects

No.	Name
0	Sigescom
1	Globalnet
2	Sisap lab
3	Virtual self-assessment system
4	Warehouse management
5	Management production (Workers)
6	Bet games – integration project
7	Task control
8	Sotester
9	Tax deductions
10	Control of materials in production

Table 3. Characterization of the target Project 4

Attribute	Value
Static or dynamic	Static/Dynamic
Type of software	Web, Mobile
Life cycle phase	Unit testing, component integration testing, system testing, system integration testing, acceptance testing
Scalability	>3001
Environment-language	C#
Inputs	Source code, executable program, execution scenarios, execution logs, application requirements, higher and lower values of variables
Knowledge	Application domain, scenario identification, object-oriented paradigm, coding, test script generation, teamwork, leadership
Experience	Some experience with coverage tests, requirements testing, web testing
Outputs	Test cases, coverage information, flaws, test scenarios
Types of defects	interface, algorithm, function, timing/serialization, documentation
Applicability in tasks	Code coverage testing, requirements testing, system testing
Availability of tools	Commercial

Recommendations. The three experts recommended three testing techniques for each target project and three testing techniques were recommended too by the system. In Appendix A, we present both experts (E1, E2, E3) and system (RS) recommendations for the 11 target projects.

Table 5 shows the degree of agreement between one expert and another, obtaining a general average of 24%. Grade of agreement is between zero and one, where zero indicates no agreement and one indicates that both experts recommended the same three techniques, without taking in account the ranking position.

Table 4. Preferences for recommendation -Target project 4

Attribute	Weight
Completeness	20
Efficacy	20
Test suite size	08
Interaction	08
User guidance	08
Sources of information	05
Comprehensibility	15
Subjective satisfaction	06
Effort	10

Table 5. Agreement between experts

TP	E1 vs E2	E1 vs E3	E2 vs E3	Mean
0	0.33	0.67	0.33	0.44
1	0.33	0.33	0.33	0.33
2	0.33	0.67	0.67	0.56
3	0.00	0.33	0.33	0.22
4	0.00	0.33	0.33	0.22
5	0.00	0.33	0.00	0.11
6	0.33	0.33	0.33	0.33
7	0.33	0.33	0.00	0.22
8	0.33	0.00	0.00	0.11
9	0.00	0.00	0.33	0.11
10	0.00	0.00	0.00	0.00
Mean	0.18	0.30	0.24	0.24

As it is shown in Table 6, as a result for the 11 target projects (TP), the recommendation System (RS) has a degree of agreement with the experts (E) of 19%. Agreement with expert 1 (E1) is 15%, with both expert 2 (E2) and expert 3 (E3) is 21%. These results are close to the agreement between any pair of human experts. We have found that the agreement between experts is not more than 24%, and the agreement of recommendation System respect to the experts is not more than 21%; so, we can say the system behavior is similar to human.

These initial results are good signal and allow us for try system with more data and other evaluation approaches and metrics where is necessary a major effort and time given the nature of items to recommend.

Table 6. Agreement of recommender system vs. experts

TP	RS vs E1	RS vs E2	RS vs E3	Mean RS vs E
0	0.00	0.33	0.00	0.11
1	0.00	0.00	0.00	0.00
2	0.00	0.00	0.00	0.00
3	0.33	0.33	0.33	0.33
4	0.67	0.33	0.33	0.44
5	0.00	0.00	0.00	0.00
6	0.00	0.33	0.00	0.11
7	0.33	0.00	0.67	0.33
8	0.00	0.33	0.67	0.33
9	0.00	0.33	0.33	0.22
10	0.33	0.33	0.00	0.22
Means	0.15	0.21	0.21	0.19

4.2   Threats to the Validity

The quality of the recommendations can be affected by the quality of the initial data feed to the repository. To improve that quality, the initial feeding was restricted to practitioners with at least three years of experience in software testing.

The validation, using the contrast of the recommendations made by experts from the industry and those performed by the system, might not be completely objective in the sense that there might be uncontrolled variables that affect the human experts' recommendations, such as their experience time, the diversity of projects in which they have worked, among others. To minimize this threat, an expert profile considering at least 10 years of experience in software testing and experience in at least three different organizations was determined. In future work, the validation of the recommendation system based on its own results (trust, accuracy, use intention, user subjective satisfaction, etc.), after applying the recommendations made in target projects, will be carried out.

It is possible that the characterization scheme used by the recommendation method, does not completely describe the context of the target project, however, the intention of this work (at this stage) is to determine the basic feasibility of the method. The validation and improvement of the characterization scheme is left for future research.

5   Conclusions

A content-based method for recommendation of software testing techniques has been proposed. Its repository or knowledge base is collaboratively and dynamically fed with the experience of industry practitioners and it is possible to offer recommendations of software testing techniques for target projects.

A software tool has been built for the implementation of the proposed method; with an architecture that makes it scalable, it is possible to perform future tests with different characterization schemes. It is even possible to reuse it for applications in other domains.

The proposed method offers similar recommendations to those of the human experts, although before generalizing these results, it is necessary to feed the repository with more instantiations of testing techniques so that it is possible to have a more records about testing technique instantiated into the repository.

A Recommendations for Target Projects (TP)

TP	Recommender	Recommendation	TP	Recommender	Recommendation
0	E1	DT, ECP, FI	6	E1	FGN, NL, FT
0	E2	DCC, DT, DCH	6	E2	OA, FT, DTC
0	E3	DT, STD, FI	6	E3	CL, BC, FT
0	RS	BC, DCC, CW	6	RS	DCC, GM, OA
1	E1	DT, STD, APT	7	E1	DT, CW, SC
1	E2	DTC, DT, OA	7	E2	FC, DTC, DT
1	E3	ECP, DT, DCC	7	E3	BC, NL, CW
1	RS	CW, BC, FC	7	RS	BC, DCC, CW
2	E1	DTC, ECP, BVA	8	E1	DTC, DT, FI
2	E2	OA, DTC, DT	8	E2	DT, DCC, GM
2	E3	ECP, DTC, DT	8	E3	CL, BC, CW
2	RS	DCC, GM, BC	8	RS	BC, DCC, CW
3	E1	FGN, CW, FT	9	E1	ECP, CW, STD
3	E2	BC, ECP, SL	9	E2	OA, BC, DTC
3	E3	BC, APT, FT	9	E3	DCC, OA, DCH
3	RS	CW, OA, BC	9	RS	BC, DCC, FC
4	E1	BC, ECP, CW	10	E1	CL, FT, DCC
4	E2	DCC, FT, DT	10	E2	BC, FC, BVA
4	E3	ECP, DCC, FI	10	E3	DT, SC, FI
4	RS	CW, BC, DCC	10	RS	BC, OA, DCC
5	E1	CL, FI, STD			
5	E2	DTC, SC, BVA			
5	E3	DT, NL, FI			
5	RS	OA, DCC, CW			

Legend

.DT: Decision Tables	OA: Orthogonal Arrays
ECP: Equivalence Class Partitioning	FC: Function Coverage
FI: Formal Inspections	BVA: Boundary Value Analysis
DCC: Decision/Condition Coverage	GM: Graph Matrices
DCH: Desk checking	FGN: Flow Graph Notation
STD: State Transition Diagrams	FT: Fuzz Testing

(continued)

(continued)

BC: Branch coverage	SL: Simple Loops
CW: Code walkthrough	CL: Concatenated loops
APT: All Pairs Technique	SC: Statement coverage
DTC: Deriving Test Cases	NL: Nested loops

References

1. Eldh, S., Hansson, H., Punnekkat, S., Pettersson, A., Sundmark, D.: A framework for comparing efficiency, effectiveness and applicability of software testing techniques. In: Testing: Academic and Industrial Conference-Practice and Research Techniques TAIC PART 2006, pp. 159–170 (2006)
2. Khan, M.E.: Different forms of software testing techniques for finding errors. Int. J. Comput. Sci. Issues 7(3), 11–16 (2010)
3. Luo, L.: Software testing techniques. Institute for Software Research International Carnegie Mellon University, Pittsburgh, PA, vol. 15232, no. 1–19, p. 19 (2001)
4. Farooq, S.U., Quadri, S.M.K.: Empirical evaluation of software testing techniques – need, issues and mitigation. Softw. Eng. Int. J. 3(1), 41–51 (2013)
5. Vegas, S., Basili, V.R.: A characterization schema for software testing techniques. Empir. Softw. Eng. 10(4), 437–466 (2005)
6. Vos, T., Marín, B., Panach, I., Baars, A., Ayala, C., Franch, X.: Evaluating software testing techniques and tools. In: Proceedings of XVI JISBD, A Coruña, pp. 531–536 (2011)
7. Vos, T.E.J., et al.: A methodological framework for evaluating software testing techniques and tools. In: 12th International Conference on Quality Software, QSIC 2012, Xi'an, Sha, pp. 230–239 (2012)
8. Brosse, E., Bagnato, A., Vos, T.E.J., Condori-Fernandez, N.: Evaluating the FITTEST Automated Testing Tools in SOFTEAM : An Industrial Case Study, May 2014
9. Condori-Fernández, N., Kruse, P.M., Vos, T.E.J., Brosse, E., Bagnato, A.: Combinatorial testing in an industrial environment - analyzing the applicability of a tool. In: Proceedings of 2014 9th International Conference on the Quality of Information and Communications Technology QUATIC 2014, pp. 210–215 (2014)
10. Cotroneo, D., Pietrantuono, R., Russo, S.: Testing techniques selection based on ODC fault types and software metrics. J. Syst. Softw. 86(6), 1613–1637 (2013)
11. Felfernig, A., Jeran, M., Ninaus, G.: Toward the next generation of recommender systems: applications and research challenges. In: Tsihrintzis, G., Virvou, M., Jain, L. (eds.) Multimedia Services in Intelligent Environments. SIST, vol. 24, pp. 81–98. Springer, Heidelberg (2013). https://doi.org/10.1007/978-3-319-00372-6_5
12. Bobadilla, J., Ortega, F., Hernando, A., Gutiérrez, A.: Recommender systems survey. Knowl.-Based Syst. 46, 109–132 (2013)
13. Bridge, D., Göker, M.H., Mcginty, L., Smyth, B.: Case-based recommender systems. Knowl. Eng. Rev. 20(3), 315–320 (2005)
14. Ohsugi, N., Tsunoda, M., Monden, A., Matsumoto, K.-i.: Effort estimation based on collaborative filtering. In: Bomarius, F., Iida, H. (eds.) PROFES 2004. LNCS, vol. 3009, pp. 274–286. Springer, Heidelberg (2004). https://doi.org/10.1007/978-3-540-24659-6_20

15. Tsunoda, M., Kakimoto, T., Ohsugi, N., Monden, A., Matsumoto, K.-I.: Javawock: a Java class recommender system based on collaborative filtering. In: Proceedings of the 17th International Conference on Software Engineering and Knowledge Engineering (SEKE 2005), pp. 491–497, July 2005

16. Borg, M., Runeson, P.: Changes, Evolution and Bugs - Recommendation Systems for Issue Management. In: Robillard, M.P., Maalej, W., Walker, R.J., Zimmermann, T. (eds.) Recommendation Systems in Software Engineering, pp. 477–509, Springer, Heidelberg, 2014. https://doi.org/10.1007/978-3-642-45135-5_18

17. Dias-Neto, A.C., Travassos, G.H.: Supporting the combined selection of model-based testing techniques. IEEE Trans. Softw. Eng. **40**(10), 1025–1041 (2014)

18. Pilar, M., Simmonds, J., Astudillo, H.: Semi-automated tool recommender for software development processes. Electron. Notes Theor. Comput. Sci. **302**, 95–109 (2014)

19. Engström, E., Runeson, P., Skoglund, M.: A systematic review on regression test selection techniques. Inf. Softw. Technol. **52**(1), 1–35 (2010)

20. Vegas, S.: Characterization schema for selecting software testing techniques. Ph.D. Thesis. Facultad de Informática, Universidad Politécnica de Madrid, February 2002

21. Dias-Neto, A.C., Travassos, G.H.: Model-based testing approaches selection for software projects. Inf. Softw. Technol. **51**(11), 1487–1504 (2009)

22. Zaidan, A.A., Zaidan, B.B., Al-Haiqi, A., Kiah, M.L.M., Hussain, M., Abdulnabi, M.: Evaluation and selection of open-source EMR software packages based on integrated AHP and TOPSIS. J. Biomed. Inform. **53**, 390–404 (2015)

23. Nidhra, S.: Black box and white box testing techniques - a literature review. Int. J. Embed. Syst. Appl. **2**(2), 29–50 (2012)

24. Jovanovic, I.: Software testing methods and techniques. IPSI BgD Trans. Internet Res. **5**(1), 30–41 (2009)

Crowdsourcing High-Quality
Structured Data

Harry Halpin[1,2](✉) and Ioanna Lykourentzou[1,2](✉)

[1] Inria, 2 Rue Simone Iff, Paris, France
harry.halpin@inria.fr
[2] Utrecht University, Princetonplein 5, Utrecht, Netherlands
i.lykourentzou@uu.nl

Abstract. One of the most difficult problems faced by consumers of semi-structured and structured data on the Web is how to discover or create the data they need. On the other hand, the producers of Web data do not have any (semi)automated way to align their data production with consumer needs. In this paper we formalize the problem of a *data marketplace*, hypothesize that one can quantify the value of semi-structured and structured data given a set of consumers, and that this quantification can be applied on both existing data-sets and data-sets that need to be created. Furthermore, we provide an algorithm for showing how the production of this data can be crowd-sourced while assuring the consumer a certain level of quality. Using real-world empirical data collected via data producers and consumers, we simulate a crowd-sourced data marketplace with quality guarantees.

Keywords: Crowdsourcing · Structured data · Resource allocation · Human computation

1   Introduction

Given there are few things more valuable in the information economy than having access to the right data at the right time, we find it likely that accurate and well-maintained data has monetary value, even though currently most efforts on producing structured data for the Web have been so far focusing on public open data that is created at public cost and published for anyone to use. Therefore, our first hypothesis is that consumers of data can use *financial incentives* to attract domain experts to produce and update structured data. In this way, the lack of structured data could be corrected by a *data marketplace*, a service that matches the consumers of structured data to producers of such data. Current research on crowd-sourcing focuses on asking pools of (usually untrained) unknown users to solve micro-tasks that do not require specialized knowledge and can be solved effectively via optimizing for cost [13] or even for fun [18]. However, the problem of crowd-sourcing structured data is very different insofar as it requires asking pools of domain experts to solve potentially very difficult

© Springer Nature Switzerland AG 2019
J. A. Lossio-Ventura et al. (Eds.): SIMBig 2018, CCIS 898, pp. 304–319, 2019.
https://doi.org/10.1007/978-3-030-11680-4_29

tasks, where optimization is for quality rather than cost. Yet crowd-sourcing literature in general has only begun to investigate mechanisms that optimize for quality [10]. For example, a qualified domain expert in medicine that creates reliable structured data about the interactions of drugs is worth much more than that of a data-set created by an amateur, and thus the domain expert can reasonably ask for much more of a financial reward. Thus, our second hypothesis is that in order to optimize for quality, some **crowd co-ordination mechanism** is necessary that takes expertise into account, and directs experts to producing domain-specific data-sets that require expert knowledge in an efficient manner.

2 Literature Review

Coordination mechanisms for human crowds is an active field of research. State-of-the-art research has started to increasingly examine such mechanisms, mainly in optimizing crowdsourcing for cost in online labor markets like Amazon Mechanical Turk [9]. Such markets work by outsourcing small "human intelligence tasks" (such as phrase translation or "captcha" recognition) to several people in a "crowd" for a specific monetary remuneration. In this context, Bernstein et al. [3] view crowd co-ordination from a queue theory perspective and solve for optimal solutions analytically with the goal to minimize cost. More similar to our proposed work, Shahaf and Horvitz focus more on micro-task quality. Their work proposes an algorithm which estimates the capacity of each potential worker and develops a global coordination strategy of allocating workers to micro-tasks with the goal to maximize quality while respecting capacity constraints [15], but do not investigate the problem in terms of structured quality data or in terms of quality assurance with regards to domain expertise. Other methods, such as mechanism design [14] and game theory [6] are also being investigated as a means to optimize the allocation of workers to tasks in a cost-effective manner. An entire stream of work has explored the effect of monetary incentives and non-monetary incentives [8], showing that in general financial incentives can systematically modify tradeoffs in speed vs. quality in task completion [12], with payment-per-task leading to highest task completion.

Most crowd-sourcing studies have engaged workers with relatively simple tasks that can be accomplished using wide-spread cognitive abilities. However, while there is a large amount of work on expert-finding (in domains ranging from social networks [17] to enterprise environments [2]) and the optimal creation of teams involving differing levels and kinds of expertise [5], there is little work on crowd-sourcing from pools of experts. Unlike traditional crowd-sourcing that assumes a large pool of workers who pass a baseline qualification in order to accomplish large amounts of similar tasks (such as Amazon Mechanical Turk HITs), the task of crowd-sourcing data-sets requires pools of domain experts whose availability may be scarce to accomplish creating high quality data-sets about varying domains. This is a topic that has only begun to be studied in crowd-sourcing research, with current scheduling results based on workers with the same level of knowledge being scheduled in order to maximize productivity [16]. The objective of proposing a "smart" crowd scheduling mechanism for the data marketplace is to ensure the production of as many as possible datasets of good quality, while not exceeding the pricing requirements of the consumers and using the available workforce of producers as effectively possible.

3   Crowd-Sourcing Optimization as Resource Scheduling

3.1   Structured Data Terminology

A *dataset* is a general term to describe a group of items of structured data in a specific domain of knowledge. Data consumers need specific datasets in order to fulfill an information need, and data producers have some domain and technical expertise that allows them to create and modify datasets. The typical functionality of the data marketplace is as follows: A consumer enters the marketplace and creates a request for a dataset, accompanied by specific requirements. These requirements may include characteristics such as the description of the domain of the data, and the maximum cost that the consumer is willing to pay for the creation of the dataset. On the other side, producers of data-sets with particular domain expertise enter the system, and they select the micro-tasks they can complete given their domains of expertise.

The objective of the proposed crowd-coordination mechanism for the data marketplace is to ensure the production of as many as possible data-sets while not exceeding the pricing requirements of the consumers and using the available workforce of producers as effectively possible. To achieve the above, the coordination mechanism seeks to effect the micro-task selection process by recommending producers to contribute to specific micro-tasks so that overall the requirements of as many consumers and producers are met as possible. Given there are in general more requests for data-sets than available experts (as demonstrated by empirical data in Sect. 4), we can treat the problem of which micro-producer should be recommended which micro-task as a crowd-souring *resource scheduling* problem. We illustrate the approach in Fig. 1 and map the terms from structured data crowd-sourcing to a resource allocation framework in Sect. 3.1. Given that no crowd-sourcing systems for producing quality data-sets through a market exist, we will use simulation with a non-profit and non-coordinated baseline, although we will parameterize our simulation from an existing crowd-sourced data-set.

3.2   Resource Scheduling Algorithm

For each producer that enters the marketplace, the resource scheduling algorithm identifies the domain in which the producer has the highest expertise. If this domain has a micro-task that can pay higher than the producer's minimum accepted wage in the domain, then the algorithm allocates the specific micro-task to the producer. In case the producer accepts, which is when the price paid by the micro-task is indeed higher than their minimum accepted wage, the producer undertakes the micro-task and the algorithm schedules the next producer. In case the micro-producer rejects, the algorithm suggests another

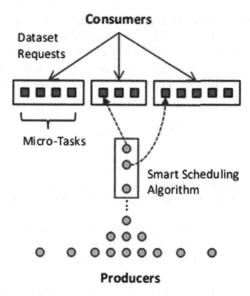

Fig. 1. A resource scheduling algorithm

micro-task. The process ends when: (i) the micro-producer accepts to contribute to a recommended micro-task, (ii) the micro-producer leaves the marketplace or (iii) there are no more micro-tasks. Figure 2 illustrates the functionality of the smart scheduling algorithm in a pseudocode format.

3.3   Model

The simulation of the marketplace is implemented according to definitions given below.

3.4   Producer

– **Domain expertise** e_d: The expertise e that the producer has in each knowledge domain d of the data marketplace. It is modeled as a vector with length equal to the total number of domains of knowledge in the data marketplace. The value of the expertise of the producer in each domain is randomly initialized at the beginning of the simulation according to a uniform probability distribution with the range [0,1], with zero being equivalent to no expertise in the domain and one being the expertise necessary to create a data-set of highest possible quality. It remains unchanged for all the interactions of the producer with the marketplace. In our simulation we assume accurate knowledge of their domain expertise by the data marketplace.
– **Minimum accepted wage** w_d: The minimum accepted wage (w_d) per micro-task is a minimum price threshold, such that the producer will not accept to contribute to micro-tasks that pay less than their threshold. It is modeled as a

vector with length equal to the number of domains of the marketplace, i.e. we assume a different minimum wage per domain of knowledge. Since a producer is more likely to want to be paid more for domains where their expertise is higher, the value of the minimum accepted wage of a user is assumed to be a linear function of their expertise (e_d) in each specific domain d, and the minimum accepted wage per micro-task in our simulation receives values e [0,1] range.

3.5 Data-Sets

A simulated dataset models a dataset request made by a consumer of the marketplace. Each dataset has the following modeled characteristics:

- **Knowledge domain** d: Models the domain of knowledge that the dataset belongs to. Each dataset is randomly assigned to exactly one domain $d \in D$ at the beginning of the simulation, with D being the total number of domains of the simulated marketplace.
- **Number of micro-tasks** m: The total number of micro-tasks that the dataset consists of. Each of the micro-tasks of the dataset can be selected and completed by exactly one producer. In this simulation we consider a fixed number of micro-tasks per dataset.
- **Quality** q: The quality of the dataset is the average of the quality of its micro-tasks. The quality of each micro-task is initially equal to zero and it changes as soon as the micro-task is completed by a producer. In the specific simulation we consider that the quality of the micro-task is linearly analogous to the expertise of the producer that completed it.
- **Total upper cost** C: The total cost that the consumer is willing to pay for the creation of the dataset. It is initialized randomly over a cost range distribution at the beginning of the simulation and is considered fixed for the specific data-set. It is used to determined the price of each micro-tasks of the dataset where p_i is the price of each micro-task and m is the total number of micro-tasks of the dataset.

3.6 Modeled Systems

To test our hypothesis, we model three data marketplace systems. The first two test the differences between profit and non-profit systems, and the third to test our hypothesis that a directed crowd-sourcing algorithm is necessary in order to crowd-source high-quality data-sets.

Non-profit System. In this system, the producer enters the system and selects an available micro-task at random that is in their domain of highest expertise. If there are no available microtasks in the domain, they will choose a micro-task that matches their next highest domain of expertise, and continue in descending order of their expertise. They will then exit and may re-enter the system.

1. For every producer u_i that arrives:
2.   Create array M containing all unallocated micro-tasks m_j per knowledge domain d: $M = m_j^d, d \in D$
3.   Sort micro-tasks in descending price order per domain: $m_{j_1}^d.price > m_{j_2}^d.price, \forall d \in D$
4. Create array A containing the estimated expertise e_i of u_i in every knowledge domain d: $A = e_i^d, d \in D$
5.   Sort array A in decending estimated expertise value order: $e_i^{d_1} > e_i^{d_2}$
6.   Select next element of A: $e_i^{d_1}$ and identify its domain: d_1
7.   if M is not empty in domain d_1: $m_j^{d_1}$ not empty
8.   if price the micro-task pays is higher than the producer's minimum wage in domain d_1: $m_{j_1}^{d_1}.price > u_i.w_{d_1}$
9.   allocate micro-task $m_{j_1}^{d_1}$ to producer u_i
10.   else return to line 6
11.   exit

Fig. 2. Smart profit directed crowd-sourcing algorithm

Simple Profit System. When a producer enters the system, the producer selects a micro-task that will bring them the greatest profit. Each producer will select among the available micro-tasks the one that has the highest price and, at the same time, exceeds their personal modeled minimum wage requirement. As soon as the production of the micro-task is finished, the producer exits and may re-enter the system.

Smart Profit System. The crowd-coordinated system is similar to the simple profit system, except that producers select the micro-tasks that they will contribute to amongst the recommendations given by the resource scheduling algorithm presented in Sect. 3.2.

4 Empirical Parameterization

In order to make our simulation realistic, we empirically determined our parameters from The Data Hub, a website that features a 4,826 public datasets.[1] Datasets are added and kept up-to-date by the voluntary contributions of users, with no financial incentives in place. In order to be concise, we abbreviate The Data Hub as CKAN, given that it was built using the Comprehensive Knowledge Archive Network (CKAN). The site has been crowd-sourcing contributions of structured data-sets since June 2006 and is probably the largest source of open public structured data-sets on the Internet.[2] The website features vastly different sizes and kinds of data-sets, such as Canada's Open Government data, bibliographic data, biological data such as BioPortal. Data-sets are given domains in a "bottom-up" fashion by tagging. The data-sets come in a variety of formats ranging from XML to RDF to CSV. User contributions and modifications of

---

[1] http://datahub.io/.

[2] While Infochimps claims to be larger (approximately 9,000 data-sets) than The Data Hub, approximately half of its data-sets are APIs rather than structured data and thus cannot be queried, and no history of users and revisions are available as Infochimps does not use crowd-sourcing.

data-sets are recorded and available as metadata for each data-set. Currently, the Data Hub has 3,700 users, although only 1,654 have actually contributed to data-sets. On November 18th 2012, we crawled and used the CKAN API to get statistics for all CKAN data-sets, including their number of tags, creation date, revisions, and users. We use the data as the basis of the parameterization of our simulation. The dataset extracted from CKAN covers a timespan of 67 months, and so we set the simulation time equal to 67 simulation units, with each unit being equivalent to one month, so that we can compare accurately to an empirical baseline, the CKAN data-set. The rest of the extracted parameters used to run the simulation are given in Table 1.

The arrival rate of users in CKAN is shown in Fig. 3. Based on this empirical data, the number of total users in the system grows exponentially with time, according to the following equation, with $\alpha = 34.01$, $\beta = 0.05898$ and goodness of fit $R^2 = 0.9922$:

$$f(t) = \alpha \cdot e^{\beta \cdot t}$$

From the derivative of $f(t)$, we find the user arrival rate λ:

$$\lambda = \frac{df}{dt} = \alpha \cdot \beta \cdot e^{\beta \cdot t}$$

From the above, the inter-arrival time can be found by inverting previous equation:

$$\lambda = \frac{1}{\alpha \cdot \beta \cdot e^{\beta \cdot t}}$$

We identify the inter-arrival time of the requested datasets, as given by Fig. 3. From the CKAN data we have that the total number of requested datasets also grows exponentially with time, with $\gamma = 139.1$, $\sigma = 0.05328$ and $R^2 = 0.9926$:

$$g(t) = \gamma \cdot e^{sigma \cdot t}$$

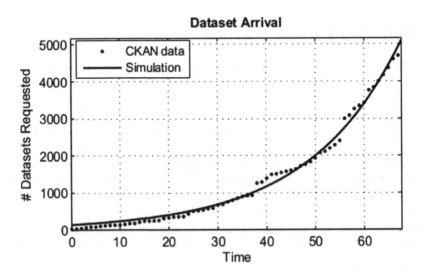

Fig. 3. Data-set arrival rate from the data-hub and parameterized simulation

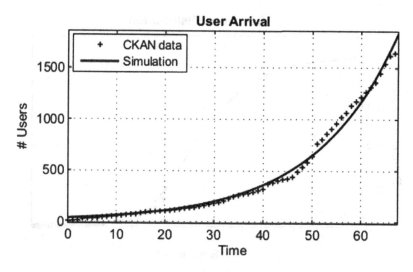

Fig. 4. User arrival rate from the data-hub and parameterized simulation

To determine the number of thematic categories that may be found in CKAN, we consider the tags in relation to the number of datasets that each tag appears, which appear to follow an exponential distribution with semi-long tail. This distribution is not technically a power-law, but an exponential distribution that seems to be in process of evolving into a power-law distribution, so there is a large amount of data-sets with a very low number of tags although these datasets are too few to be the majority of the distribution. After eliminating the tail of tags that appear in less than 4% of the total number of CKAN datasets (the semi-long tail), we find 20 main thematic categories. Based on this, the number of domains of knowledge is set to 20 (Fig. 4).

To determine the number of micro-tasks that we will model each job to have, the distribution of datasets in CKAN against the number of revisions that each dataset has is given in Fig. 5. Similarly to the number of tags, there is an exponential distribution with the beginning of the development of a long tail. Again, after removing the long tail (data-sets with revisions of 1), we again the weighted average of the number of microtasks per dataset is 3.

Finally, we need to determine the user expertise distribution and the distribution of cost per dataset. These parameters cannot be found through the data mining performed in CKAN, since the CKAN system does not employ any method of determining expertise and is a non-profit system. Thus expertise is modeled as a beta distribution, with shape coefficients equal to $\alpha = 0.3, \beta = 1$. A beta distribution was chosen because it is a double bounded function with the range of [0,1]. The specific selection of shape parameters gives a user population that follows an exponential expertise distribution, with most users having little or average knowledge on a given thematic category and a small number of users having high expertise on the subject [2]. Finally since currently there is no

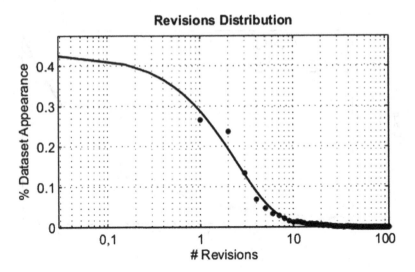

Fig. 5. Distribution of revisions per dataset

research into the potential cost distribution of high-quality structured data, we define the value of each dataset in our simulations as a random number from a uniform distribution in the range [0,3], due to the average number of micro-tasks being 3.

Table 1. Parameters of model

Name	Value
Simulation time	67
Users	1654
Domains	20
Micro-tasks per job	3
Dataset inter-arrival time	$\frac{1}{\lambda} = \frac{1}{g'(t)} = \frac{1}{\gamma \cdot \sigma \cdot e^{\sigma \cdot t}}$, with $\gamma = 139.1$, $\sigma = 0.05328$
User inter-arrival time	$\frac{1}{\lambda} = \frac{1}{f'(t)} = \frac{1}{\alpha \cdot \beta \cdot e^{\beta t}}$, with $\alpha = 34.01$, $\beta = 0.05898$
User expertise distribution	Normal $(\bar{x} = 0.5, \sigma = 0.3)$
Cost distribution	Beta ([1000, 10000])

5   Results

Our simulation was run using the empirically-derived and estimated parameters given in Table 1, and the results are examined in detail to test our hypotheses, namely that financial incentives will increase the quality of dataset production and that a crowd-sourcing co-ordination mechanism (formulated and solved by us as a resource-scheduling problem) will increase the quality of dataset production. The performance of the three algorithms in terms of average dataset quality (the maximization of which is the objective of the resource scheduling algorithm) is given in Fig. 6 by a quality histogram of the produced datasets. In this diagram, the x axis represents the quality in the range $[0,10]$, with 0 representing no quality at all and 10 the highest possible quality. The figure has two y axes, with the left corresponding to the datasets produced by the "simple profit" algorithm and the right axis illustrating the datasets produced by the "non-profit" and the "smart profit" ones, as the "simple profit" algorithm produces a high amount of low quality datasets off the scale of the other two algorithms.

As can be observed in Fig. 6, the smart scheduling algorithm achieves much higher quality compared to both the non-monetary and the simple profit algorithms. This can be attributed to the fact that the algorithm identifies the domain each user is mostly expert at and "guides" the user's contribution towards a micro-task from this domain. It is also straightforward to observe that the simple profit system performs very poorly in terms of quality, because users in this system have no incentive to make good contributions and they rather select the micro-task they will undertake based only on the price it pays. Interestingly enough, perhaps the above rationale can be used to partially explain the low quality results for which purely-profit based systems such as Amazon Mechanical Turk have been often criticized for.

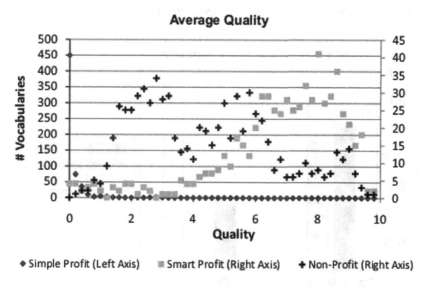

Fig. 6. Quality of dataset per crowd-sourcing algorithm

Figure 7 compares the three algorithms in terms of production efficiency, defined in terms of the ratio of completed and partially completed datasets. The term "completed" refers to the datasets whose micro-tasks have all been allocated and completed by producers and the term "partially completed" refers to the datasets with one or more completed micro-task that have not been fully completed. Figure 8 shows the production process of these datasets in relation to time. The simple profit system produces a similar number of partially completed datasets as the non-profit system. The smart scheduling algorithm does not produce many partially complete datasets as it produces mostly completed data-sets. Strangely enough, the non-profit system produces more completed data-sets than the simple profit systems. This is likely because in the non-profit system, the users attempt to self-allocate according to their domain expertise, but do not do it as efficiently when compared to the smart scheduling algorithm. In contrast, the simple profit system produces many partially completed results of low quality as users are optimizing for profit but ignoring their own domain of expertise. Overall, the contributions of the producers are more focused when the smart scheduling algorithm is used, and they tend to be more dispersed when users self-select the tasks that they will undertake (with this dispersion being higher in the case of the simple profit system due to the over-riding of profit in comparison to domain knowledge).

Therefore, our first hypothesis is incorrect. *Financial incentives by themselves do not produce higher-quality data-sets, but instead skew the creation of data-sets towards a lower-quality due to optimization of cost over quality by the contributors.* However, our second hypothesis is correct. *The combination of financial incentives with a smart scheduling algorithm that directs producers of data-sets to tasks in their domain of expertise produces higher-quality datasets.*

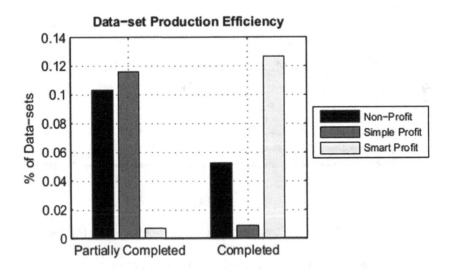

Fig. 7. Dataset completion rates

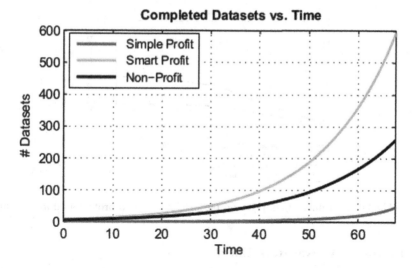

Fig. 8. Completion of datasets over time

5.1 Price Effect

Next, we examine the effect that the price per micro-task and producer has on the performance of the proposed algorithm. Figure 9 explores the effect of price variability on the performance of the two monetary incentive-based algorithms, the simple and the smart profit ones. The x axis shows the cost per micro-task, in a range between 0, 1, and 20 monetary units. The left y axis shows the number of datasets completed for each price level in a scale between 0 and 700 datasets. The right y axis shows the average quality achieved, in the range of 0 (no quality) to 10 (best quality). The dark grey lines correspond to the simple profit system and the light grey lines to the smart profit algorithm. Of course, as observed earlier, the smart scheduling algorithm performs better than the simple alternative at all price ranges, both in terms of number of completed datasets and in terms of quality achieved. Yet what is interesting as regards to the number of datasets completed is that the simple profit system performs very poorly, especially when the offered micro-task price drops. In fact it takes very high monetary rewards (approximately 20 monetary units per micro-task) to approach the performance of the smart profit algorithm. In parallel, when examining the two systems in terms of quality, the smart profit system manages to achieve and maintain high quality levels regardless of the offered price, while the simple profit algorithm's quality is analogous to the offered price. This result can be attributed to the fact that the smart scheduling algorithm targets tasks to the expertise of the producer. Thus, we can conclude that the smart scheduling algorithm allows consumers of datasets to offer their task at a lower cost and expect the same or better quality as a simple system based on financial incentives, and increasing the task price does not produce a large increase in quality if smart scheduling is used.

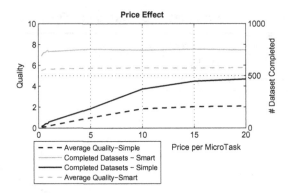

Fig. 9. The effect of micro-task price on performance of two-profit based algorithms

5.2 Producer Arrival Rate Effect

The final parameter to examine is the number of producers arriving in the data marketplace, which would be considered the "resource supply" in terms of resource scheduling and allocation algorithms. The present producer arrival rate, as extracted from the CKAN data, is low in comparison to the number of dataset requests. As it can be intuitively expected, the growth of the arrival of producers may increase if the system starts offering monetary incentives. So understanding the response of the crowdsourcing system under differentiated conditions of producer arrival is essential if one is to implement a working system. To simulate different growth levels of the producer arrival rate, we modify the shape factor β of the producer arrival rate function with β values in the range [1,2], i.e. from the CKAN levels ($\beta = 1$) to an almost double growth of the arrival rate ($\beta = 2$) in $f(t) = \alpha \cdot e^{\beta \cdot t}$.

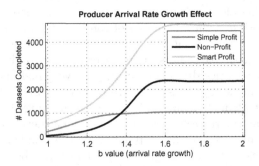

Fig. 10. The effect of the growth in the producer arrival rate for the three algorithms

Figure 10 illustrates the response of the three systems in regards to production efficiency based on increasing the arrival of the number of producers in terms of number of completed datasets. As shown, the smart profit algorithm outperforms the other two for all of the simulated producer arrival rates, managing to keep directing the producer contributions and to prioritizing them towards dataset completion, in the same manner as observed in earlier simulation work. It is very interesting to observe the surprising behavior of the other two algorithms. The non-profit system performs better than the simple profit one, for user arrival rates augmented up to 20% ($\beta = 1.2$). After this value, the non-incentive system is outperformed by both the two monetary incentive-based systems. Thus, one can argue that for systems with small amount of producers, a non-profit system makes more sense than a profit-driven one unless a crowd-sourcing system is used. However, our first hypothesis is correct under certain conditions: Assuming a high arrival rate of new users, a simple crowd-sourcing based system based on monetary incentives, even without a smart directed crowd-sourcing algorithm, will outperform a crowd-sourcing system without financial incentives.

6   Conclusions and Future Work

Although the vision of a data marketplace is not without precedence (e.g. see [4] for an initial vision of the Web as an "Information Marketplace"), much of the data produced to-date is erroneous and not well maintained [7]. Nevertheless, the attempts for financially-powered data marketplaces have so far been very rare, mainly due to the absence of performance guarantees for consumers, and the fact that research on computational methods for gathering data from people in a systematic manner and with performance guarantees is still in its infancy (indicatively see [1,11]). In this paper we argue for a data marketplace system where various actors can produce and consume datasets for financial contributions. Our first hypothesis was that financial incentives inside such a marketplace would lead to the production of greater amounts of high-quality structured data than systems without such incentives. This hypothesis was shown to be only partially correct: Only with high arrivals of producers into the system does such a simple data market-place succeed. However, there is an important caveat: our second hypothesis, that crowd-sourcing with financial incentives would perform better in terms of quality if a directed crowd-sourcing algorithm is used (as given in our example by a smart resource scheduling algorithm) was shown to be correct.

However, our simulations made a number of simplifying assumptions in order to test our hypotheses under the strictest possible conditions, and future work should test these parameters. For instance, we assume that data-sets do not vary in any way systematically (instead, they vary randomly) over quality and regarding their knowledge domain. These assumptions are based on the fact that there are no studies of macro-scale quality effects in structured data. However, adding a ranking and feedback system (such as "one to five stars") would allow us to judge quality. We could then experiment with algorithms that estimated both the difficulty of the task and expertise of the producers and then used a non-beta distribution to model these parameters. Furthermore, we could allow datasets to belong to more than one domain. Also, it is highly likely that the decomposition of particular datasets into micro-tasks varies considerably, and thus we could also examine varying the number of micro-tasks as well. Finally, much more sophisticated pricing schemes could be used that let the consumers set minimum and maximum dataset price ranges, allow producers to negotiate the price for the micro-tasks they undertake, and then the mechanism could attempt to optimize the crowd-sourcing system for both quality and price.

Finally, in the current research we are forced to model a future incentive-based system for structured data out of empirical parameters gathered from a non-profit based system. The goal of this stage of the simulation is to gain a general understanding of the effect that parameters such as the dataset arrival rates and producer arrival rates' have on the quality of structured data. Given our hypothesis that directed crowd-sourcing with financial incentives outperforms non-profit based systems, the next step of our research beyond simulation is to create an implemented data market-place for real humans initially using the parameters derived from this work. The implementation of such a system will then no doubt lead to more empirical results that then can be used to optimize further iterations of a data marketplace design.

References

1. Amsterdamer, Y., Grossman, Y., Milo, T., Senellart, P.: Crowd mining. In: SIG-MOD Conference, pp. 241–252 (2013)
2. Balog, K., Azzopardi, L., de Rijke, M.: Formal models for expert finding in enterprise corpora. In: Proceedings of the 29th Annual International ACM SIGIR Conference on Research and Development in Information Retrieval, SIGIR 2006, pp. 43–50. ACM, New York (2006)
3. Bernstein, M.S., Karger, D.R., Miller, R.C., Brandt, J.: Analytic methods for optimizing realtime crowdsourcing. CoRR, abs/1204.2995 (2012)
4. Dertouzos, M., Gates, B.: What Will Be: How the New World of Information Will Change Our Lives. HarperCollins, New York City (1998)
5. Dorn, C., Dustdar, S.: Composing near-optimal expert teams: a trade-off between skills and connectivity. In: Meersman, R., Dillon, T., Herrero, P. (eds.) OTM 2010. LNCS, vol. 6426, pp. 472–489. Springer, Heidelberg (2010). https://doi.org/10.1007/978-3-642-16934-2_35
6. Ghosh, A., Hummel, P.: Implementing optimal outcomes in social computing: a game-theoretic approach. In: Proceedings of the 21st International Conference on World Wide Web, WWW 2012, pp. 539–548. ACM, New York (2012)
7. Halpin, H., Hayes, P.J., McCusker, J.P., McGuinness, D.L., Thompson, H.S.: When owl:sameAs Isn't the same: an analysis of identity in linked data. In: Patel-Schneider, P.F., et al. (eds.) ISWC 2010. LNCS, vol. 6496, pp. 305–320. Springer, Heidelberg (2010). https://doi.org/10.1007/978-3-642-17746-0_20
8. Ho, C.-J., Slivkins, A., Suri, S., Vaughan, J.W.: Incentivizing high quality crowd-work. In: Proceedings of the 24th International Conference on World Wide Web, International World Wide Web Conferences Steering Committee, pp. 419–429 (2015)

9. Huang, E., Zhang, H., Parkes, D.C., Gajos, K.Z., Chen, Y.: Toward automatic task design: a progress report. In: Proceedings of the ACM SIGKDD Workshop on Human Computation, HCOMP 2010, pp. 77–85. ACM, New York (2010)

10. Ipeirotis, P.G., Provost, F., Wang, J.: Quality management on amazon mechanical turk. In: Human Computation Workshop (KDD-HCOMP 2010) (2010)

11. Lykourentzou, I., Vergados, D.J., Naudet, Y.: Improving wiki article quality through crowd coordination: a resource allocation approach. Int. J. Semantic Web Inf. Syst. **9**(3), 105–125 (2013)

12. Mao, A., et al.: Volunteering versus work for pay: incentives and tradeoffs in crowd-sourcing. In: First AAAI Conference on Human Computation and crowdsourcing (2013)

13. Mason, W., Watts, D.J.: Financial incentives and the "performance of crowds". In: Human Computation Workshop (HComp2009) (2009)

14. Nath, S., Zoeter, O., Narahari, Y., Dance, C.: Dynamic mechanism design for markets with strategic resources. In: Proceedings of the Twenty-Seventh Conference Annual Conference on Uncertainty in Artificial Intelligence (UAI 2011), pp. 539–546. AUAI Press, Corvallis (2011)

15. Shahaf, D., Horvitz, E.: Generalized task markets for human and machine computation. In: National Conference on Artificial Intelligence (2010)

16. Shen, H.Y.Z., Fauvel, S., Cui, L.: Efficient scheduling in crowdsourcing based on workers. In: 2017 IEEE International Conference on Agents (ICA), pp. 121–126. IEEE (2017)

17. Smirnova, E.: A model for expert finding in social networks. In Proceedings of the 34th International ACM SIGIR Conference on Research and Development in Information Retrieval, SIGIR 2011, pp. 1191–1192. ACM, New York (2011)

18. Von Ahn, L.: Human computation. Ph.D. thesis, Carnegie Mellon University, Pittsburgh, PA, USA (2005). AAI3205378

Ethical and Socially-Aware Data Labels

Elena Beretta[1,3]($\boxtimes$) ⬤, Antonio Vetrò[1,2] ⬤, Bruno Lepri[3] ⬤,
and Juan Carlos De Martin[1] ⬤

[1] Nexa Center for Internet and Society, DAUIN, Politecnico di Torino, Turin, Italy
{elena.beretta,antonio.vetro,demartin}@polito.it
[2] Future Urban Legacy Lab, Politecnico di Torino, Turin, Italy
[3] Fondazione Bruno Kessler, Trento, Italy
lepri@fbk.eu

Abstract. Many software systems today make use of large amount of personal data to make recommendations or decisions that affect our daily lives. These software systems generally operate without guarantees of non-discriminatory practices, as instead often required to human decision-makers, and therefore are attracting increasing scrutiny. Our research is focused on the specific problem of biased software-based decisions caused from biased input data. In this regard, we propose a data labeling framework based on the identification of measurable data characteristics that could lead to downstream discriminating effects. We test the proposed framework on a real dataset, which allowed us to detect risks of discrimination for the case of population groups.

Keywords: Data ethics · Automated decisions · Data quality

1 Introduction

The availability of large-scale data, often regarding human behavior, is profoundly changing the world in which we live. The automated flow and analysis of this type of data offers an unprecedented opportunity for actors in both public and private sectors to observe human behaviors for a large variety of purposes: to provide insights to policy-makers; to build personalized services like automated recommendations on online purchases; to optimize business value chains; to automate decisions; etc. However, the way data are collected, tested and analyzed poses a number of risks and questions related to the context of use [3]. Many researchers, in fact, identified a number of ethical and legal issues where the application of software automated techniques in decision-making processes has led to intended and unintended negative consequences, and especially disproportionate adverse outcomes for disadvantaged groups [1,12]. Recent scandals such as the one involving Cambridge Analytica and Facebook[1] or the study conducted by ProPublica of the COMPAS Recidivism Algorithm[2], are two well-known examples of the relevance of these issues for our societies. Recent research

---

[1] https://bit.ly/2Hoa2q7.
[2] See https://bit.ly/1XMKh5R.

ⓒ Springer Nature Switzerland AG 2019
J. A. Lossio-Ventura et al. (Eds.): SIMBig 2018, CCIS 898, pp. 320–327, 2019.
https://doi.org/10.1007/978-3-030-11680-4_30

efforts have focused on the data collection and data exploitation issues of software systems (e.g., in the field of machine learning [7] or, more in general, in software-related conferences [11]). We place the problem in the context of software engineering practice proposing a data labeling framework, the Ethical and Socially-Aware Data Labels (EASAL), to identify data input properties that could lead to downstream potential risks of discrimination towards specific population groups. The beta version of the framework, presented here, relies on three building blocks, each one supported by previously published evidence drawn from different disciplines. We describe our data labeling in Sect. 2, and we show and discuss the results from testing EASAL on a real dataset in Sect. 3. We conclude by summarizing our contribution and providing indications for future research work in Sect. 4.

2   Ethically and Socially-Aware Data Labeling (EASAL)

Many software systems today rely on statistical techniques and prediction models, fed by large amount of available data. Such data is used for training algorithms whose scope is to recognize patterns and find relationships in data. A problem that characterizes automatic approaches that rely solely on data and algorithms is that they miss the human capability to perform important tasks, among which the context-aware interpretation of the results, the elaboration of explanations and cause-effect relationships, the recognition of biases (and possibly their correction). Regarding the latter, which is the focus of our work, we report a statement made by the mathematician Cathy O'Neal in her book "Weapons of Math Destruction" [13]:

> if the admission models to American universities had been trained on the basis of data from the 1960s, we would probably now have very few women enrolled, because the models would have been trained to recognize successful white males.

This observation entails an important fact: not only data processes such collection and analysis have ethical consequences, but also input data properties are connected to important ethical issues. In fact, some characteristics of the collected data involve ethical issues, and those problems propagate throughout all subsequent phases of the data life-cycle in software systems, until affecting the output, i.e. the decisions or recommendations made by the software. Our hypothesis is that certain data characteristics may lead to discriminatory decisions and therefore it is important to identify them and show the potential risks.

Moved by these motivations, we defined the Ethically And Socially-Aware Labeling (EASAL) framework, which is a way of labeling datasets using measures of certain input data characteristics (e.g., uneven distribution in gender balance, co-linearity of attributes, etc.) that represent a risks of discrimination if used in decision making (or decision support) systems. We believe that this information will be useful to software engineers to be more aware of the risks of discriminations and to use the dataset in an more ethically and socially-aware manner. In addition, it could be used by third parties to certify such risks on a given dataset.

To the best of our knowledge, labelling approaches for ethical purposes are being investigated in two other ongoing research initiatives. The first one is a collaboration between the Berkman Klein Center at Harvard University and the MIT Media Lab, which led to "The Dataset Nutrition Label Project"[3]. The project aims to avoid that incomplete, misunderstood or problematic data have an adverse impact on artificial intelligence algorithms. The second research is conducted by Gebru *et al.* [6], who propose "Datasheets for Datasets": with respect to our proposal, this approach is towards more discursive technical sheets to encourage better communication between creators and users of a dataset. These approaches are not mutually exclusive, instead they can be seen as mutually reinforcing. Herein we describe the building blocks of EASAL.

2.1 Disproportionate Datasets

Most of today software-automated decisions are based on the analysis of historical data. This is very often done with machine learning models. It has been proven that problems of fairness and discrimination inevitably arise, mainly due to disproportionate datasets [13]. Disproportionate datasets lead to disproportionate results, generating problems of representativity when the data are sampled - thus leading to an underestimation or an overestimation of the groups - and of imbalance when the dataset used has not been generated using random probabilistic sampling methods. Many of the datasets used today have not been generated using these methods, but are rather selected through non probabilistic methods, which do not provide to each unit of the population the same opportunity to be part of the sample; this means that some groups or individuals are more likely to be chosen, others less. For this reason, it is essential to keep this aspect under control in non-probabilistic samples. In general, solutions relating to demographic or statistical parity are useful in cases where there is no deliberate and legitimate intention to differentiate a group considered protected, which would otherwise be penalized [4]. It should therefore be borne in mind that the solutions vary according to both the nature and use of the data. Take as an example a type of analysis that includes in its attributes individual income. If the choice to include in the sample only individuals with a high income is voluntary, no representativity problems arise, since the choice of a given group is based on the purposes of the analysis. However, if the probability of being included in the sample is lower as the income is lower, then the sample income will on average be higher in the overall income of the population.

---

[3] See https://datanutrition.media.mit.edu.

2.2   Correlations and Collinearity

In statistics two variables x_1 and x_2 are called collinear variables when one is the linear transformation of the other and therefore there is a high correlation. In general there are always relationships between variables that involve a certain degree of linear dependence, but it is essential to keep this aspect under control to avoid negative effects: in fact, in case of collinearity, small variations in the data may correspond to significant variations in the estimated values. Since the analysis of collinearity reveals the presence of redundant connections between variables, it is useful in those areas more sensitive to the risk of discrimination. To prevent this effect some researchers adopt a naïve approach that precludes the use of sensitive attributes such as gender, race, religion and family information, but in some cases may not be effective. The use of geographic attributes, for example, is shown to be unsuitable when the use of protected data is to be foreclosed, because it easily leads to tracing protected attributes, such as race [12]. Hardt [7] points out that the condition of non-collinearity requires that the predictor $(\hat{Y})$ and the protected attribute (A) are independent conditional on Y: e.g., if *income* has to be predicted, it must be independent of *gender*. Another common error is *"mistake correlation with causation"* [8]; cause-effect ratios are often confused with correlations when features are used as proxies to explain variables to be predicted. For example, the IQ test is a test that measures logical-cognitive abilities, but if used as a proxy to select the smarter students for admission to a university course, it would almost certainly reveal itself as an imperfect proxy, since intelligence is a too broad concept to be measured by a number only. As a consequence, although there is a correlation between the test value of the IQ and the predicted variable, it is not sufficient to explain it. As Friedler *et al.* [5] remark, *"determining which features should be considered is a part of the determination of how the decision should be made"*. In light of the problems mentioned in the previous paragraphs, we expect EASAL synthetically summarizes the analysis of collinearity and correlation between protected attributes in order to avoid possible discriminatory results.

2.3   Data Quality

In computer science, "garbage in, garbage out" (GIGO) is a popular sentence to identify where "flawed, or nonsense input data produces nonsense output"[4]. The GIGO principle implies that the quality of the software is affected by the quality of the underlying data. As a consequence, computer generated recommendations or decisions are affected by poor input data quality. For these reasons, we include data quality as third building block of EASAL. The ISO/IEC standard 25012 [9] defines 15 data quality characteristics, operationalized by 63 metrics defined in the ISO/IEC 25024 [10]. Recent research efforts (e.g., [2,14,15]) showed that a measurement approach is effective in revealing data quality problems, especially for the inherent quality dimensions, that are also more effective for our purposes,

---

[4] See https://en.wikipedia.org/wiki/Garbage_in,_garbage_out.

because they are not affected by the context of use (e.g., hardware and software environment, computer-human interface). We propose the ISO/IEC 25012 and 25024 standards models as a reference for quantitatively assessing the quality of data input and the consequential confidence of the decision made out of that data. In particular, we refer to the inherent quality dimensions: accuracy, completeness, consistency, credibility, currentness[5].

3   Testing EASAL on Real Datasets

We tested the EASAL on Credit Card Default dataset, that *contains information on default payments, demographic factors, credit data, history of payment, and bill statements of credit card clients in Taiwan from April 2005 to September 2005*[6]. The field of creditworthiness often appears in the literature alongside issues related to ethical decisions. Recently, some studies have shown that access to credit for black people is modulated by certain attributes such as race, rather than by information about the payer's status[7]. The dataset that we use does not contain the protected attribute *race*, but contains other personal information that can be used in a discriminatory way if applied to assess creditworthiness, such as gender and level of education.

Disproportion. Figure 1 reports an example of visualization of a disproportionate dataset: the histogram shows that the frequency distribution of the age attribute is highly skewed, and the group most represented is that of 25 to 40 years. The analysis of the frequencies for the other protected attributes included in the dataset, shows that: 60% of individuals are women; 46.7% of individuals have attended university; married and single individuals are equally represented. Although we do not have information neither on the real frequencies of protected

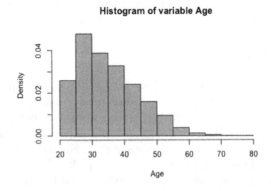

Fig. 1. Frequency of variable *age*

---

[5] For the definitions of inherent quality measures see [10].

[6] https://www.kaggle.com/uciml/default-of-credit-card-clients-dataset.

[7] See https://bit.ly/2NyNVPx.

attributes in the source population nor on the sampling method used (if any), the results of the analysis of disproportions suggests to use the age variable with caution: in fact the variable age shows a more considerable disproportion compared with the other protected attributes, exposing a potential risk of discrimination (e.g., if the dataset is used to automate decisions or recommendations on the capability to repay a debt, and attribute age is one of the predictors).

Correlation and Collinearity. We perform the analysis for each protected attribute in the Credit Card default dataset, in relation to *default payment* (1 = yes, 0 = no). We report on Fig. 2 an example of mosaic plot11 for the attribute *education*: blue indicates cases in which there are more observations in that cell than would be expected under the null model of independence between attribute *education* and attribute *default payment*; red means there are fewer observations than would have been expected; eventually, grey indicates that observations are coherent with the assumption of independence. Figure 2 shows that *default payment* is highly correlated to the *education* level, for all its levels. The test has been performed also on the other protected attributes included in the dataset, and showed that the correlation between the protected attributes and the *default payment* variable is significant for the *gender* variable (both male and female), and it is significant for the marital status variable in correspondence with the *default payment group* = *yes*. In addition, Pearson residuals show that the most correlated categories are: the *education* variable and the male, both in correspondence with *default payment* = *yes*. As a consequence of the analysis, the identified correlations should be taken into account when using the dataset in an algorithm that supports or automate decisions.

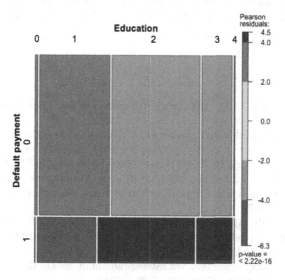

Fig. 2. Conditional mosaic plot for conditional independence of variable *education*. *Legend: 0 = na; 1 = graduate school; 2 = university; 3 = high school; 4 = others (Color figure online)*

Inherent Data Quality. The nature of dataset allows to test only two of the five inherent quality dimensions, *accuracy* and *completeness*; of these, five metrics are selected from the ISOIEC 25024, namely: *Acc-I-4: Risk of data set inaccuracy, Com-I-1: Record completeness, Com-I-2: Attribute completeness, Com-I-4: Data values completeness, Com-I-5: Empty records in a data file*[8]. The test provides the extreme positive value of the index, therefore it is not necessary to report the values obtained.

4   Conclusions

We presented a theoretical framework for labeling input data used in decision making software and for identifying risks of discrimination towards specific population groups. The Ethically and Socially Aware Labels (EASAL) are composed of three building blocks: measures for assessing disproportion; measures for assessing correlation and collinearity involving protected attributes; measures for assessing data quality. The building blocks have been identified on the base of literature studies and authors experience. We intend to address our future work along the following directions: test and specification of the use of correlation and collinearity metrics for different types of statistical variables; graphical design of an intuitive label that could help software engineers in quickly understanding the discriminatory risks of using a dataset; automation of label creation and source code freely available to allow replication studies and improvements. We also invite the software engineering community to contribute to this initial work by improving the building blocks measures, by identifying new building blocks and by applying EASAL for benchmarking purposes. This would facilitate an increase of awareness of software practitioners regarding the ethical implications of the data-driven systems that they design, build, and to which are probably subject, at least in some scenarios.

References

1. Barocas, S., Selbst, A.D.: Big data's disparate impact. Calif. Law Rev. **104**(3), 671–732 (2016)
2. Corrales, D.C., Corrales, J.C., Ledezma, A.: How to address the data quality issues in regression models: a guided process for data cleaning. Symmetry **10**(4), 99 (2018). https://doi.org/10.3390/sym10040099. https://bit.ly/2xOLVzN
3. Doshi-Velez, F., et al.: Accountability of AI under the law: the role of explanation. Berkman Center Research Publication Forthcoming, Harvard Public Law Working Paper 18(07) (2017)
4. Dwork, C., Hardt, M., Pitassi, T., Reingold, O., Zemeln, R.: Fairness through awareness. In: Proceedings of the 3rd Innovations in Theoretical Computer Science Conference, pp. 214–226. ACM (2012)
5. Friedler, S.A., Scheidegger, C., Venkatasubramanian, S.: On the (im) possibility of fairness. arXiv preprint arXiv:1609.07236 (2016)

---

[8] Details on how to compute each metric can be retrieved from [10].

6. Gebru, T., et al.: Datasheets for datasets. arXiv:1803.09010 (2018)
7. Hardt, M., Price, E., Srebro, N.: Equality of opportunity in supervised learning. In: Advances in Neural Information Processing Systems (2016)
8. Hosni, H., Vulpiani, A.: Forecasting in light of big data. Philos. Technol. **13**, 1–13 (2017)
9. ISO-IEC: ISO/IEC 25012:2008 Software engineering - Software product Quality Requirements and Evaluation (SQuaRE) - Data quality model. Standard, International Organization for Standardization, Geneva, CH, December 2008
10. ISO-IEC: ISO/IEC 25024:2015 - Systems and software engineering - Systems and software Quality Requirements and Evaluation (SQuaRE) - Measurement of data quality. Standard, International Organization for Standardization, Geneva, CH, October 2015
11. Karim, N.S.A., Ammar, F.A., Aziz, R.: Ethical software: integrating code of ethics into software development life cycle. In: 2017 International Conference on Computer and Applications (ICCA), pp. 290–298, September 2017. https://doi.org/10.1109/COMAPP.2017.8079763
12. Lepri, B., Staiano, J., Sangokoya, D., Letouzé, E., Oliver, N.: The tyranny of data? The bright and dark sides of data-driven decision-making for social good. In: Cerquitelli, T., Quercia, D., Pasquale, F. (eds.) Transparent Data Mining for Big and Small Data. SBD, vol. 11, pp. 3–24. Springer, Cham (2017). https://doi.org/10.1007/978-3-319-54024-5_1
13. O'Neil, C.: Weapons of Math Destruction: How Big Data Increases Inequality and Threatens Democracy. Crown Publishing Group, New York (2016)
14. Torchiano, M., Vetrò, A., Iuliano, F.: Preserving the benefits of open government data by measuring and improving their quality: an empirical study. In: 2017 IEEE 41st Annual Computer Software and Applications Conference (COMPSAC), vol. 1, pp. 144–153, July 2017. https://doi.org/10.1109/COMPSAC.2017.192
15. Vetrò, A., Canova, L., Torchiano, M., Minotas, C.O., Iemma, R., Morando, F.: Open data quality measurement framework: definition and application to open government data. Gov. Inf. Q. **33**(2), 325–337 (2016). https://doi.org/10.1016/j.giq.2016.02.001

Shadow Removal in High-Resolution Satellite Images Using Conditional Generative Adversarial Networks

Giorgio Morales$^{(\boxtimes)}$, Samuel G. Huamán , and Joel Telles

National Institute of Research and Training in Telecommunications (INICTEL-UNI),
National University of Engineering, San Luis 1771, 15021 Lima, Peru
{gmorales,shuaman,jtelles}@inictel-uni.edu.pe

Abstract. In satellite image processing, obscure zones that were affected by shadows are normally discarded from further processing. Nevertheless, for specific applications, such as surveillance, it is desirable to remove shadows despite the fact that reconstructed zones do not necessarily have real reflectance values. In that sense, we propose a shadow removal method in high-resolution satellite images using conditional Generative Adversarial Networks (cGANs). The generator network is trained to produce shadow-free RGB images with condition on a satellite image patch altered with artificial shadows and concatenated with its respective binary shadow mask, while the discriminator is adversely trained to discern if a given shadow-free image comes from the generator or if it is an original RGB image without artificial alteration. The method is tested in the proposed dataset achieving an error ratio comparable with the state of the art. Finally, we confirm the feasibility of the proposed network using real shadowed images.

Keywords: Generative Adversarial Networks · Shadow removal ·
Satellite imagery

1 Introduction

Satellite images are normally affected by shadows caused by clouds or elevated mountains. For some applications it is enough to detect shadowed areas and avoid any further processing over them due to their reflectance values have been modified. Nevertheless, for other applications, such as surveillance, in which it is involved human visual inspection, it is useful to remove the shadow effect although the reflectance of the new images will not exactly correspond to reality.

Shadow removal in high-resolution satellite images is a challenge, as such images have limited spectral bands and details should be kept after processing. Typical approaches relies on multi-temporal analysis for any kind of satellite images [1–3]. However, it is not always possible to have multiple satellite images from the same scene; thus, other methods propose shadow removal from single

© Springer Nature Switzerland AG 2019
J. A. Lossio-Ventura et al. (Eds.): SIMBig 2018, CCIS 898, pp. 328–340, 2019.
https://doi.org/10.1007/978-3-030-11680-4_31

images multiplying the affected zones by constants to standardize the scene illumination [4–6] or using gradient based methods [7,8], which do not take into account the shadow variations inside the umbra region. Besides, other methods propose the use of color transformations techniques such as histogram matching and color transfer [5,9,10]. Alternatively, information theory based techniques is proposed in [11] but this approach either require user assistance or careful parameter selection. In addition, another method that needs user interaction is [12], which register the penumbra to a normalised frame which allows to efficiently estimate non-uniform shadow illumination changes. It is worth to mention that, since many of the aforementioned methods are designed to work with natural images, such as photographs, they can assume that the texture remains approximately the same across the shadow boundary, which means they will not necessarily work well with high-resolution satellite images, where it is possible to find drastic terrain changes inside a shadowed region.

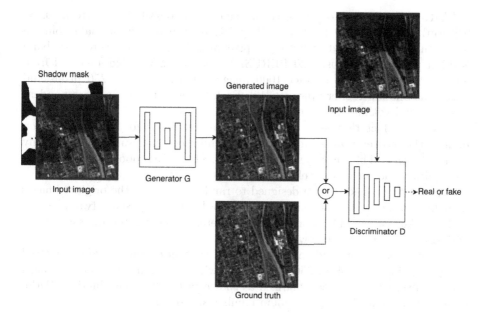

Fig. 1. Proposed methodology scheme.

In recent years, many deep learning techniques have been developed thanks to the availability of large datasets and computational resources (such as dedicated GPUs), resulting in a landmark change in the computer vision community. However, to the best of our knowledge, very few deep learning based techniques have been proposed to tackle the problem of shadow detection and removal. Owing to these reasons, we propose a new method to remove shadows in high-resolution satellite images from PERUSAT-1, a Peruvian satellite managed and supervised by the Space Agency of Peru (CONIDA), using conditional Generative Adversarial Networks (cGANs). This method consist on training a Generator network with condition on a satellite image patch concatenated its respective shadow mask to produce shadow-free images. Adversarially, a Discriminator network is trained to distinguish between real shadow-free images and those produced by the Generator, as shown in Fig. 1. In this way, we train the networks using the proposed ShadowfreePeru dataset conformed by shadow-free images and images altered with artificial shadows, thus expecting that the trained network will generalize well when it comes to remove real shadows in satellite images. We use fake shadows to train the networks because it is too difficult to capture the shadow-free and the shadowed satellite image of the completely same scene (e.g. with no vegetation or water body changes) and with the same light conditions.

2   Proposed Method

2.1   ShadowfreePeru Dataset

A PERUSAT-1 scene has four spectral bands: red (0.63–$0.7\,\mu$m), green (0.53–$0.59\,\mu$m), blue (0.45–$0.50\,\mu$m) and NIR (0.752–$0.885\,\mu$m). The spatial resolution of the multispectral bands is 2.8 m per pixel and that of the panchromatic band is 0.7 m per pixel. We used 20 PERUSAT-1 scenes of variable area and from different geographies to extract 1000 image patches of 256×256 pixels and create the ShadowfreePeru dataset [13]. Each image patch has a correspondent image altered with artificial shadows.

Given the fact that shadow removal is mainly used for visual inspection because the recovered zones do not have trustful reflectance values, we take only the red, green and blue channels of the images. Thus, the proposed method could be applicable not only to satellite images, but any RGB image.

The artificial shadows are designed to randomly darken the original images with different degrees of intensity and noise, so that the proposed network trained with this dataset would be more robust to different types of alterations and could perform well with real shadows.

In first place, the random shapes of the synthetic shadows are generated using a Perlin noise [14] of 256×256 pixels with a random seed value between two and five, which is then binarized as shown in Fig. 2a. The binarized Perlin noise ($P1$) is multiplied by a random number according to:

$$P2 = P1 \times k, \quad \text{s.t.} \quad k \sim U(0.4, 0.8), \tag{1}$$

where k is a random variable with uniform distribution, $U(a, b)$, whose probability density function $u(k)$ is formally defined as:

$$u(t) = \begin{cases} 1, \text{if } t \in [a, b] \\ 0, \text{otherwise} \end{cases}. \tag{2}$$

Then, $P2$ is used to create three masks ($P3_1$, $P3_2$ and $P3_3$) that will individually alter the red, green and blue channels of the original images. Hence, as it is desirable to generate shadows with different intensities, the values of $P2$

will be slightly increased or decreased. Besides, the effect of the shadows is not necessarily the same for the three channels, so a small aleatory variation is added to each of them, as in Eq.2. Figure 2c shows one generated $P3$ mask.

$$P3_1 = P2/u, \quad \text{s.t.} \quad u \sim U(0.85, 1.15)$$
$$P3_2 = P2/(u+v), \quad \text{s.t.} \quad v \sim U(-0.02, 0.02) \tag{3}$$
$$P3_3 = P2/(u+r), \quad \text{s.t.} \quad r \sim U(-0.02, 0.02).$$

Next, we invert the $P3$ masks and apply a median filter with a 3×3 - pixel kernel in order to reduce noise. After that, we get the $P4$ masks (Fig. 2d) applying a mean filter with a disk kernel of diameter d $(d \sim U(1,8))$, so we have shadows with different types of blurring of their edges (penumbra). Finally, we get the artificially shadowed image (Fig. 2e) altering its red, green and blue channels using the element-wise products shown in Eq.3. Figure 3 shows different shadows generated with this method.

$$R' = R \circ P4_1,$$
$$G' = G \circ P4_2, \tag{4}$$
$$B' = B \circ P4_3.$$

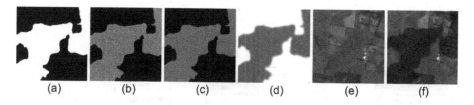

Fig. 2. Shadow generation. (a) Binarized Perlin noise. (b) $P2$ mask. (c) $P3_1$ mask. (d) Filtered $P4_1$ mask. (e) Original image. (f) Altered image with artificial shadows. (Color figure online)

Fig. 3. Sample shadowed images from the ShadowfreePeru dataset. (a)(c) Smooth shadows with blurred edges. (b)(d) Dark shadows without blurred edges.

In order to avoid overfitting problems when training a model, we use data augmentation to increase the dataset size. Thus, we rotated each patch 90°, 180° and 270° so that we have a total of 4000 patches. We split 90% of the data to create the training set, 5% to the validation set and 5% to the test set.

2.2   Conditional Generative Adversarial Network

Conditional Generative Adversarial Networks (cGANs) [15] are generative models that learn a mapping from observed image x and random noise vector z to y, $G : \{x, z\} \rightarrow y$. In this work, the generator G is trained to produce outputs that cannot be distinguished from real shadow-free high-resolution satellite images by an adversarially trained discriminator, D. Therefore, the objective of a cGAN can be expressed as:

$$\mathcal{L}_{cGAN}(G, D) = \mathbb{E}_{x,y}[log(D(y, x))] + \mathbb{E}_{x,z}[log(1 - D(G(x, z), x))], \quad (5)$$

$$G^* = arg \min_{G} \max_{D} \mathcal{L}_{cGAN}(G, D), \quad (6)$$

where the discriminator D maximizes the expected log-likelihood of correctly distinguishing real samples from fake ones, while the generator G maximizes the expected error of the discriminator by trying to synthesize better images.

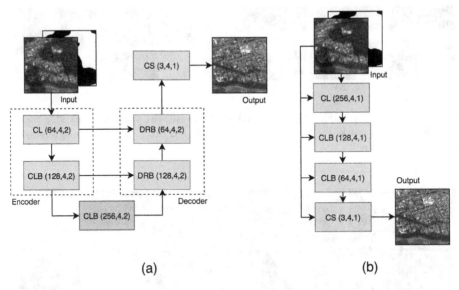

Fig. 4. Generator architectures. (a) cGAN1: U-NET architecture. (b) cGAN3: continuous condition concatenation architecture.

For the generator, we compared the use of a network with continuous condition concatenation (Fig. 4a) [17] and a U-NET [18], a network composed by an encoder and a decoder with skip connections (Fig. 4b). The motivation of adding skips between separate layers is that structure in the input is aligned with structure in the output and so it is convenient to share information from low levels of the network to the higher ones. The encoder-decoder structure of the U-NET tends to extract global features of the inputs and generate new representations by this overall information, while the continuous condition concatenation offers a detailed spatial local guidance through all the layers, which seems to be more suitable for this local color transformation task.

For the discriminator, a decoder network called PatchGAN [16] that works at the scale of local images patches of $N \times N$ pixels, trying to classify if each patch in an image is real or fake. We chose $N = 16$, which means that each pixel from the discriminator's output corresponds to the believability of a 94×94 patch of the input image.

Figures 4 and 5 show in detail the architecture of the generator and discriminator. The layers of Convolution, Deconvolution, Batch Normalization, ReLU, LeakyReLU and Sigmoid are represented by C, D, B, R, L and S, respectively. Numbers in parentheses indicate the number, size and stride of the convolution and deconvolutional filters. Both networks were trained using an Adam optimizer [19] with a learning rate of 0.00005, a momentum term β_1 of 0.9, a momentum term β_2 of 0.999 and a mini-batch size of 32 during 100 epochs due to the limited computational capability.

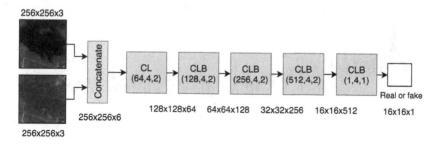

Fig. 5. Discriminator architecture.

We compared four different network configurations using 200 images from the validation dataset to select the optimal architecture of the generator network. Table 1 shows the comparison of metrics of the four networks considering the Root Mean Square (RMSE) as a metric; the first two networks (cGAN1 and cGAN2) have a U-NET architecture but differs in the number of layers inside the encoder and the decoder, while the last two networks have a continuous

Table 1. Comparison of metrics between different architectures

Network	Layers		Root Mean Square Error (RMSE)
	Encoder	Decoder	
cGAN1	2	2	0.5141
cGAN2	3	3	0.5894
cGAN3	4		0.5056
cGAN4	3		0.6248

condition concatenation architecture and differs in the number of convolutional layers. From this comparison, we choose the cGAN3 network (shown in Fig. 4b) because it presents the best performance. In addition, Fig. 6 shows the evolution of the RMSE value computed on both the training and validation sets over training time.

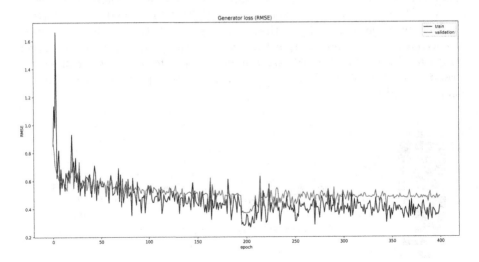

Fig. 6. Epoch vs. RMSE

3   Results

3.1   Fake Shadows Removal

The proposed algorithm was trained using Python 3.6 on the High Performance Computational Center of the Peruvian Amazon Research Institute (IIAP), with an Intel CPU Xeon E5-2680 v4 and a NVIDIA Tesla K80 GPU. We used 200 images from the test set to evaluate the performance of the proposed method comparing the shadow removal results with ground truth images using the ShadowfreePerudataset.

In order to consider the size of the shadows and the fact that some shadows may be darker than others, we quantitatively compare the shadow removal results from different methods computing the error ratio $E_r = E_n/E_0$, where E_n is the Root Mean Square Errors (RMSE) between the RGB intensities of the ground truth (original non-shadowed image) and the shadow removal result, and E_0 measures the same type of error between the ground truth and the shadowed image. The standard deviation of this measure (S_r) is also computed. These metrics are shown in Table 2, where the non-stared columns (E_r and S_r) indicate the error score using all pixels in the image, and the stared columns (E_r^* and S_r^*) indicate the error within the shadowed areas only. Figure 7 shows a comparison between all this methods. The mean time required to process an image is

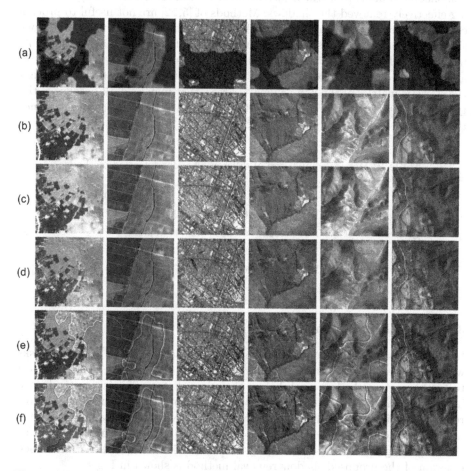

Fig. 7. Fake shadow removal in satellite images. (a) Artificially shadowed images. (b) Original images. (c) Our proposed method. (d) Gong et al. [12]. (e) Zigh et al. [5]. (f) Deb et al. [6].

calculated after evaluating each method in a PC with Intel CPU Xeon E5-2690 without GPU. All methods are tested using CPU only in order to compare the processing times using the same framework.

From Table 2, it seems that the method of [5] has a good performance and works faster than the others, but as can be seen in Fig. 7, it always presents a whitish border corresponding to the penumbra area because it multiplies each channel of the image by constants without taking into account the possible variations inside the shadowed area, just like [6]. The method of [12] has the best performance, but requires user interaction and take long time to process each image, while the proposed method consist on an end-to-end learning approach that is already trained to work in many geographical contexts. It is worth mentioning that, while the other methods alter the shadowed areas exclusively, the Convolutional Neural Network process the entire image and slightly modifies the non-shadowed areas too, which increases the overall error ratio, in spite of having effectively removed the shadows. Methods of [5,12] are not useful to process completely shadowed images because they need as input both shadow and non-shadow references. What is more, it is possible that they will fail when the whole non-shadowed areas and the shadowed areas correspond to low reflectance soil and high reflectance soil, respectively; i.e., these methods will attempt to assign the statistic metrics of the non-shadowed areas, that may be dark because of the type of soil, to the shadowed areas that correspond to high reflectance soil and, therefore, are supposed to have high pixel values.

Table 2. Metrics comparison of different shadow detection methods

Method	E_r	S_r	E_r^*	S_r^*	Time per image (sec)
Deb et al. [6]	0.6538	0.5305	0.4817	0.4012	0.054
Zigh et al. [5]	0.3837	0.2220	0.3280	0.2371	0.012
Gong et al. [12]	0.3003	0.1512	0.2891	0.1460	11.260
Proposed method	0.3154	0.1559	0.3804	0.1647	1.369

3.2 Real Shadows Removal

In the previous case, shadows were artificially created so that shadow masks were known a priori. However, when dealing with real images, it is necessary to firstly create a shadow mask before the removal step. This is a critical issue in satellite images due to the fact that shadows can easily be mistaken for low reflectance soil or water. That is why we use an adversarially trained generator network with a U-NET architecture, as proposed in [20], to semantically segment shadows using the four spectral bands of the satellite images. Hence, the block diagram of the proposed shadow removal method is shown in Fig. 8.

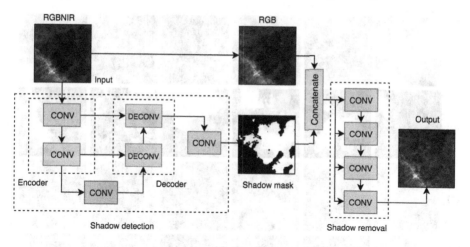

Fig. 8. Block diagram for remove shadows in real images.

Figure 9 shows a visual comparison of all mentioned methods. We compare the results visually due to the fact that we do not have the ground truth of the shadow-free images, like in the previous case, and therefore we cannot compute such metrics as the RMSE or the error ratio. It can be seen that the methods of [6] and [5], despite having a good performance removing fake shadows, are not suitable for this type of task because of the intensity variations inside the shadowed areas. In addition, the method of [12] shows to be effective removing real shadows and producing realistic results; nevertheless, it could fail in situations such as the second, fourth and fifth cases of Fig. 9c because it copies to the shadowed regions the color and texture of the surroundings areas, which results inappropriate in the presence of clouds. Our method proves that, although it was trained to remove fake shadows, which were defined as random localized obscurations of the image, it can generalize well and effectively remove real shadows, though some shadow borders are still distinguishable like in the second, third and fourth cases of Fig. 9b. It is worth mentioning that if shadows are incorrectly segmented, that would cause degeneration of color in non-shadowed areas. Nevertheless, the results reported in [20] show a high value of precision (95.82%), which means that it does not present too many false positives. In that case, false positives are mainly caused by leftovers along borders of the detected shadows and not by incorrectly detected shadows.

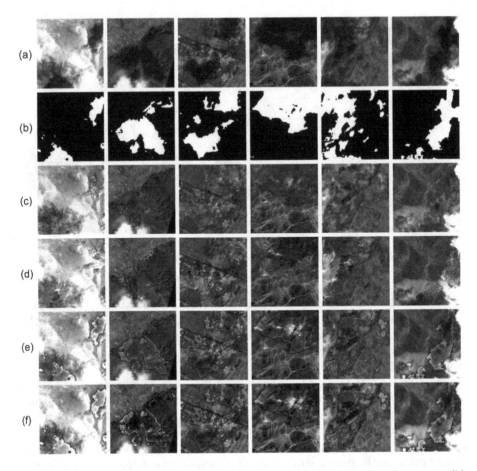

(a)

(b)

(c)

(d)

(e)

(f)

Fig. 9. Real shadow removal in satellite images. (a) Real shadowed images. (b) Detected shadows. (c) Our proposed method. (d) Gong et al. [12]. (e) Zigh et al. [5]. (f) Deb et al. [6]. (Color figure online)

4   Conclusions

In this paper, we have presented a novel end-to-end trainable deep neural network for addressing the problem of removing shadows in satellite images. The proposed model consists of a generator network with continuous condition concatenation architecture based on adversarial training. The method shows similar results to the state of art and improves some aspects, such as the time processing, with respect to that of the most efficient of the compared methods, and the fact that colors and textures of the reconstructed areas are not exclusively dependent on the surrounding non-shadowed areas.

In addition, it has been shown that a network that has been trained using a dataset of shadow-free and artificially shadowed images, can generalize well when dealing with real shadows. That means that if the artificial shadows are more carefully modeled to have a closer behavior to real shadows, the performance of the neural network can be improved.

Acknowledgement. The authors would like to thank the National Commission for Aerospace Research and Development (CONIDA) and the National Institute of Research and Training in Telecommunications of the National University of Engineering (INICTEL-UNI) for the support provided. The training of all the networks was carried out by the High Performance Computational Center of the Peruvian Amazon Research Institute (IIAP). For more information please visit http://iiap.org.pe/manati.

References

1. Wang, B., Ono, A., Muramatsu, K., Fujiwara, N.: Automated detection and removal of clouds and their shadows from landsat TM images. IEICE Trans. Inf. Syst. **E82–D2**, 453–460 (1999)
2. Sah, A.K., Sah, B.P., Honji, K., Kubo, N., Senthil, S.: Semi-automated cloud/shadow removal and land cover change detection using satellite imagery. Int. Arch. Photogramm. Remote. Sens. Spat. Inf. Sci. **39–B7**, 335–340 (2012)
3. Candra, D.S., Phinn, S., Scarth, P.: Cloud and cloud shadow removal of landsat 8 images using Multitemporal Cloud Removal method. In: 6th International Conference on Agro-Geoinformatics, Fairfax, VA, USA (2017)
4. Murali, S., Govindan, V.K.: Shadow detection and removal from a single image using LAB color space. Cybern. Inf. Technol. **13**(1), 95–103 (2013)
5. Zigh, E., Belbachir, M.F., Kadiri, M., Djebbouri, M., Kouninef, B.: New shadow detection and removal approach to improve neural stereo correspondence of dense urban VHR remote sensing images. Eur. J. Remote. Sens. **48**(1), 447–463 (2015)
6. Deb, K., Suny, A.H.: Shadow detection and removal based on YCbCr color space. Smart Comput. Rev. **4**(1), 23–33 (2014)
7. Finlayson, G.D., Hordley, S.D., Lu, C., Drew, M.S.: On the removal of shadows from images. TPAMI **28**(1), 59–68 (2006)
8. Liu, F., Gleicher, M.: Texture-consistent shadow removal. In: Forsyth, D., Torr, P., Zisserman, A. (eds.) ECCV 2008. LNCS, vol. 5305, pp. 437–450. Springer, Heidelberg (2008). https://doi.org/10.1007/978-3-540-88693-8_32
9. George, G.E.: Cloud shadow detection and removal from aerial photo mosaics using light detection and ranging (LIDAR) reflectance images. Ph.D. thesis, The University of Southern Mississippi, Mississippi, USA (1996)
10. Khan, S.H., Bennamoun, M., Sohel, F., Togneri, R.: Automatic shadow detection and removal from a single image. IEEE Trans. Pattern Anal. Mach. Intell. **6**(1), 431–446 (2015)
11. Kwatra, V., Han, M., Dai, S.: Shadow removal for aerial imagery by information theoretic intrinsic image analysis. In: IEEE International Conference on Computational Photography, pp. 1–8. IEEE, Seattle (2012)
12. Gong, H., Cosker, D.: Interactive shadow removal and ground truth for variable scene categories. In: Proceedings of the British Machine Vision Conference. BMVA Press, Nottingham (2014)

13. ShadowfreePeru Dataset. http://didt.inictel-uni.edu.pe/dataset/datasetshadow-correction.hdf5
14. Perlin, K.: Improving noise. ACM Trans. Graph. **21**(3), 681–682 (2002)
15. Mirza, M., Osindero, S.: Conditional generative adversarial nets. arXiv preprint arXiv:1411.1784 (2014)
16. Isola, P., Zhu, J.Y., Zhou, T., Efros, A.: Image-to-image translation with conditional adversarial networks. In: IEEE Conference on Computer Vision and Pattern Recognition (CVPR), Honolulu (2017)
17. Cao, Y., Zhou, Z., Zhang, W., Yu, Y.: Unsupervised diverse colorization via generative adversarial networks. arXiv preprint arXiv:1702.06674v2 (2017)
18. Ronneberger, O., Fischer, P., Brox, T.: U-Net: convolutional networks for biomedical image segmentation. In: Navab, N., Hornegger, J., Wells, W.M., Frangi, A.F. (eds.) MICCAI 2015. LNCS, vol. 9351, pp. 234–241. Springer, Cham (2015). https://doi.org/10.1007/978-3-319-24574-4_28
19. Kingma, D., Ba, J.: Adam: a method for stochastic optimization. In: International Conference on Learning Representations (ICLR), San Diego (2015)
20. Morales, G., Arteaga, D., Huamán, S., Telles, J.: Shadow detection in high-resolution multispectral satellite imagery using generative adversarial networks. In: 2018 IEEE XXV International Conference on Electronics, Electrical Engineering and Computing (INTERCON), pp. 1–4. IEEE, Lima (2018). https://doi.org/10.1109/INTERCON.2018.8526416

A Mixed Model Based on Shape Context and Spark for Sketch Based Image Retrieval

Willy Puenternan Fernández[✉] and César A. Beltrán Castañón[✉]

Department of Engineering, Artificial Intelligence Research Group (IA-PUCP),
Pontificia Universidad Católica del Perú, Av. Universitaria 1801,
San Miguel, Lima 32, Peru
{willy.puenternan, cbeltran}@pucp.pe

Abstract. Nowadays, information is not limited to textual representation but takes several other forms such as sketch-based image retrieval, where the user draws a query, and the system retrieves the most similar images. In this work we present a mixed approach combining shape context and Spark features, previously we had applied the Bag-of-Features strategy to select regions of interest, achieving significant improvement in effectiveness of the retrieval task. Our method works as a local strategy for key-points detection. Results are very auspicious, and we show different experiments conducted to demonstrate our proposed methodology. The highlight of this paper is the step-by-step description of the methodology to create a framework for sketch-based image retrieval.

Keywords: SBIR · Bag-of-features · Shape context · Spark feature

1 Introduction

Information has changed, now it is represented not only by textual documents but by the video and graphics as well [1, 2]. Now it has abstract ways of representation that cannot be accessed using the old fashion textual method as input. It is necessary to find ways to retrieve that kind of data and accordingly it is that content based image retrieval (CBIR) mechanisms arise to meet this increasing demand to help people make complex searches of videos or images [3]. Those kind of searches are possible but until now implied having a prior model similar to the one that the user wanted to find [4]. In many cases, the user does not have that, and yet it is quite illogical searching for an image that you already have.

The user usually does not have the image so is forced to do complicated text-based searches or (in the worst case) to make largely manual searches one by one inside a big set of images [5]. On the other hand, the use of sketches has been present in the history of humanity since remote times. Our prehistorical ancestors have used it to represent several kinds of knowledge and what is more, since very early ages [6]. So now, with this perspective, we propose a new way to communicate to the browser the image we want to find. Why not use the sketches as an input parameter into the browser's search engine? Naturally, lots of difficulties will have to be overcome. Some issues are solved and new problems raised.

J. A. Lossio-Ventura et al. (Eds.): SIMBig 2018, CCIS 898, pp. 341–348, 2019.
https://doi.org/10.1007/978-3-030-11680-4_32

As you can see, there is a lot of work to do in the field of visual information retrieval, but the spectrum of opportunities derived from its development is immense and not only for use with web browsers. There are many applications in several fields like didactic teaching, video games and technological security [8]. Precisely, it is this kind of diversity that is the project's most valuable source of inspiration, to foster the base for the development of new technologies through the people's mind language: the sketch [6]. The present work shows a strategy for the retrieval of images based on the input of drawings. Some steps were implemented of an algorithm capable of obtaining useful results.

2 Proposed Methodology

The methodology used in the present project was Bag-of-Features derived from the widely used Bag-of-Words in the fields of information retrieval and classification. Both methods are similar in several aspects, but they differ in that Bag-of-Features handles the processing of images rather than words [9].

With Bag-of-Words, each document is represented according to the frequency of the words it contains and without consideration of the syntactic representation between words. On the other hand, the methodology of Bag-of-Features characterizes each image as a set of regions of interest which describes its appearance but ignores its structure in the global image [10].

Another tool used in the project is the image descriptor Shape Context [11], which is used to compare figures based on their shapes. It establishes a set of random points inside the sketch's strokes to create vectors that connect each point with the others. The set of those vectors allows us to have a detailed description of the context of an image. This descriptor is especially useful when we work with sketches [7]. Figure 1 shows the visual representation with Shape Context.

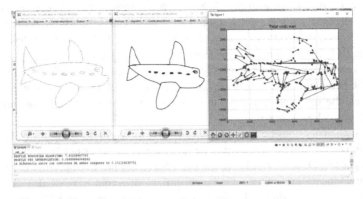

Fig. 1. Visual representation of how the descriptor shape context works when are compared two similar images.

Finally, the work used another descriptor which performs similarly to Shape Context, called the Spark feature. It consists in the detection of multiple strokes in the images by first obtaining some points located in regions free of strokes. After we get those points, Spark feature generates various straight lines from the mentioned points, recording those that cross any stroke from the image [7]. Figure 2 shows a visual representation of the Spark algorithm.

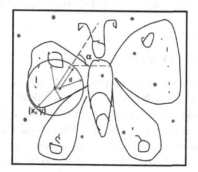

Fig. 2. Visual representation of how the descriptor spark feature works.

To implement the framework our strategy followed two steps: preprocessing, to select images, clean noise, and detect edges; and processing of the image, to extract features and create the key points representation.

3 Image Pre-processing

To transform images into sketches, we first defined the image database which will be used with the project. We manually selected one by one the best free images we could use from the ImageNet web. Those images had to be free of visual noise and imperfections to later obtain the best results we could. However, there was the possibility that we could have selected some images with considerable visual distortions, for which we developed an automatic noise cleaning. We chose a total of 320 images divided into four categories of 80 images: airplane, alarm clock, angel, and ant. Next, we apply the Canny algorithm through OpenCV 2.4.1 [12], a computer vision library. To simplify the complexity of that procedure, we decided to work only with the resultant binary images, giving the appearance to be working just with sketches and not with colorful photos from the database. One advantage of the Canny algorithm is that it preprocesses images to delete any visual noise.

It is important to say that not all the elements from our database were initially colorful pictures. We wanted to make possible getting sketches through another sketch, so we purposely added some sketches into the database and also applied them the preprocessing described with the Canny algorithm. Another important point is the interface for the user to sketch the image he wanted to retrieve created using several

OpenCV libraries. In that interface, the user inserted their sketch into a black panel, which would capture and invert the binary sketch's colors just for esthetic visualization.

4 Image Processing

In this section we describe the main process of the work that start with the feature extraction, detection of strokes, define the index structure and retrieve of images based on sketches.

4.1 Extraction of Local Features

With the image database established, we proceeded with the extraction of local characteristics from each image, which consists of the random selection of small regions with the most critical strokes from an image (in this case, a sketch). Each image was divided into ten local square regions, with side's length equal to the 25% of the image's diagonal. To get the ten most representative local regions per image, only those with a minimum of 2% of black pixels over the rest of its pixels were selected. Through this mechanism, obtaining local regions with none or inadequate quantity of information were avoided. Figure 3 shows the visual process for patches selection.

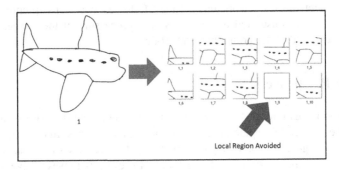

Fig. 3. Example of the visual patches selection applied to the images database.

Given the significant amount of images and their local regions, the use of a server Centos 7 GPU Nvidia Grid and an algorithm based on the use of threads was implemented to enhance the computational processing reducing time from 8 to 3 h approximately. With the local regions of all the images in our database, the next step generates a vector of features which allow us to differentiate a local area from other.

4.2 Detection of Strokes

Now with the local regions, we have to determine a set of patterns that allows us to associate each of these with a specific image. Creating a feature vector of coordinates where there are strokes in the local region for each image will be useful. There are plenty of feature descriptors, but for this work, we used the Spark feature [7] because it was more practical working with sketches than others like SIFT or specialized histograms. Spark features allow better visualization of the points in the local regions detected by the descriptor and to save computational cost processing each image. The more coordinates we get in the local area, the more robust its feature vector will be. The present work used a total of 500 coordinates per local region. These points were chosen randomly and put inside an index structure. In Fig. 4 we can see graphically the result, where the green dots are the coordinates where a stroke was detected.

Fig. 4. Visualization of the results obtained using Spark feature. The green dots are the coordinates where a stroke was detected. (Color figure online)

Once we have the set of 500 coordinates per local region saved into an index structure (a feature vector), the next step was sorting those points from the lowest to the highest point to obtain a simulation of displacement into the image, first by the abscissas and then by the ordinates.

4.3   Index Structure

We implemented a structure to store the location of every image extracted from our database and the information associated with those like their ten local regions and the features vectors from each of them. This structure is our codebook, and it consists of 320 elements where each of them corresponds to an image from our database. Each of these elements contains a list of 10 items to store the local regions from the images referenced by the codebook, and this will be our list of features vectors lists. We also said that we would work with 500 coordinates per image, that's why every element from the list of features vectors will reference another list of 500 elements; this will be our list of features vectors (Fig. 5). Finally, each item from the last list mentioned will have a feature vector of two components for X and Y coordinates. Through this structure, we can map out all the essential characteristics from our database images. Similarly to this structure, we will have another one exclusive to the user's sketch.

4.4   Comparing Image Database vs. Sketch

Now, with all the previous steps done, we can proceed to compare index structures to find the most similar images in our database to the sketch entered by the user. In

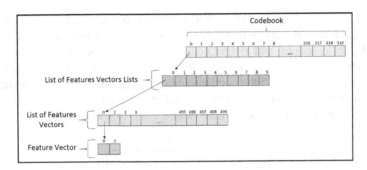

Fig. 5. Index structure used in the work to store information associated with the image database.

summary, we can compare their coordinates and try to retrieve the photos with the significant quantity of equivalences or similarities based on their numerous characteristics.

We search for the top 10 most similar images to the user's sketch and we decided to use the Shape Context descriptor as an additional filter which will help our framework to reduce the universe of possible candidates and to reduce the computational cost for processing every image.

The results obtained were promising and made it possible for the project to return not only the most similar image from the database but also some others with some similarity to the user's sketch. The comparison in detail between both structures was made very simple to avoid misunderstandings with those readers who want to replicate it.

Two points are similar if their coordinates have as the Euclidean distance value less than or equal to 50. For the local region comparison, two of them (one for the sketch and the other for the database image) are similar if their percentage of similar points is higher than or equal to the 40% of coordinates. Finally, both images are similar if they have at least one local region considered equivalent on the previously described filters.

In Fig. 6 we can see results of alarm clock query showing the input image and the top-ten most similar images.

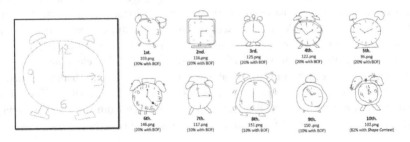

Fig. 6. A query sample, for the alarm clock class, showing the top ten most similar images.

5   Evaluation

The results obtained thanks to the fusion between our comparing algorithm and the descriptor Shape Context were successful, getting in the majority of cases the most similar images with the user's sketch considering the 3-top retrieved images. To show that, we used a precision-recall metric where we can visually see the results obtained with 80 samples. Better scores resulted with images belonging to the alarm clock category, followed by the categories angel, ant, and airplane (Fig. 7).

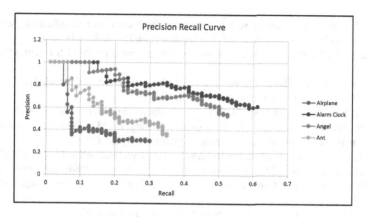

Fig. 7. Precision-recall curve using 80 samples.

Based on the experimentation, it is possible to say that the results obtained in each of the categories are mainly due to the degree of detail captured when we randomly extracted the most representative local regions in each image. The significant aspect the captured region has, the higher is the hit rate of the framework, for our experiment, images of alarm clocks in the database presented the higher score. In this case, the location of the strokes in the alarm clock sketches are always over the clock hand's silhouettes. The rest of the categories were not stroked similar to those, making the alarm clock category easy to identify, but not the others. For this reason, if we want to expand the range of correct images returned by the framework at the top of its results, it would be enough to adapt a better descriptor.

6   Conclusions

If there is the necessity to work with a large number of pictures, it is indispensable to execute the framework with hardware with high processing capacity, especially for the generation of the codebook and its previous steps. The present work used a server Centos 7 with GPU Nvidia Grid.

A high quality set of images is essential to obtain better results in comparing an image with an input sketch image. The development of a framework for the retrieval of images based on the image information is a useful tool because it allows increasing the range of complex models that the user wants to find.

There is the possibility to improve the framework's results through the use of a more influential figure descriptor. There are multiple ways to generate the codebook. In the present work, the codebook was made in a way to make it simple and easy to understand.

References

1. Choi, Y.: Analysis of image search queries on the web: query modification patterns and semantic attributes. J. Am. Soc. Inf. Sci. Technol. **64**(7), 1423–1441 (2013)
2. Mysen, C.C., Verma, N., Chen, J.: Ranking custom search results. US Patent 8,930,359, 6 Ene. 2015
3. Veltkamp, R.C., Burkhardt, H., Kriegel, H.P.: State-of-the-Art in Content-Based Image and Video Retrieval, vol. 22. Springer, New York (2013)
4. Lew, M.S.: Principles of Visual Information Retrieval. Springer, London (2013)
5. Cao, Y., Wang, H., Wang, C., Li, Z., Zhang, L., Zhang, L.: MindFinder: interactive sketch-based image search on millions of images, Firenze, Italy (2010)
6. Yu, Q., et al.: Sketch-a-net that beats humans. arXiv preprint arXiv:1501.07873 (2015)
7. Hildebrand, K., Eitz, M., Boubekeur, T., Alexa, M.: Sketch-based image retrieval: benchmark and bag-of-features descriptors. IEEE Trans. Vis. Comput. Graph. **17**, 1624–1636 (2011)
8. López, J.C.: Xataka, 7 Abril 2016. http://www.xataka.com/robotica-e-ia/
9. Feijo, S.P.: Aplicación del Modelo Bag-of-Words al Reconocimiento de Imágenes. Universidad Carlos III de Madrid, Madrid, España (2009)
10. Rosado Rodrigo, P., Figueras Ferrer, E., Planas Roselló, M., Ferran Reverter, C.: La imagen toma la palabra: Construcción de un vocabulario visual. In: de 2nd Art, Science, City International Conference ASC 2015, Valencia, España (2015)
11. Belongie, S., Malik, J., Puzicha, J.: Shape context: a new descriptor for shape matching and object recognition. In: Advances in neural information processing systems, pp. 831–837 (2001)
12. Doxygen: Open Source Computer Vision, 22 de Agosto de 2016. http://docs.opencv.org/3.1.0/da/d22/tutorial_py_canny.html

Continuous Detection of Abnormal Heartbeats from ECG Using Online Outlier Detection

Yuhang Lin[1,3P], Byung Suk Lee[1(✉)], and Daniel Lustgarten[2]

[1] Department of Computer Science, University of Vermont, Burlington, VT, USA
{Yuhang.Lin,Byung.Lee,Daniel.Lustgarten}@uvm.edu
[2] Department of Medicine, University of Vermont, Burlington, VT, USA
[3] Department of Computer Science, North Carolina State University,
Raleight, NC, USA
ylin34@ncsu.edu

Abstract. Detecting abnormal heartbeats from an electrocardiogram (ECG) signal is an important problem studied extensively and yet is a difficult problem that defies a viable working solution, especially on a mobile platform which requires computationally efficient and yet accurate detection mechanism. In this project, a prototype system has been built to test the feasibility and efficacy of detecting abnormal ECG segments from an ECG data stream targeting a mobile device, where data are arriving continuously and indefinitely and are processed online incrementally and efficiently without being stored in memory. The processing comprises three steps: (i) segmentation using R peak detection, (ii) feature extraction using discrete wavelet transform, and (iii) outlier detection using incremental online microclustering. Experiments conducted using real ambulatory ECG datasets showed satisfactory accuracy. In addition, comparing personalized detection (tuned separately for each patient's ECG datasets) and non-personalized detection (tuned aggregated over all patients' datasets) confirms a definite advantage of personalized detection for ECG.

Keywords: ECG · Anomaly detection · Outlier detection · Data stream

1 Introduction

There has been a significantly large body of work on automatically detecting abnormal segments from electrocardiogram (ECG) signal. Different methods have been used for different work with different objectives, and in this project the objective is real-time online incremental detection with a lightweight computational algorithm. Ideally, the computation overhead should be light enough to run on a mobile platform such as a smartphone. The method chosen with this objective in mind is online outlier detection based on microclustering. Additionally the following choices have been made to support the objective: (1) only

J. A. Lossio-Ventura et al. (Eds.): SIMBig 2018, CCIS 898, pp. 349–366, 2019.
https://doi.org/10.1007/978-3-030-11680-4_33

one lead ECG is used as opposed to the full 12-lead ECG, and (2) clustering is performed on features extracted from ECG segments to be computationally efficient ($O(N)$) and resilient to errors [14]. Our goal in this paper is to examine the method in light of the objective.

The project started on Android smartphone platform, but then migrated to a laptop platform and stayed there until now. Yet, with the real-time processing expectation in mind, the algorithm chosen worked incrementally over incoming ECG data stream with instantaneous processing, i.e., without having to store the data in memory. The processing was done in three steps: (1) segmentation of the ECG data, (2) feature extraction from the ECG segments, and (3) online outlier detection from the features. Segmentation relied on R peak detection equipped with false R peak removal. Feature extraction used Haar discrete wavelet transform. Outlier detection used incremental microclustering [22].

The outcome was evaluated in terms of the detection accuracy using MIT-BIH arrhythmia ECG datasets. When the parameters for outlier detection were tuned personalized to individual patients' datasets, the sensitivity, specificity, and accuracy on average were 83%, 88%, and 92%, respectively, and when aggregated over all patients' datasets using the average parameter values, they were 56%, 87%, and 82%, respectively. In addition, the accuracy was higher when there was a clearer majority between normal or abnormal segments, that is, when the ECG segments were skewed in their distribution of abnormality. These results demonstrated the feasibility and efficacy of the detection method employed and strongly indicated the need for *personalized* detection.

Main contributions of this paper can be summarized as follows: (i) to the best of our knowledge, this is the first project using online outlier detection mechanism to detect abnormal segments from an ECG signal; (ii) comprehensive evaluations on the accuracy of abnormal segment detection presents a new insight into the behavior of the online outlier detection mechanism and an empirical perspective on the merit of personalized anomaly detection as opposed to non-personalized.

Following this Introduction, Sect. 2 describes the ECG datasets used in the project, Sect. 3 discusses the steps of anomaly detection process, Sect. 4 reports the anomaly detection accuracy in the experiment results, Sect. 5 discusses related work, and Sect. 6 concludes the paper.

2   ECG Datasets

Electrocardiogram (ECG) is an electrical signal manifesting the heartbeat over time. It is a sequence of segments, one segment per heartbeat. Figure 1a shows a raw (i.e., unfiltered) ECG signal with noise in it. Figure 1b illustrates the composition of an ECG segment – each segment consists of a P wave, a QRS complex, and a T wave.

The ECG datasets used in this project were downloaded from MIT-BIH Arrhythmia Database, which contains 48 half-hour excerpts of two-channel ambulatory ECG recordings obtained from 47 subjects (i.e., patients) studied by

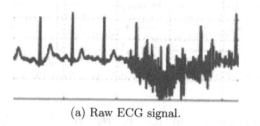
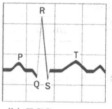

(a) Raw ECG signal.   (b) ECG segment.

Fig. 1. ECG signal.

the BIH Arrhythmia Laboratory from 1975 to 1979. All recordings were gathered using the sampling rate of 360 samples per second per channel. In this project, we used the ECG signal on the first channel (or lead) in each dataset, namely MLII, for all patients.

The ECG datasets are annotated with the codes denoting the normality or abnormality of segments at each R peak location. A complete list of those annotation codes can be found at the PhysioBank Annotation web site [19]. Table 1 shows the annotation labels used in this project. Database code is the R peak type annotated in the database, and AHA code is the type categorized by the American Heart Association (AHA). We used the AHA code in this project, and, thus, the database codes 'V', 'F', '!', 'E', 'P', 'f', 'p', 'Q', '/', and – (empty) were considered abnormal. (The code '!' is a heart beat code but incorrectly listed as a non-beat code, and was corrected in this project.)

Table 1. List of ECG annotation codes used in this project.

Database code	AHA code	Description
N	Normal	Normal beats
L	Normal	Left bundle branch block beat
R	Normal	Right bundle branch block beat
A	Normal	Atrial premature beat
S	Normal	Supraventricular premature or ectopic beat (atrial or nodal)
!	Normal	Ventricular flutter wave
V	Abnormal	Premature ventricular contraction
j	Abnormal	Nodal (junctional) escape beat
F	Abnormal	Fusion of ventricular and normal beat
f	Abnormal	Fusion of paced and normal beat
E	Abnormal	Ventricular escape beat
Q	Abnormal	Unclassifiable beat
/	Abnormal	Paced beat
–	Abnormal	No annotation found within this segment

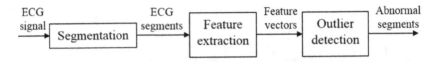

Fig. 2. Overview of the approach to continuous anomaly detection from ECG signal.

The downloaded ECG datasets have already been filtered through a bandpass filter to retain only the frequency range 0.1 Hz to 100 Hz and digitized at 360 Hz using hardware built at the MIT Biomedical Engineering Center and at the BIH Biomedical Engineering Laboratory [18]. Some ECG datasets still contained significant noise, which had adverse effect on the resulting accuracy, while most others were stable enough to be used without such an effect (see Sect. 4.2).

3   Approaches

This section discusses the specific approaches used in this project.

3.1   Overview

Figure 2 shows a high-level overview of the steps for detecting abnormal ECG segments from an incoming ECG signal. First, the signal is divided into consecutive segments (or heartbeats). Then, in the feature extraction step, each segment is transformed to a feature vector, which is mapped to a point in a feature vector space. The outlier detection algorithm then picks out those points farther than a threshold from other points. These outliers are considered abnormal ECG segments. In effect, the detection is done as an incremental unsupervised binary classification of each segment, i.e., either normal or abnormal. Each step is discussed in more detail in the rest of this section.

3.2   Segmentation

Segmentation comprises three steps: R peak detection, segment extraction, and false R peak removal.

R Peak Detection. There are different algorithms used to detect R peaks from raw ECG data. In this project, Chen and Chen's *moving average based filtering algorithm* [6] was used for its good performance and low computation overhead. This algorithm performs three steps over a moving average of consecutive ECG samples: (1) linear high-pass filtering, (2) nonlinear low-pass filtering, and (3) decision making with adaptive threshold.

In the decision-making (step 3), an adaptive threshold T is updated in each moving window using the formula below

$$T = \alpha\gamma P + (1 - \alpha)T$$

where P is the local maximum newly detected in the waveform, α is the forgetting factor, and γ is the weighting factor to determine the contribution of peak values to threshold adjustment.

It is suggested in their algorithm that the moving average window size M can be 5 or 7 samples, α can be chosen from the range of 0.01 to 0.1, and γ can be 0.15 or 0.2. In this project, we set M to 5, α to a random number from 0.01 to 0.1 at each run, and γ to 0.17 (as its showed higher R peak location accuracy than 0.15 or 0.2).

Segment Extraction. After detecting R peaks, the next step is to extract segments from the ECG data. We adopted the following formula, introduced in Veeravalli et al.'s work [23]:

$$P_{window} = QR_{max} + 0.2 * RR_{prev} + 0.1$$

$$T_{window} = 1.5 \times QTc_{max} \times \sqrt{RR_{prev}} - QR_{max}$$

where QR_{max} ($= 0.08$) is half of the maximum of QRS duration and QTc_{max} ($= 0.42$) is the maximum value for the QT coefficient in Bazett's formula [3] shown below.

$$QT_cB = \frac{QT}{\sqrt{RR}}$$

The extracted segment spans the P_{window} and the T_{window}.

Figure 3 shows an example ECG segment extracted using the formula. The yellow dot marks the R peak; on its left is the P window, and on its right is the T window.

False R Peak Removal. While the adopted R peak detection and segment extraction algorithms worked adequately for most segments, there were quite a number of segments that contained two R peaks, where the first one was a true peak and the second one was a false peak. We, therefore, added one more step to remove the false second R peak from the segment. Specifically, if any extracted segment has two R peaks and if the second R peak is within the T_{window}, it is detected as a false R peak, and the end of the segment is cut 15

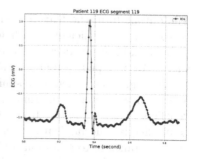

Fig. 3. A sample ECG segment.

samples before it. This reduction length of 15 was chosen as a result of manually checking the results for different reduction lengths ranging from 0 to 25 at the increment of 5.

3.3 Feature Extraction

Discrete wavelet transform (DWT) was used to transform each ECG segment to a feature vector. This step is, in effect, reducing the dimensionality of an ECG segment of approximately 300 samples to a feature vector of 32 coefficients. Daubechies wavelets and Haar wavelets were compared, and Haar was chosen for its better signal restoration ability and faster speed.

Haar wavelet transform takes a pair of consecutive numbers from the input sequence, calculates their pairwise average and puts the result in the first half, and calculates their pairwise difference and puts the result in the second half. Then, by taking the first half, which contains only the pairwise averages, we can approximate the original signal, and repeating this process, we can reduce the number of coefficients to half each time. We continued repeating until we had 32 coefficients in the first half, and then extracted the first 32 coefficients as the feature vector. In this project, it typically took 3 iterations to finish the process.

Symmetric Padding: Given the recursive two-way division performed by Haar wavelet transform, the input length (i.e., number of data elements) should be a power of 2. Since an ECG segment length (i.e., number of ECG samples in it) is not a power of 2 for most segments, a certain number of data elements should be added to make it a power of 2. There are several different ways to do it [21], and we chose the *symmetric padding*, which mirrors the data to increase the length to the nearest next power of 2. For example, if the segment has n samples $x_1 x_2 ... x_{n-1} x_n$, it is mirrored on both sides to $...x_2 x_1 | x_1 x_2 ... x_{n-1} x_n | x_n x_{n-1} ...$ symmetrically until the resultant length is a power of 2.

3.4 Outlier Detection

The output from the feature extraction step is a continuous stream of feature vectors of 32 coefficients. Each feature vector is mapped to a point in a 32-dimensional feature space. In feature extraction, ECG segments that have similar shapes are mapped to points at similar coordinates in the feature space. Thus, normal segments, which have similar shapes, are mapped to similar coordinates and form a cluster. Abnormal segments, on the other hand, are mapped to "outliers", i.e., points far from other points in a cluster. So, outlier detection is an effective mechanism to identify a point farther off from others, and such outliers translate to abnormal ECG segments.

We adopted an outlier detection method called *"Microcluster-based Continuous Outlier Detection (MCOD)"* [8,14]. This method is one of popularly used distance-based outlier detection algorithms [22], and works well as long as there is a majority between normal or abnormal segments. The algorithm requires the following three parameters to detect outliers.

- r: maximum allowed radius from a point
- k: minimum number of points required within the radius
- w: size (i.e., number of points) of a moving window

Microcluster-Based Continuous Outlier Detection (MCOD). MCOD is a distance-based outlier detection (DBOD) algorithm over a data stream, enhanced from Continuous Outlier Detection (COD) [14]. Figure 4 illustrates the distance-based outlier detection approach. The point a is not an outlier because there are 3 points b, c, and e within distance r from e. In contrast, the point e is an outlier because there is only one (i.e., less than 3) point b within the distance r from a [12,13].

COD is computationally efficient in handling two cases – a new point entering the window and an old point leaving the window. In the former case, it checks if any existing outliers should become inliers after the addition of the new point. In the latter case, it checks if any existing inliers should become outliers after the removal of the old point. To handle these two cases, COD supports a range query to find points within the distance r and uses an event-based queue to check if an inlier becomes an outlier because of a point removed from the window.

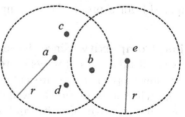

Fig. 4. Distance-based outlier detection when $k = 3$ (source: [8]).

Executing a range query can be expensive, especially when the dimensionality of points is high. MCOD can greatly reduce the number of range queries performed, thereby improving the overall performance of the algorithm. A microcluster can be thought of as a small sphere in the data space. The algorithm requires a microcluster to be of radius $r/2$ and contain more than k points at all time. Any point that belongs to a microcluster is never an outlier because there always exist more than k points within the range r in the same microcluster. In contrast, any point that does not belong to any microcluster is very likely to be an outlier. Every time a new point arrives in a stream, if the window is not full, then no point is removed. Otherwise, the oldest point is removed and, if it belongs to a microcluster, the number of members in that microcluster is reduced by one and, if the resulting number drops below $k + 1$, the microcluster is removed and for all its members, their lists of nearest microcluster centers within a distance of $3r/2$ is updated.

MCOD may label an ECG segment as an outlier when it enters a window and then later change it to inlier. In this project, an ECG segment is considered an outlier only if it is labeled as an outlier throughout from the time it enters the sliding window till the time it leaves the window.

3.5   Complexity Analysis

The outlier detection algorithm MCOD can tell if an ECG segment is normal or abnormal only after the feature point mapped from the segment passes through the window completely. Thus, the complexity can be expressed in terms of the segment size s and the window size w. The three-step approach – comprising segmentation, feature extraction, and outlier detection – requires $O(s + wk)$ memory space, where s is the largest ECG segment size, and takes $O(sw)$ run time, as discussed below.

Space Complexity. The segmentation step requires linear processing of the incoming ECG data samples, and memory buffer large enough to hold the largest ECG segment suffices to support this processing. So, the space complexity is $O(s)$. The feature extraction step processes each segment at a time, and for each segment the run time is proportional to the segment size. Therefore, it requires only the buffer space to hold the largest segment, hence $O(s)$. The outlier detection step requires $O(wk)$ space. Readers are referred to the MCOD paper [14] for a proof of this complexity.

Time Complexity. Since the segmentation step processes ECG data samples linearly, it takes $O(sw)$ to generate w feature points in the window of the outlier detection step. The feature extraction step to generate the w feature points in the window is $O(sw)$ as well because for each segment DWT takes linear time with the segment size and each segment generates one feature point in the window. The outlier detection step takes $O(l \log w) + O(m)$, where l is the average number of times feature points are re-labeled as outliers as a result of an old point removed from the window of size w and m is the average number of feature points within the maximum allowed radius r (discussed in the MCOD paper [14]). While the theoretical worst case time complexity of this step could be $O(w \log w)$, in practice it is near $O(\log w)$ because both $l \ll w$ and $m \ll w$ hold. Thus, the total run time complexity for all three steps is $O(sw) + O(\log w)$, which asymptotically equals $O(sw)$.

4 Evaluations

This section presents the setup, results, and analysis of the experiments performed to evaluate the accuracy of detecting abnormal segments from an ECG data stream.

4.1 Experiment Setup

Development Platform: The main development platform was Windows 10 laptop with 2.6 GHz dual core CPU, 8 GB RAM, and 240 GB SSD. In addition, a virtual Ubuntu Server with 2.2 GHz single core CPU, 1 GB RAM and 25 GB SSD, running MySQL and PHPMyAdmin on DigitalOcean cloud server was used to train and test algorithms and store the experiments results.

Datasets: MIT-BIH Arrhythmia datasets described in Sect. 2 were used in the experiments. (Due to space cosntraint, tables in this section show results from 20 randomly selected datasets. Results from all 48 datasets are available at https://github.com/yuhang-lin/ECGAD_extended_result.) Each ECG dataset was divided into training and testing datasets with 60%–40% split. Table 2 shows the number and ratio of abnormal segments in training dataset and testing dataset, respectively, for each dataset. Different patients show different abnormal segment ratios (i.e., ratio of abnormal segments over all segments), and the

Table 2. Ratios of abnormal segments in training and testing.

Patient number	Training dataset			Testing dataset		
	Number of abnormal segments	Total number of segments	Ratio of abnormal segments	Number of abnormal segments	Total number of segments	Ratio of abnormal segments
106	224	1212	18.48%	296	808	36.63%
114	58	1138	5.10%	4	759	0.53%
116	73	1438	5.08%	38	959	3.96%
118	19	1383	1.37%	19	922	2.06%
119	232	1192	19.46%	213	796	26.76%
122	0	1485	0.00%	0	990	0.00%
200	536	1688	31.75%	466	1126	41.39%
201	135	1173	11.51%	65	782	8.31%
202	21	1281	1.64%	2	855	0.23%
205	39	1591	2.45%	41	1061	3.86%
207	307	1444	21.26%	426	963	44.24%
208	927	1760	52.67%	433	1174	36.88%
213	370	1950	18.97%	212	1300	16.31%
217	1125	1324	84.97%	838	883	94.90%
219	39	1291	3.02%	27	862	3.13%
221	281	1452	19.35%	116	969	11.97%
222	3	1500	0.20%	17	1001	1.70%
228	332	1387	23.94%	237	925	25.62%
231	2	942	0.21%	0	628	0.00%
233	497	1841	27.00%	342	1228	27.85%

(Ratios of all 48 datasets are available at https://github.com/yuhang-lin/ECGAD_extended_result/blob/master/abnormal_segment_ratio.md.)

ratio varies widely. Notably, the patent 122 is a healthy patient with no heartbeat anomaly, and the patients 222 and 231 are in good shape as well, with only a few abnormal heartbeats. In contrast, the patient 217 is in a very poor shape, with approximately 90% heartbeats abnormal.

Performance Measures: We used the traditional performance measures – sensitivity, specificity, and accuracy. In light of detecting abnormal segments, (i) true positive means detecting an abnormal segment as abnormal, (ii) true negative means detecting a normal segment as normal, (iii) false positive means detecting a normal segment as abnormal, and (iv) false negative means detecting an abnormal segment as normal.

Outlier Detection Parameter Tuning: The set of parameter values that maximizes the anomaly detection performance was found using a random search iterated 1000 times for each ECG dataset. 1000 iterations is more than enough, and it gives 99.996% probability of achieving near optimum within 1% of the true optimum. (A random search of n iterations has $1 - (1 - \epsilon)^n$ probability of finding parameter values achieving near-optimum within the error ϵ from the true optimum [7].) The ranges of each MCOD parameter used in the experiments are $[0.1, 3.0)$ for the radius r, $[2,80]$ for k, and $[25,100]$ for window size w. After the training process of parameter tuning through random search, we picked the set of parameter values that maximized the accuracy, subject to the constraint that minimum 80% was required for both sensitivity and specificity. (In case none met the constraint, the lower bound was lowered progressively until one was found.) When more than one set of parameter values gave the same accuracy, then the one that had the smaller window size was picked because a smaller window can output the outlier quicker and can be more robust when the input data is smaller.

4.2 Experiment Results and Analysis

The results are presented in two different scenarios. One is *personalized*, where the parameters are tuned for individual patients as discussed in Sect. 4.1. The other is *aggregated*, where the average of the individual optimal parameter values are used as generic parameter values for all patients.

Personalized Results. Table 3 summarizes the accuracy, sensitivity, and specificity obtained for each patient's ECG dataset when the outlier detection parameters were optimized for each dataset separately. This case reflects personalizing the anomaly detection for individual patients.

Overall, the performance using personalized parameters is good. 37 out of 48 datasets achieved accuracy higher than 90%. Note from Table 2 that ECG datasets of the patients 122 and 231 have no abnormal segments in the testing data, so the sensitivity for them is not applicable (N/A).

For the ECG datasets of patients 207, 208, 213, and 228, the accuracy was lower than 90%, as low as 65% for the patient 207. There are a few reasons we believe can explain these lower accuracies. The first reason is the noise in the filtered dataset. The ECG datasets are from ambulatory devices, which cause significant noises such as baseline drifts, motion artifacts, and powerline noise. Although filtered, some datasets still show significant noise. (Figure 5 illustrates typical noisy segments from dataset 207.) Further removing noise from pre-filtered data would require sophisticated signal processing, and was beyond the scope of the project. The second reason is the change of statistics between training dataset and testing dataset during the performance evaluation. This in part can be reflected by the different abnormal segment ratio between training and testing as shown in Table 2. In this project, the performance tuning is not adaptive to such a change (called "concept drift") and, therefore, the algorithm may not be able to react to unexpected changes by adjusting the tuned parameter values.

Table 3. Best testing performance on each patient using personalized parameters (sensitivity N/A for zero abnormality ratio).

Patient #	Sensitivity	Specificity	Accuracy	Optimal parameter values		
				Window	K	Radius
106	0.98	0.97	0.98	97	45	2.19
114	0.67	0.95	0.95	80	14	0.92
116	0.97	0.97	0.97	87	13	3.05
118	0.74	0.94	0.93	65	4	1.65
119	1.00	0.89	0.92	31	21	2.58
122	N/A	1.00	1.00	28	3	2.53
200	0.92	0.94	0.93	74	40	2.10
201	0.94	0.96	0.95	73	36	1.42
202	1.00	0.93	0.93	85	5	0.91
205	1.00	1.00	1.00	98	65	1.33
207	0.28	0.97	0.65	59	25	1.90
208	0.96	0.58	0.72	75	26	1.67
213	0.81	0.87	0.86	80	17	1.73
217	1.00	0.00	0.94	99	80	1.08
219	0.68	0.97	0.97	60	5	3.04
221	0.96	1.00	0.99	44	19	2.12
222	0.06	1.00	0.98	25	4	2.07
228	0.60	0.91	0.83	74	33	2.09
231	N/A	1.00	1.00	25	14	2.87
233	0.97	0.94	0.95	27	10	3.07

Average from 48 datasets: sensitivity 0.83, specificity 0.88, accuracy 0.92. (Results for all 48 datasets are available at https://github.com/yuhang-lin/ECGAD_extended_result/blob/master/personalized_result.md.)

Abnormal Segment Ratio and Accuracy Measures: Figure 6 shows the trend of accuracy, sensitivity, and specificity for datasets sorted by the abnormal segment ratio in the testing dataset. Note that the distribution of abnormality ratios in the datasets is skewed to approximately 45% or lower and approximately 90% or higher. The achieved accuracy is in a fairly consistent range across the two skewed ranges of abnormality ratio, which indicates robustness of the employed outlier detection mechanism to the ratio. The sensitivity shows a similar trend, but it drops very low when the ratio is near zero (<1%). It makes sense because lower ratio means fewer abnormal segments (i.e., true positives) and, hence, lower statistical significance. In contrast, the specificity drops very low when the ratio is near 1 (>90%). It makes sense because higher ratio means fewer normal segments (i.e., true negatives) and, hence, lower statistical significance.

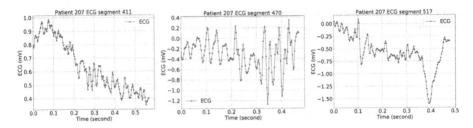

Fig. 5. Example noisy segments from the ECG of the patient 207. (The ECG dataset of patient 207 in particular shows the lowest accuracy overall among all datasets. Looking into the dataset in detail, we found that segments in this dataset have several different types of normal segments, such as 1457 left bundle branch block beats (L), 86 right bundle branch block beats (R), and 107 atrial premature beats (A), as well as several different types of abnormal segments such as 105 premature ventricular contractions (V), 472 Ventricular flutter waves(!) and 105 Ventricular escape beats (E) (see Table 1 for different annotation codes of ECG segments). Indeed, this dataset is mentioned as "an extremely difficult record" [20] in PhysioBank.)

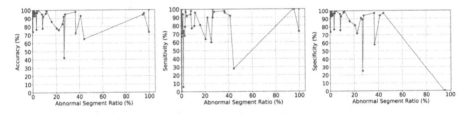

Fig. 6. Accuracy measures for different abnormal segment ratios.

Aggregate (i.e., Non-personalized) Results. Based on the optimum parameter values determined in the personalized anomaly detection experiment (see Table 3), we calculated their mean values as the generic parameter values used commonly for all 48 patients, namely, non-personalized. The mean values are 62 for w, 17 for k, and 1.8 for r. Table 4 shows the resulting performances.

The accuracy achieved using the non-personalized approach is lower than that of the personalized approach for 31 out of 48 datasets, although 24 datasets still achieved accuracy higher than 90%. Overall there were significant degradation of accuracy. The datasets for patients 217 and 219 in particular sustained the biggest degradation – from 94% to 33% for the patient 217 and from 97% to 17% for the patient 219.

4.3   More on Personalized Versus Non-personalized

The histograms in Fig. 7 show the number of ECG datasets in each 10% range of accuracy when the anomaly detection was personalized and not personalized. It is visually evident that personalized detection by far outperforms non-personalized detection. Numerically, the chi-squared distance of the personalized histogram from the non-personalized is 10.2.

Table 4. Performance on each patient when not personalized (window $w = 62$, $k = 17$, and radius $r = 1.8$ for all patients' datasets; sensitivity N/A for zero abnormality ratio.)

Patient #	Sensitivity	Specificity	Accuracy
106	0.68	0.96	0.86
114	0.33	1.00	0.99
116	1.00	0.81	0.82
118	1.00	0.87	0.88
119	0.30	0.78	0.65
122	N/A	0.98	0.98
200	0.37	0.98	0.72
201	0.03	0.99	0.90
202	0.00	0.94	0.94
205	0.98	1.00	1.00
207	0.21	0.97	0.61
208	0.82	0.75	0.78
213	0.84	0.83	0.83
217	0.34	0.16	0.33
219	1.00	0.14	0.17
221	0.97	0.99	0.99
222	0.18	0.99	0.98
228	0.60	0.91	0.83
231	N/A	1.00	1.00
233	1.00	0.55	0.68

Average from 48 datasets: sensitivity 0.56, specificity 0.87, accuracy 0.82. (Results for all 48 datasets are available at https://github.com/yuhang-lin/ECGAD_extended_result/blob/master/nonpersonalized_result.md.)

The fairly large difference in the accuracy performance is understood when the distribution of the optimal sets of outlier detection parameters (i.e., w, k, r) are examined, as shown in the scatter plot in Fig. 8. It shows the MCOD parameters tuned personalized for each patient's ECG dataset (see Table 3) and also the aggregated mean values of them (i.e., $w = 62, k = 17, r = 1.8$) used in the non-personalized case. The parameter values tuned for different datasets are widely spread in the parameter space, as indicated by their standard deviations 24, 21, and 0.85 for w, k, and r, respectively. These observations confirm that ECG varies a lot for individual patients and, therefore, personalized detection is much desired.

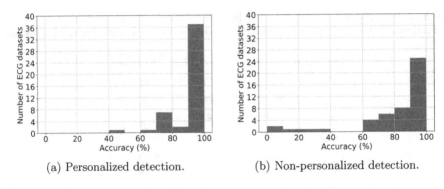

Fig. 7. Number of ECG datasets in different accuracy ranges (bin size = 10%).

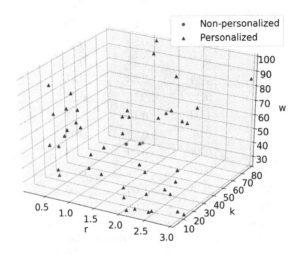

Fig. 8. Scatter plot of personalized MCOD parameters.

5   Related Work

There is a large body of work done on anomaly detection from ECG. In this section, we discuss briefly what we believe are a representative sample reflecting the state of the art in three aspects of this project: (a) machine learning methods used for automatic detection of abnormal ECG segments, (b) feature extraction methods to reduce ECG segments to feature vectors, and (c) distance-based outlier detection from a data stream.

Machine Learning Methods: Various machine learning methods have been used for ECG classification, such as decision tree [4], support vector machine [24], artificial neural network [26], and their ensemble [15]. These methods, however, are geared for offline classifications and are not necessarily handling individual ECG segments separately. In contrast, some recent work are far more suitable for online real-time classification of ECG segments as done in this project. For example, Veeravalli et al. [23] used dynamic time warping (DTW) based similarity calculation, personalized to individual patients by obtaining the normal ECG segment through clustering (K-means). For another example, Chauhanv and Vig [5] used long short-term memory (LSTM) recurrent neural networks (RNN) as a predictive model trained with normal ECG segments to detect abnormal segments. Both methods have the advantage of working well with continuously arriving ECG segments. To the best of our knowledge, there is no prior work that examined using outlier detection based on online clustering, which was the goal in this project.

Feature Extraction Methods: As mentioned in Sect. 3.3, feature extraction in this project is for dimensionality reduction from an ECG segment to a feature vector. There have been two different ways the extracted features are used. One way is as a synopsis of an ECG segment characterizing a certain anomaly [9]. For example, a P-R interval can be used to detect premature ventricular contraction, an R-R interval to detect premature atrial heartbeat, and the QRS duration to detect a ventricular premature complex [17]. Another way is as an input model of the segment to a subsequent machine learning algorithm. In this project, it is the discrete wavelet transform (DWT), chosen for its efficiency and resilience to noise. There are several others, such as principal component analysis (PCA) [16], rank correlation coefficient (RCC) [11], and B-splines [10]. They all extract dominant features that represent the ECG signal approximately but differ in the specific sense of dominance. Specifically, DWT selects the first 2^n, where $n = 5$ in our work, coefficients as the dominant features; PCA selects dominant linear components that, when linearly combined, approximates the input signal; RCC selects a subset of ECG data samples whose RCC values are the highest, where RCC is a measure of the correlation based on the ranks of data values; B-splines are used as bases that are linearly combined to fit ECG signal "curve" lines, and the resulting "knots" and parameters of the B-splines are used as the features.

Outlier Detection Methods: In addition to MCOD [8], the distance-based outlier detection method used in this project, there are other methods that can be used for outlier detection. For example, MOA supports the following ones that we believe are in the mainstream of online outlier detection over data streams: ExactSTORM and ApproxSTORM (STORM stands for "Stream Outlier Miner") [1], Abstract-C [25], COD [14] and MCOD, and additionally AnyOut [2]. All of these methods except AnyOut are distance-based methods developed progressively for improvements. (AnyOut is a method enabling the detection of an outlier "any time" the time expires, and is orthogonal to the detection mechanism (e.g., distance-based, density-based)). As shown in the comprehensive experiments conducted by Tran et al. [22], MCOD performs best among all distance-based methods.

6 Conclusion

The feasibility and accuracy of detecting abnormal segments from an ECG data stream using distance-based online outlier detection have been demonstrated in this project. Combined with features extracted using Haar discrete wavelet transform, the microcluster-based continuous outlier detection algorithm successfully detected abnormal ECG segments with higher than 90% accuracy for a majority of datasets. The accuracy performance compared between personalized and non-personalized anomaly detection scenarios showed that personalized showed by far higher accuracies.

There are several issues in the employed algorithms that still warrant further work. First, the outlier detection mechanism was implemented as a binary classifier to normal versus abnormal, without distinguishing among different types of abnormality. It is suggested that the mechanism is extended to a multi-class classifier that can label the anomaly type of abnormal segments. Second, the R peak detection algorithm used in this project has a significant room for improvement so it will not result in multiple R peaks in the same segment as happened in this project. There are more advanced techniques, and one of them should be adopted for better results. Third, the online outlier detection algorithm used in this paper works well when the ECG data stream is stationary, and as a result, the accuracy performance was somewhat inadequate for some ECG datasets. It would be desired to enhance the algorithm to be adaptive to the change of the ECG segment statistic, such as abnormal segment ratio, to adjust the outlier detection parameters according to the change of statistic.

As mentioned in Introduction, the project initially started out on an Android smartphone platform. The project will continue to migrate the program codes of all steps into the Android platform. Then, a performance profile (i.e., the elapsed time of individual steps of the processing algorithm) will be built to assess the real-time "fitness' of the method used in this project.

References

1. Angiulli, F., Fassetti, F.: Distance-based outlier queries in data streams: the novel task and algorithms. Data Min. Knowl. Discov. **20**(2), 290–324 (2010)
2. Assent, I., Kranen, P., Baldauf, C., Seidl, T.: AnyOut: anytime outlier detection on streaming data. In: Lee, S., Peng, Z., Zhou, X., Moon, Y.-S., Unland, R., Yoo, J. (eds.) DASFAA 2012. LNCS, vol. 7238, pp. 228–242. Springer, Heidelberg (2012). https://doi.org/10.1007/978-3-642-29038-1_18
3. Bazett, H.C.: An analysis of the time-relations of electrocardiograms. Heart **7**, 353–370 (1920)
4. Bensaid, A.M., Bouhouch, N., Bouhouch, R., Fellat, R., Amri, R.: Classification of ECG patterns using fuzzy rules derived from ID3-induced decision trees. In: Proceedings of the Conference of the North American Fuzzy Information Processing Society, pp. 34–38, August 1998
5. Chauhan, S., Vig, L.: Anomaly detection in ECG time signals via deep long short-term memory networks. In: Proceedings of the IEEE International Conference on Data Science and Advanced Analytics, pp. 1–7, October 2015

6. Chen, H.C., Chen, S.W.: A moving average based filtering system with its application to real-time QRS detection. In: Computers in Cardiology, pp. 585–588, September 2003

7. Firebug: Practical hyperparameter optimization: Random vs. grid search (2016). https://stats.stackexchange.com/q/209409. Accessed 18 April 2018

8. Georgiadis, D., Kontaki, M., Gounaris, A., Papadopoulos, A.N., Tsichlas, K., Manolopoulos, Y.: Continuous outlier detection in data streams: an extensible framework and state-of-the-art algorithms. In: Proceedings of the ACM SIGMOD International Conference on Management of Data, pp. 1061–1064, June 2013

9. Houghton, A.R., Gray, D.: Making Sense of the ECG. CRC Press, Boca Raton (2007)

10. Karczewicz, M., Gabbouj, M.: ECG data compression by spline approximation. Signal Process. **59**(1), 43–59 (1997)

11. Khare, S., Bhandari, A., Singh, S., Arora, A.: ECG arrhythmia classification using spearman rank correlation and support vector machine. In: Deep, K., Nagar, A., Pant, M., Bansal, J.C. (eds.) Proceedings of the International Conference on Soft Computing for Problem Solving December, 2011. AISC, vol. 131, pp. 591–598. Springer, New Delhi (2012). https://doi.org/10.1007/978-81-322-0491-6_54

12. Knorr, E.M., Ng, R.T.: Algorithms for mining distance-based outliers in large datasets. In: Proceedings of the 24th International Conference on Very Large Data Bases, pp. 392–403 (1998)

13. Knorr, E.M., Ng, R.T., Tucakov, V.: Distance-based outliers: Algorithms and applications. VLDB J. **8**(3–4), 237–253 (2000)

14. Kontaki, M., Gounaris, A., Papadopoulos, A.N., Tsichlas, K., Manolopoulos, Y.: Continuous monitoring of distance-based outliers over data streams. In: Proceedings of the IEEE International Conference on Data Engineering, pp. 135–146, April 2011

15. Macek, J.: Incremental learning of ensemble classifiers on ECG data. In: Proceedings of the 18th IEEE Symposium on Computer-Based Medical Systems, pp. 315–320, June 2005

16. Monasterio, V., Laguna, P., Martinez, J.P.: Multilead analysis of t-wave alternans in the ECG using principal component analysis. IEEE Trans. Biomed. Eng. **56**(7), 1880–1890 (2009)

17. Patel, A.M., Gakare, P.K., Cheeran, A.N.: Real time ECG feature extraction and arrhythmia detection on a mobile platform. Int. J. Comput. Appl. **44**(23), 40–45 (2012)

18. PhysioNet: MIT-BIH Arrhythmia Database Directory (Introduction) (2010). https://physionet.org/physiobank/database/html/mitdbdir/intro.htm. Accessed 29 May 2018

19. PhysioNet: PysioBank Annotation (2016). https://www.physionet.org/physiobank/annotations.shtml. Accessed 25 May 2018

20. PhysioNet: MIT-BIH Arrhythmia Database Directory (Records) (2018). https://www.physionet.org/physiobank/database/html/mitdbdir/records.htm#207. Accessed 14 June 2018

21. PyWavelet: Signal extension modes PyWavelets Documentation (2018). http://pywavelets.readthedocs.io/en/latest/ref/signal-extension-modes.html. Accessed 2 June 2018

22. Tran, L., Fan, L., Shahabi, C.: Distance-based outlier detection in data streams. Proc. VLDB Endow. **9**(12), 1089–1100 (2016)

23. Veeravalli, B., Deepu, C.J., Ngo, D.H.: Real-time, personalized anomaly detection in streaming data for wearable healthcare devices. In: Khan, S.U., Zomaya, A.Y., Abbas, A. (eds.) Handbook of Large-Scale Distributed Computing in Smart Healthcare. SCC, pp. 403–426. Springer, Cham (2017). https://doi.org/10.1007/978-3-319-58280-1_15

24. Venkatesan, C., Karthigaikumar, P., Paul, A., Satheeskumaran, S., Kumar, R.: ECG signal preprocessing and SVM classifier-based abnormality detection in remote healthcare applications. IEEE Access **6**, 9767–9773 (2018)

25. Yang, D., Rundensteiner, E.A., Ward, M.O.: Neighbor-based pattern detection for windows over streaming data. In: Proceedings of the 12th International Conference on Extending Database Technology: Advances in Database Technology, pp. 529–540 (2009)

26. Yang, T.F., Devine, B., Macfarlane, P.W.: Artificial neural networks for the diagnosis of atrial fibrillation. Med. Biol. Eng. Comput. **32**(6), 615–619 (1994)

Do Public and Government Think Similar About Indian Cleanliness Campaign?

Aarzoo Dhiman$^{(\boxtimes)}$ and Durga Toshniwal

Department of Computer Science and Engineering,
Indian Institute of Technology Roorkee, Roorkee 247667, Uttarakhand, India
{aarzoodhiman.dcs2017,durgefec}@iitr.ac.in

Abstract. With the growth of internet, social networks has become primary source for people to present their views on different topics. The data collected from social media are considered enough as well as reliable to be processed and gather insights on the perceptions of people towards any topic. In this research work, an empirical study of the Twitter data (i.e. around 400,000 tweets) collected for the period of December 1, 2017 to March 31, 2018, pertaining to Indian Cleanliness Campaign called *Swachh Bharat Abhiyan (SBA)*, which focuses on improving the cleanliness situation in the country, has been done. Here, a demographic distribution of the Twitter data has been generated by augmenting *partial keyword matching* along with *Named Entity Recognition* for geoparsing the tweets. This will help to study the involvement of the people in different areas of the country. Furthermore, *Sentiment Analysis* of the tweets has been performed to gather the perception of people towards the campaign. Also, to assure the integrity of the campaign, the tweets have been segregated into *public and government generated tweets* and the respective sentiments have been compared to determine the difference in perception of public and government in different areas of the country. This work can be considered of interest because there has not been any research work which focuses on analyzing the awareness and perception of people on SBA in detail.

Keywords: Indian cleanliness campaign · Swachh Bharat Abhiyan · Social media data analysis · Twitter · Sentiment analysis · Demographic analysis

1 Introduction

India is the 2nd largest populated country and has been ranked 125th (out of 183) in life expectancy all over the world. There may be many reasons behind such low ranking which are rather interconnected. For example, records have shown that a vast number of water-borne, air-borne and food-borne diseases are caused by poor sanitation, sewage and hygiene conditions of a place as proved by Snow [12] during Cholera outbreak, London in 1854.

© Springer Nature Switzerland AG 2019
J. A. Lossio-Ventura et al. (Eds.): SIMBig 2018, CCIS 898, pp. 367–380, 2019.
https://doi.org/10.1007/978-3-030-11680-4_34

Current Indian government has been laying out many efforts to improve the health standards of their citizens. This includes the launch of many government policies, health programs and services that are related to female health, mother and child care, infrastructure building, cleanliness and financial support [7] etc. In October 2014, government launched a national level cleanliness campaign, named *Swachh Bharat Abhiyan (SBA)* [11] which aims to clean cities, roads, public places and transportation services in urban as well as rural areas of the country. One primary aim of this campaign is to make India free from open defecation and achieve 100% scientific solid waste management by October 2019.

With the inception of SBA, the Indian government has promoted it on a very large scale. The primary focus of the Indian government has been on proper sanitation and waste management in the country. Government has been providing budgets to build toilets, better roads, deploy dustbins in all the public places, teach public about its importance in urban and rural areas of the country and take up yearly surveys [13] to measure its performance in different areas of the country. However, there are not enough statistics provided by the government which can ensure the level of involvement and belief of citizens in SBA. This knowledge is important because it can help to classify places based on their performance such that more attention may be given to the locations which are lagging behind.

In this research work, *web data* collected from Twitter has been processed to first, monitor the involvement of people in SBA by studying the demographic distribution of the tweets by geoparsing the tweets for *Named Entities* with help of *Named Entity Recognition (NER)* and *partial keyword matching* and second, study the perceptions of common public towards SBA by using off-the-shelf *Sentiment Analysis* tool. Further, analysis of the demographic characteristics of some *active* and *inactive* places have been performed to find some pattern. Also, a manual selection of the twitter handles has been performed by using some keywords to segregate the twitter data into *public generated tweets* and *government generated tweets* and determine the difference between the overall sentiments of public and government tweets. Currently, there is no research work done which focuses on studying the awareness/activeness and perception of common public and the difference in perception of government and common public towards SBA using the Twitter data.

The rest of the paper is organized as follows. Section 2 contains related work, which discusses some of the previous work done in this field. Section 3 contains proposed methodology, which discusses the details on data collection, pre-processing and its analysis. Section 4 contains results and discussion, highlighting important findings of our analyses. Finally, the paper ends with conclusion and references.

2   Related Work

The primary process of surveying the effects of any campaign includes gathering public data using some *direct or indirect approaches* and then processing it to uncover some hidden patterns and trends. However, these data collection and processing measures are very time consuming and laborious and thus researchers have moved their focus towards data available online i.e. web data. The most notable contribution to predict spread of a disease using web data was made by Google in 2009. Here, Ginsberg et al. introduced Google Flu Trends (GFT) [6], a web service that helps in early detection of spread of a disease in a given population. They monitored the daily queries related to Influenza like Illness (ILI) posed by users on online search engine i.e. Google.

Later, many other online data sources such as Twitter, Yahoo! and Google were explored to perform the outbreak detection. In 2009, De Quincey and Kostkova [3] studied the potential of Twitter as an indirect source of data collection for Epidemic Intelligence. They collected tweets related to Swine Flu Pandemic and affirmed that this data can be used in conjunction with traditional data sources to predict outbreaks. In 2010, Culotta [2] tried to replicate the results of GFT using the Twitter data. From his analysis, the author concluded that simple regression techniques are not enough for Twitter data because of its unstructured nature. Better pre-processing and classification tools can help in increasing accuracy of the results, while using Twitter data. Later, researchers applied many different techniques on the Twitter data to improve accuracy of results for example, regression techniques like least absolute shrinkage and selection operator (LASSO) [8] and BOLASSO [9], Classifiers like Support Vector Machine (SVM) [1] etc.

In this paper, we focus on tracking the impact of Swachh Bharat Abhiyan (SBA) on people using Twitter data. SBA was first launched in 2014 and since then very less research work has been done related to it. In 2015, Raj and Kajla [10] collected tweets related to SBA and performed simple sentiment analysis to find out perception of Indian citizens towards SBA. Later in 2017, Tayal and Yadav [14] tested their sentiment analysis tool Senti-Meter on Twitter data related to SBA. They processed 1200 tweets collected for the period of January 2016 to March 2016. They performed manual tagging to calculate the accuracy of their tool. Both of the works related to SBA worked on very less number of tweets and did not consider any other demographic details of the places in India.

3   Proposed Work

In this study, we track the involvement in and influence of SBA on common people in different regions of the country. We also study the difference between the sentiments of public and government generated tweets. To perform this study, we used the data available from Twitter for the period of December 1, 2017 to March 31, 2018. Details related to dataset is given further in the section.

3.1   Dataset Description

We use Twitter as the source of social media data because of the word limit of 150 words. Further, availability of Twitter Streaming API facilitates the data collection process as depicted in retrieval module of Fig. 1. The API helped us to collect around 400 thousand tweets for the period of December 1, 2017 to March 31, 2018. A sample description of the keywords and tweets is given in Table 1.

Table 1. Description of Twitter data collected using keywords related to SBA for the duration of Dec 2017 to Mar 2018

Description	Hashtags examples	Tweet example	Number of tweets
General	#SwachhBharat Abiyan # MyCleanIndia	@marineravin: @tavleen_singh anything you wanna more to add abt swachh Bharat Abhiyan ... @sanjayuvacha @amitmehra	322,287
Toilet related	#Open Defecation #MyCity-MyPride	@paramiyer_: Congratulations to Team @swachhbharat. Tirunelveli district in Tamil Nadu has been declared # OpenDefecationFree	26,846
Cities related	#SwachhUP #SwachhJhar	RT @lezlietripathy: Participated in Cleaning #Vesave Beach Today. An initiative by @AfrozShah1 Supported by @Dev_Fadnavis @AUThackeray Today. #SwachhBharat #Swachhmaharashtra #swachhversova	24,736
Rural area related	#ZSBP #SbmZSBP	@kishanganjzsbp: Morning follow up and pit digging in Gachpada Panchayat #ZSBP #SwachhBharat #SwachhBihar #SBMGramin @SwachhBihar @LSBA_Bihar @swachhbharat	86,868

To determine the demographic distribution of the tweets in different regions of the country, we applied partial keyword matching along with the NER tool for geoparsing. To perform the partial keyword matching, we used a set of names of different locations in India. However, India is a vast country of around 4000

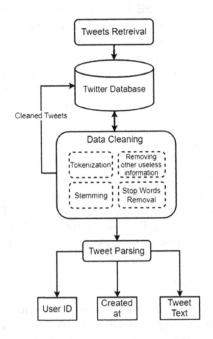

Fig. 1. Steps of data retrieval and preprocessing

Table 2. Sample list of cities chosen for analysis

Zones	Tier 1	Tier 2	Tier 3
Central India	-	Raipur, Bhopal	Mungeli, Ujjain
Eastern India	Kolkata	Patna, Asansol	Kishangarh, Brahmapur
Northern India	Delhi	Chandigarh, Faridabad	Solan, Leh
North-Eastern India	-	Guwahati	Silchar, Pasighat
Southern India	Bangalore, Hyderabad	Warangal, Mangalore	Tirupati, Kavaratti
Western India	Ahmedabad, Mumbai	Surat, Nagpur	Silvassa, Diu

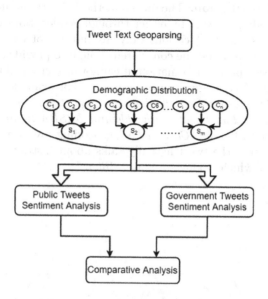

Fig. 2. Steps of data analysis process

cities and towns. So, to simplify our experimentation, we use representative cities from different zones[1] by applying the standard HRA classification [15] where cities are classified into three *Tiers* (Tier 1 being the highest) on the basis of their population. A sample description is given in Table 2. We have also classified urban and rural areas on the basis of this HRA classification, where cities belonging to Tier 1 and upper Tier 2 are regarded as urban areas and rest of the cities and towns as little backward areas of the country.

Data Preprocessing. Thousands of tweets related to SBA are published daily on Twitter. For our study, we are primarily concerned with the tweet text date and time of tweet post and user ID, but it is surrounded by other information such as hyperlinks, images and videos etc. Also, the text is subjected to many impurities due to the use of informal language by the users. To overcome

---

[1] https://en.wikipedia.org/wiki/Administrative_divisions_of_India.

these issues and improve the quality of results, the tweets are first cleaned as illustrated in pre-processing module of Fig. 1. The data cleaning process includes tokenization, stemming, removing stop words and other useless information such as hyperlinks and videos and images etc. The resultant cleaned tweets are then parsed and analyzed to determine the names of cities and states from the tweet texts. Details on further analysis are given in the following section.

3.2   Proposed Methodology

Demographic Distribution. The formal method of getting information about any place is geocoding a tweet by either through geo-locations or user's profile information. In case of geo-locations only around 1% of tweets contain geo-locations and this data cannot be considered enough to provide accurate results. In case of using users profile information, it cannot be confirmed that one's tweet is about the same place as given in one's profile. For example, for tweets such as *"RT @Swacchbegusarai: Door to door visit and triggering to increase uses of toilets in Fatehpur Panchayat @swachhbharat @LSBA_Bihar # ZSBP ht..."* posted by the Twitter handle *"CleanupTN"*, geocoding gave us the name of place *"TamilNadu"* whereas the tweet actually talks about district *"Begusarai"* and village *"Fatehpur"* which belongs to state *"Bihar"*.

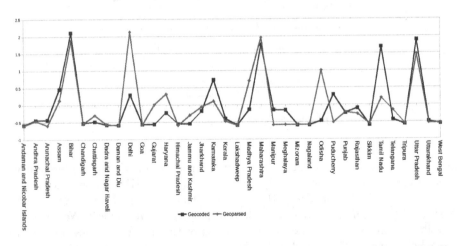

Fig. 3. Normalized number of tweets collected by using geocoding and geoparsing for all states and Union Territories in India

Thus, in our study we geoparsed [5] the tweets to find the number of tweets for different towns, cities, UTs and states as shown in Fig. 2. Here, we geocode the named entities recognized by using NER and partial keyword matching. To perform NER, we used off-the-shelf Python NLTK libraries and to perform partial keyword matching, we calculate the partial match between the tweet text token and the city or state name and then geocode only those entities for which the partial match comes out to be greater than 95%. To perform geocoding we used the off-the-shelf *geopy Python Library*. By using this technique, we were able to collect results even for very small cities and towns which were not mentioned in user's profile or geo-locations. The number of tweets collected per city and state by using geoparsing were also found to be higher than by using geocoding the tweets. We found a correlation coefficient of 0.7408 between the number of tweets collected by using the two techniques. To provide a pictorial depiction, we normalized the number of tweets collected by geocoding and geoparsing by using z-score normalization. Figure 3 gives a line plot for the number of tweets for different states on the sample data for the time period of March 1 to March 15, 2018.

Table 3. Twitter handles and keywords used for segregating the government and public tweets

Sr. no.	Keywords	Handles example
1	Swachh	@swachhbharat,@swachhfaridabad
2	Gov	@CEOMyGovIndia
3	Clean	@CleanupTN, @CleanIndia
4	BJP	@BJP4India, @bjpsamvad
5	MoHUA_India	@MoHUA_India
6	SBM	@sbmodfindia
7	MC	@AiMCHaryana
8	CMO	@CMO_Odisha, @cmohry

Following paragraph gives the mathematical formulations of the steps involved. Let S be the set of name of states and union territories in India. So, $S = \{S_1, S_2, \ldots S_n\}$, where n is the number of states and union-territories. Let C be the set of name of cities and towns in India. So, $C = \{C_1, C_2, \ldots C_m\}$, where m is the number of cities and towns in India. We used around 1500 of the city names for our analysis. A dictionary CS consists of the names of cities and their corresponding state name. So, $CS = \{C_i : S_j\}, 1 \leq i \leq n, 1 \leq j \leq m$ such that $CS[C_i] = [S_i]$.

Let T_i be the twitter corpus and frequency score of a city $F(C_i)$ is calculated by using the formula given in (1). A tweet t_j in the corpus is given a score 1 if name of city C_i is present in that tweet and 0 otherwise as given in (2). We took care of the partial matching of names in the tweets as well as given in (3).

$$F(C_i) = \Sigma\Sigma F(C_i, t_j), \forall i \forall j \tag{1}$$

$$where, F(C_i, t_j) = 1 \quad if\ C_i \in t_j, Otherwise\ 0 \tag{2}$$

$$PartialMatch = \left(\frac{2M * 100}{T}\right) \tag{3}$$

where, M is the number of matches and T is the total number of elements in the string. Similarly, the frequency score of a state is calculated by the using the frequency score of the state and its corresponding cities. More formally, frequency score of a state S_j is equal to the sum of frequency score of the state S_j and sum of frequency scores of all the cities C_i for which $CS\,[C_i] = S_j\ \forall i$ as given in (4) and (5).

$$F\,(S_i) = \Sigma\Sigma F\,(S_i, t_j) + \Sigma F\,(C_k)\,, \forall i \forall j \forall k \text{ and } SC\,[C_k] = S_i \qquad (4)$$

$$\text{where, } F\,(S_i, t_j) = 1 \quad \text{if } S_i \in t_j, \text{ Otherwise } 0 \qquad (5)$$

A city may consist of some people who are highly active on twitter and some who hardly use it. Twitter corpus when analysed using the absolute number of tweets only give cumulative results for a particular place. Also, more the population of place, more is the number of tweets collected. However, when the number of tweets is normalized by the factor of population of the place, this gives us the intensity of tweets per person in a given city C_i. We normalize the score value of a place by dividing it with the population of the place (i.e. per capita score) as given in (6).

$$N\,(C_i) = \frac{F\,(C_i)}{P\,(C_i)} \qquad (6)$$

where, $P\,(C_i)$ is the population of city C_i

Sentiment Analysis. Sentiment analysis is a supervised classification process to predict the opinion of a person through the text which is related to some topic. We used sentiment analysis on our Twitter corpus to capture the opinions and sentiments of people towards SBA. Each tweet has been classified into three opinions: positive, negative and neutral by using *Word Sense Disambiguation, Senti Word Net and word occurrence statistics using movie review corpus* [4]. We used the dedicated sentiment classification library of python for our study. If sentiment score value comes out to be greater than 0 then the sentiment is classified as positive, if it comes out to be less than 0 then the sentiment is classified as negative and otherwise neutral.

Public-Government Tweet Segregation. SBA is a national level campaign and government is making many efforts to promote it. Hence, there are many dedicated twitter accounts specific to some states or cities which promote SBA in their local communities. So, we categorize the twitter corpus into two classes on the basis of the Twitter handles, such that one class consists of all the tweets posted by using government handles and second class consists of tweets posted by common public as shown in Fig. 2. We selected the government handles manually by using some set of keywords as illustrated in Table 3. Rest of the tweets were considered to be made by common public.

A simple measure named *Public Post (PP)* percentage, which gives the percentage of tweets made by common public out of the total tweets posted for a location, is used to determine the actual activity of public in the location.

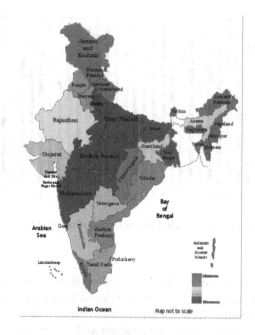

Fig. 4. Heatmap of India generated by processing tweets to study their demographic distribution.

4   Results and Discussions

In this section we have highlighted some of the important results of our experimentation. All the analyses have been done using Python 2.7.

Table 4. Description of the states which are most active on Twitter

Rank	State	Zone	Tweet (%)	PP (%)	POP (%)	LR (%)	F_LR (%)
1.	Madhya Pradesh	Central India	11.71	62.73	6.00	69.32	59.24
2.	Uttar Pradesh	North India	11.44	54.62	16.50	67.68	57.18
3.	Maharashtra	Western India	11.30	70.66	9.28	82.34	75.87
4.	Bihar	North India	10.66	53.57	8.60	61.80	51.50
5.	Delhi (Capital)	North India	9.69	80.56	1.39	86.21	80.76

The Twitter corpus was first processed to determine the demographic distribution of the tweets in different regions of the country with the help of geoparsing the named entities. Figure 4 provides the state-wise hotspot distribution over the country where red stands for maximum number of tweets and green stands for lowest number of tweets. Further the state-wise sentiment analysis of the tweets show that most of the tweets are generally neutral as shown in Fig. 5. Figure 5 shows a stacked bar graph representation of the percentage tweets per sentiment

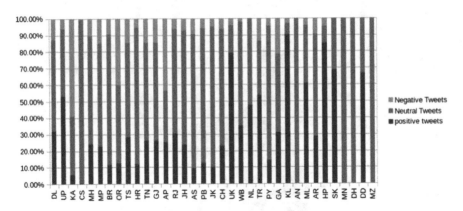

Fig. 5. Stacked bar graph representing respective sentiments for all the states and UTs in India

for all the states (abbreviated[2])and UTs, where states are arranged in ascending order of total number of tweets. This tells us that people are accepting the campaign very peacefully. In our study, a high number of tweets correspond to high awareness and activeness of the location and vice-versa.

Table 5. Description of the states which are least active on Twitter

Rank	State	Zone	Tweet (%)	PP (%)	POP (%)	LR (%)	F_LR (%)
1.	Andaman & Nicobar Islands	South-East India	0.0057	0	0.03	86.63	82.43
2.	Lakshadweep	South-West India	0.0057	90	0.01	91.85	87.95
3.	Mizoram	North-East India	0.028	87.09	0.09	91.33	89.27
4.	Sikkim	North-East India	0.039	83.52	0.05	81.42	75.61
5.	Dadra & Nagar Haveli	Western India	0.074	60	0.03	76.24	64.32

We further compare the demographic characteristics of the most active/aware and least active/aware states from the country. For this study, we first analyze the five top-most and five bottom-most states in terms of number of tweets. Tables 4 and 5 gives the names of top five states and bottom five states respectively, along with some of their demographic characteristics[3]. We would like to exempt the case of Delhi from our analysis as it is the capital of India. The analysis of these demographic characteristics show that states on top have high population and low average and female literacy rates as compared to the states in bottom. This shows that population density of a place has a very high impact on number of tweets generated as well as retweeted from that location. Through the tables, we also see that public post percentage (PP) of top five states lie between

---

[2] http://slusi.dacnet.nic.in/watershedatlas/list_of_state_abbreviation.htm.

[3] POP: Population, LR: Avergae Literacy Rate, F_LR: Female Literacy Rate, PP: Public Posts percentage, PS: Public Sentiments, GS: Government Sentiments.

Table 6. Sentiment analysis of top and bottom most states for public generated tweets and government generated tweets

	Public			Government		
	State	PS	GS	State	PS	GS
Top	UP	1.020	0.591	UP	1.020	0.591
	MH	0.543	0.595	MH	0.543	0.595
	TN	0.0435	0.692	MP	0.311	0.134
	BR	0.411	0.299	BR	0.411	0.299
Bottom	KA	0.3014	0.767	TN	0.043	0.692
	AS	0.3040	0.324	HR	0.128	0.209
	TR	0.6309	0.761	GA	0.0862	0.554
	PY	0.676	0.872	HP	1.152	0.068

50–70% (exempting the case of Delhi), which shows that people of these places are equally active as the government. In case of bottom most five states, the public post percentage (PP) have very extreme values such as 0 or 90% because there are very few number of tweets for these places.

Table 6 gives the names of top-most and bottom-most states in terms of number of government and public tweets along with their average public and government sentiment values. As illustrated from the table, Uttar Pradesh (UP), Madhya Pradesh (MP), Maharashtra (MH) and Bihar (BR) have been on top according to government tweets as well as public tweets. The sentiment analysis of all states in bottom show that the sentiments of government tweets are generally more positive as compared to public tweets. This states that government is making efforts to promote SBA in these places as well. This analysis leads to a result that people may be talking about a campaign in the different locations but they generally lack the emotions while talking about it. This means that people are just retweeting the facts and figures and tend to show very less personal experiences about SBA on Twitter. However, this also indicates that government is also making ample efforts in promoting SBA and gradually making people active to talk about the same.

Table 7. Description of lower Tier 2 and Tier 3 cities active on Twitter

Sr. no.	City	State	Tier	POP (%)	LR (%)	PP (%)	PS	GS
1	Dibrugarh	Assam	3	1.54	76.22	19	0.068	0.045
2	Durg	Chhatisgarh	2	2.68	79.06	55.2	−0.66	−0.0041
3	Kishanganj	Bihar	3	16.9	57.04	34.36	0.0077	0.015
4	Jalna	Maharashtra	3	2.86	71.52	27.74	0.039	0.079
5	Kota	Rajasthan	2	10	76.56	81.57	0.12	0.050

Table 8. Description of Tier 2 and Tier 3 cities active in terms of normalized number of tweets

Sr. no.	City	State	Tier	POP (%)	LR (%)	PP (%)	US	GS
1	Silvassa	Dadra & Nagar Haveli	3	0.98	91.01	60.4	−0.039	−0.038
2	Dibrugarh	Assam	3	1.54	76.22	19	0.068	0.045
3	Kishanganj	Bihar	3	16.9	57.04	34.36	0.0077	0.015
4	Kota	Rajasthan	2	10	76.56	81.57	0.084	0.048

Similarly, we extended our study to the cities. The analysis showed that the cities with highest number of tweets are mostly well developed and highly populated metro cities or capitals (i.e. belonging to Tier 1 and upper Tier 2 cities) e.g., Delhi, Mumbai, Hyderabad, Indore and Bhopal etc. So, we narrow down our analysis to lower Tier 2 and Tier 3 cities. Table 7 gives the names of some active Tier 2 and Tier 3 cities with other demographic details. From this table, we see that these cities have very low population and good literacy rates except for Kishanganj (Bihar). Table 7 shows that though there are high number of tweets for Dibrugarh, Kishanganj and Jalna, most of these are made by using government handles. This shows that government is making efforts in promoting the campaign in lower Tier cities as well. The public and government tweet sentiment analysis shows that government tweets are more neutral as compared to public tweets. This means that people are also taking active part in tweeting about SBA. The overall sentiment, however, has remained the same for public and government tweets. This indicates that government handles have not been making any false statements on Twitter.

To find the intensity of tweets in a place, we normalized the number of tweets by the factor of population of each place. It can be argued that the low population of a city accounts for a high normalized tweet score. This is also visible in the results as shown in Table 8. Table 8 gives the details of some of cities which are very active in terms of number of normalized tweets but are not so well known. Here, all the cities with low populations such as Silvassa and Dibrugarh have high number of normalized tweets. Also, Silvassa has a very high value of public post percentage, which tells that people are highly active there. These cities have some similarities for instance, they have good literacy rates and they belong to Tier 2 and Tier 3. So, it can be said that high literacy rate of a place leads to high number of tweets while considering normalized number of tweets.

We further analyze the corpus to identify the most active hashtags being used and the most active Twitter handles. Figure 6 gives the word cloud for the set of most active keywords, where we see that *CleanIndia*, *NarendraModi* (Prime Minister of India) and *SwachhBharat* are the most active Hashtags being used. Table 9 gives the names of most popular user handles. This table shows that most of the twitter handles that come on top are actually government handles, which indicates that government handles are the primary source of tweet generation and their spread.

Fig. 6. Word cloud depiction of most commonly used keywords and hashtags for SBA

Table 9. Description of most active Twitter handles related to SBA

Rank	Frequent user IDs	Description	Government/Personal handle
1	WeMeanToClean	Delhi-based volunteer group	Organization
2	SwachhHYD	Hyderabad, Telangana India	Government
3	Env_cleanindia	Indian open public handle	Government
4	CleanupTN	Dedicated to make Tamil Nadu a clean city	Government
5	Nitesh1901	Lead Infra Solution Architect-IBM	Personal

5   Conclusion

In this study, we used the Twitter data to capture the level of awareness and the sentiments of different towns, cities, union territories and states in India towards Indian Cleanliness Campaign, Swachh Bharat Abhiyan (SBA). Through our study, we realized that population and area of a place (i.e. population density) has a very high impact on generation of high number of tweets from any given location. However, high literacy rate has a direct impact on number of normalized number of tweets (i.e. tweets per capita). The sentiment analysis of the tweets show that most of the tweets generated are mostly neutral. This shows that most of the tweets are mostly informative and people tend to show very less emotions for SBA on Twitter. The public post percentage analysis of tweets show that in case of Tier 2 and 3 cities, which are active on Twitter, most of these tweets are made by using government handles. This indicates that government has been making efforts to promote SBA in not so developed cities as well. The public and government tweet segregation and sentiment analysis of top most and bottom most states show that people may be talking about the different events related to SBA in form of sharing some facts and figures but they lack in presenting any emotional content towards SBA. Overall, the government has been very active in spreading awareness about SBA in different places in the country. For further study, we can focus on comparing more demographic characteristics of the different locations and perform manual tagging to determine the accuracy of the sentiment analysis tool used.

References

1. Aramaki, E., Maskawa, S., Morita, M.: Twitter catches the flu: detecting influenza epidemics using Twitter. In: Proceedings of the Conference On Empirical Methods in Natural Language Processing, pp. 1568–1576. Association for Computational Linguistics (2011)
2. Culotta, A.: Towards detecting influenza epidemics by analyzing twitter messages. In: Proceedings of the First Workshop on Social Media Analytics, pp. 115–122. ACM (2010)
3. de Quincey, E., Kostkova, P.: Early warning and outbreak detection using social networking websites: the potential of Twitter. In: Kostkova, P. (ed.) eHealth 2009. LNICST, vol. 27, pp. 21–24. Springer, Heidelberg (2010). https://doi.org/10.1007/978-3-642-11745-9_4
4. Fellbaum, C.: WordNet. Wiley Online Library (1998)
5. Gelernter, J., Mushegian, N.: Geo-parsing messages from microtext. Trans. GIS 15(6), 753–773 (2011)
6. Ginsberg, J., Mohebbi, M.H., Patel, R.S., Brammer, L., Smolinski, M.S., Brilliant, L.: Detecting influenza epidemics using search engine query data. Nature 457(7232), 1012 (2009)
7. New government policies and programmes, April 2018. https://powermin.nic.in/en/content/new-government-policies-and-programmes
8. Lampos, V., Cristianini, N.: Tracking the flu pandemic by monitoring the social web. In: 2010 2nd International Workshop on Cognitive Information Processing (CIP), pp. 411–416. IEEE (2010)
9. Lampos, V., De Bie, T., Cristianini, N.: Flu detector - tracking epidemics on Twitter. In: Balcázar, J.L., Bonchi, F., Gionis, A., Sebag, M. (eds.) ECML PKDD 2010. LNCS (LNAI), vol. 6323, pp. 599–602. Springer, Heidelberg (2010). https://doi.org/10.1007/978-3-642-15939-8_42
10. Raj, S., Kajla, T.: Sentiment analysis of Swachh Bharat Abhiyan. Int. J. Bus. Anal. Intell. 3(1), 32 (2015)
11. Swachh Bharat Mission-Gramin, April 2018. http://swachhbharatmission.gov.in/sbmcms/index.htm
12. Snow, J.: On the mode of communication of cholera. Edinb. Med. J. 1(7), 668 (1856)
13. Swachh Survekshan 2018, April 2018. https://www.swachhsurvekshan2018.org
14. Tayal, D.K., Yadav, S.K.: Sentiment analysis on social campaign "Swachh Bharat Abhiyan" using unigram method. AI Soc. 32(4), 633–645 (2017)
15. Wikipedia: Classification of Indian cities (2018). https://en.wikipedia.org/wiki/Classification_of_Indian_cities.htm

Author Index

Printed in the United States
By Bookmasters